DIGITAL

Matthew Boulton College
of Further and Higher Education

LIBRARY

Classification Number	Book Number
621.3815 THI	37608

THIS BOOK MUST BE RETURNED OR RENEWED ON OR BEFORE THE LAST DATE STAMPED BELOW

08. NOV 03
03.
19. MAY 94
15. JUN 94
03. JAN 95
03. FEB 95
03. MAR 95

04. 03. 96
18. 04. 96
13. 06. 96
07. 01. 97
16. JUL 04
18 Feb 08

058661

DIGITAL TECHNIQUES
from problem to circuit

A.P. Thijssen, Delft University of Technology
H.A. Vink, Philips Research Laboratories, Eindhoven
C.H. Eversdijk, formerly Professor Delft University of Technology

Edward Arnold
A division of Hodder & Stoughton
LONDON   NEW YORK   MELBOURNE   AUCKLAND

57608
621.3815

© 1989 A. P. Thijssen, H. A. Vink and C. H. Eversdijk

First published in Great Britain 1989

Distributed in the USA by Routledge, Chapman and Hall, Inc.
29 West 35th Street, New York, NY 10001

British Library Cataloguing in Publication Data

Thijssen, A. P.
   Digital techniques
   1. Digital electronic equipment
   I. Title
   621.3815

ISBN 0-7131-3657-X

All rights reserved. No part of this publication may be reproduced or transmitted in any form or by any means, electronically or mechanically, including photocopying, recording or any information storage or retrieval system, without either prior permission in writing from the publisher or a licence permitting restricted copying. In the United Kingdom such licences are issued by the Copyright Licensing Agency: 33-34 Alfred Place, London WC1E 7DP.

Printed and bound in Great Britain for Edward Arnold, the educational, academic and medical publishing division of Hodder and Stoughton Limited, 41 Bedford Square, London WC1B 3DQ by J. W. Arrowsmith Ltd, Bristol

Preface

The present VLSI technology creates extensive possibilities for the growth of the field of applications of digital techniques. More and more application specific designs can compete with designs based on standard components. This has an impact on the design process of digital circuits. The classical trial-and-error design style leads to very high costs, because of the many and expensive redesigns. A top-down oriented design style is an automatic consequence.

For a top-down oriented design the designer must be familiar with and have some experience in many aspects of digital techniques. It is not sufficient to know the logic properties of the components. Insight into subjects such as timing of components and timing at system level is indispensible to arrive at a reliable and correctly functioning circuit. In addition, the logic designer should have some knowledge of related disciplines. Problems with transient phenomena, which may arise during the realisation, can be solved more easily when the necessary precautions have already been taken at the logic design level. The application of the clock mode view, in which a period of time is reserved for the transition phenomena before the new internal state is effected by the clock, is an example.

Within the wide field of digital techniques this book deals with the logical aspects of the design. A chapter on logic families and modelling of components has been added in order the offer some insight into the details of the realisation level. For printed-circuit board designs with standard components this will be sufficient. An integrated circuit designer, however, should have more know-how of the electronic details and of the physical aspects of components and materials.

The book can be used for digital logic design courses in undergraduate programs in computer science and electronic and electrical engineering. In order to make the book a basis for other courses in hardware orientation the necessary knowledge of other fields has been included. As such the book is suitable for self-study. It will also be useful to the IC designer at system level. A selection of 200 problems has been included.

The book has originated from our courses at the Electrical Engineering Department of the Delft University of Technology in the Netherlands. A large number of the students have contributed to this text with their work and suggestions. They have demonstrated that a structured design style has many advantages. We also wish to express our gratitude to the students of the 'Avond-HTS' at the Hague. They have frequently been used as a response group for new parts of the text. This task they have fulfilled with enthusiasm.

The authors are aware that many others have contributed to this book, with their support and their lively interest in how the project progressed. We appreciate the help of Sylvia van Winden and Gerda Duvé-Randsdorp in typing the previous editions of the manuscript and doing the everlasting corrections during the preparation of this book. With the assistance of the staff of the 'Vereniging voor Studie- en Studentenbelangen' in Delft the manuscript has been converted into this copy. The translation was done by Machiel Knol and the typesetting by John van der Elburg. The final review of the manuscript by Susan Masotty was particularly helpful. They all did a fine job.

Pijnacker/Valkenswaard A.P. Thijssen/H.A. Vink
The Netherlands
1988

Contents

PREFACE		5
INTRODUCTION		11
1	NUMBER REPRESENTATIONS AND CODES	18
	1.1 The decimal and binary number systems	18
	1.2 The octal and hexadecimal number systems	21
	1.3 Codes for alphanumeric character sets	23
	1.4 Other codes	27
2	SWITCHING ALGEBRA	30
	2.1 Digital circuits and symbolic logic	30
	2.2 Operations in symbolic logic	32
	2.3 Composition rules for formulas	36
	2.4 Switching algebra	39
	2.5 The minterm form of functions	44
	2.6 The maxterm form of functions	47
	2.7 Example of a specification	49
3	THE REDUCTION OF LOGIC FUNCTIONS	52
	3.1 Karnaugh maps	52
	3.2 Karnaugh maps and the reduction of logic functions	54
	3.3 Applications of Karnaugh maps	57
	3.4 The sum and product forms of functions	60
	3.5 Incompletely specified functions	61
	3.6 A systematic procedure to find a cover in Karnaugh maps	63
	3.7 The formulation of covering problems	65
	3.8 The Quine-McCluskey algorithm	68
4	BRANCH TYPE COMBINATIONAL CIRCUITS	73
	4.1 Contacts and relays	73
	4.2 Designing in an AND-OR structure	76
	4.3 Designing through decomposition	78
	4.4 Practical aspects	82
5	THE LOGIC DESIGN OF COMBINATIONAL LOGIC CIRCUITS	86
	5.1 Logic levels and truth values	86
	5.2 Designing with ANDs, ORs and NOTs	91
	5.3 Designing in NANDs or NORs	94
	5.4 AND-NOR gates	95
	5.5 Designing through decomposition	97
	5.6 Some practical aspects	101
6	LOGIC FAMILIES AND THEIR GATE MODELS	104
	6.1 Introduction	104

	6.2	The TTL logic family	111
	6.3	Output drivers in the TTL family	114
	6.4	The I^{2}L logic family and other bipolar families	121
	6.5	MOS field-effect transistors	124
	6.6	MOS inverters	130
	6.7	NMOS logic circuits	133
	6.8	Switching behaviour of ratio-type NMOS logic	135
	6.9	Pass transistor logic	137
	6.10	CMOS logic circuits	139
	6.11	A comparison of MOS and bipolar	142
	6.12	The gate model	143
7	MSI COMBINATIONAL LOGIC CIRCUITS		150
	7.1	Introduction	150
	7.2	The design of a 4-bit binary adder	151
	7.3	The carry problem in adders	154
	7.4	The look-ahead carry generator	158
	7.5	Comparators	161
	7.6	Data transfer	165
	7.7	Error detection and error correction	171
	7.8	Summary	176
8	PROGRAMMABLE LOGIC		179
	8.1	Read Only Memories	179
	8.2	Some applications of ROMs	183
	8.3	Programmable logic arrays	187
	8.4	Optimisation in array logic	191
9	MEMORY ELEMENTS I: LATCHES		197
	9.1	Memory elements	197
	9.2	The S-R and the $\overline{S}$-$\overline{R}$ latches	200
	9.3	The gated latch	203
	9.4	Encoding problems and race conditions.	207
	9.5	Function tables	209
	9.6	Application: the anti-bounce circuit	211
	9.7	Application: the omnibus circuit	212
	9.8	Mapping problems	215
	9.9	RAM memories	216
10	MEMORY ELEMENTS II: FLIP-FLOPS		219
	10.1	Limitations of latches	219
	10.2	The master-slave principle	221
	10.3	The level mode and the clock mode views	223
	10.4	The D flip-flop	227
	10.5	The J-K flip-flop	228
	10.6	The T flip-flop	230

10.7	S-R flip-flops	231
10.8	The determination of input signals for flip-flops	234
10.9	Designing sequential circuits	238
10.10	The state assignment	242
10.11	Sequential circuits in programmable logic	245

11 TIMING — 249

11.1	Timing of gated latches	249
11.2	Edge-triggered timing of flip-flops	255
11.3	Pulse-triggered timing of flip-flops	259
11.4	Properties of edge-triggered or pulse-triggered timing	261
11.5	Preparatory and direct acting inputs	263
11.6	Direct data transfer between flip-flops	265
11.7	Indirect data transfer between flip-flops	269
11.8	Internal and external timing	271
11.9	Data transfer at different clock stages	276
11.10	Clock skew	278
11.11	Synchronisation of external signals	283
11.12	Conclusions	286

12 REGISTERS — 290

12.1	The internal structure of registers	290
12.2	How to use programmable registers	293
12.3	Timing properties of registers	296
12.4	Applications of shift registers	299
12.5	Parallel-series and series-parallel converters	302

13 COUNTERS — 305

13.1	Introduction	305
13.2	Asynchronous binary counters	308
13.3	Synchronous binary counters	312
13.4	Modular structure of synchronous binary counters	314
13.5	4-Bit binary counters	316
13.6	4-Bit BCD counters	324
13.7	The structure of 4n-bit counters	326
13.8	The parameter specification of 4n-bit counters	331
13.9	Pseudo clock mode use of counters	335
13.10	Applications	336
13.11	Counter-control loops	339
13.12	Summary	340

PROBLEMS — 345

ANSWERS — 425

SYMBOLS AND ABBREVIATIONS — 427

INDEX — 431

Introduction

What is digital technique?

All around us we are seeing more and more digital systems come into being every day. Many kinds of equipment that have been around for a long time are increasingly being realised in a totally different technology. They are being 'digitised'. For example digital watches, meters, recorders, parking meters, etc. Also in automatic telephone systems the number of computer controlled telephone exchanges has greatly increased. The computer has changed human life to such an extent that it is quite possible that our distant descendants will call 20th century man 'homo digitalis'. The following list with some dates gives an impression of the speed of the developments which have resulted in the present penetration of digital techniques.

before 1900: Mechanical automata.
   ca. 1900: Application of electromagnetic controlled relays.
   ca. 1940: First simple computer with vacuum tubes.
       1948: Invention of the transistor (Bardeen, Schockley and Brittain).
   ca. 1960: Computers with integrated circuits.
       1971: First microprocessor (Intel 4004, 4-bit μP).
   ca. 1980: Integration of circuits for many applications (VLSI).
   ca. 1985: (Semi)Custom IC profitable in small numbers.

This list gives a picture of the developments taking place in the hardware. A similarly fast development can also be seen in computer software.

In order to explain the explosive growth in the use of digital systems and components we must first of all point to the technology of circuit integration. This technology, at first stimulated by space technology, has made it possible to produce small-size complex systems at low cost. The impulse towards the growth of digital applications, however, dates back to far before space travel. Just think of telephone exchanges built with relays. Through some simple examples we will try to trace some reasons for the fast rise in the application of digital techniques.

Example *The registration of the angular position of a spindle*
Remote registration of information (in this case the position of a spindle) is a recurring problem. Of all the means at our disposal we will restrict ourselves to those in which the information is converted into and transmitted by electrical signals.

The information is represented by an electric quantity (current or voltage). This signal is transmitted by cable. The information must then be visualised at its destination. One of the possibilities is to show it as a deflection (from 0) on a measuring instrument. Figure 0.1 shows a (rather primitive) solution to the problem.

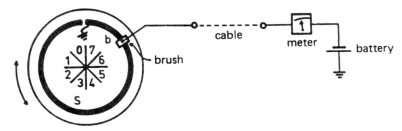

Figure 0.1 Continuous position registration by a potentiometer.

On the spindle a shaft S has been mounted, to which a strip of resistance material has been attached. The strip is grounded on one side. The distance of the brush b to the ground connection largely determines the total resistance of the measuring chain and thus the current through the meter. According to Ohm's law the value of this current is:

$$I_{meter} = \frac{U_{battery}}{R_{strip} + R_{rest}} \; A$$

with

$$R_{rest} = R_{brush} + R_{cable} + R_{meter} + R_{battery} \; \Omega.$$

If $R_{strip} \gg R_{rest}$ the deflection of the meter corresponds to the resistance R_{strip} of the chain between brush b and ground and thus with the position of the shaft. This solution is an example of a *continuous registration method*. The above solution could be rated as follows:

– the simplicity of the solution - sufficient;

– the accuracy - insufficient;

– the reliability - insufficient.

The reliability is negatively influenced by:

– the contact resistance between shaft and brush. This is highly variable as a result of wear and/or dirt.

– the cable resistance, which can usually not be neglected and which also varies with temperature.

– the direct dependence of the system on the battery supply voltage.

The accuracy is negatively influenced because close to the ground connection $R_{strip} \not\gg R_{rest}$. The influence of the resistance R_{rest} on the system is then too strong.

The device's dependence on the battery supply voltage is explained in Figure 0.2. Starting from the observation as indicated in Figure 0.2 the meter deflection is interpreted as position 5 (nominal supply voltage), as position 6 (battery supply voltage at 80% of the nominal value) or as position 4 (battery supply voltage too high). Continuous adjustments in the system are necessary. Another disadvantage of the above solution is that the relation between the position of the spindle and the current/meter deflection is strongly *non-linear*.

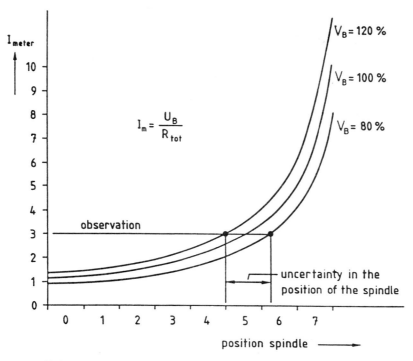

Figure 0.2 Relation between meter deflection and angular position.

The reliability and the accuracy of the system could be improved, for example, by a stabilised power supply, although this would make the system more expensive. A different approach is shown in Figure 0.3. A grounded metal strip whose length is one eighth of the circumference is fitted on the spindle. Eight brushes with fixed positions determine which lamp indicating the spindle position is switched on.

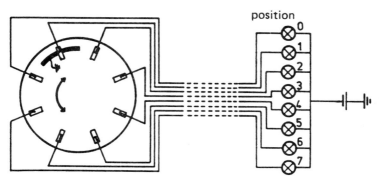

Figure 0.3 Discrete position registration, parallel solution.

In Figure 0.3 the information about the position of the spindle has been determined by a lamp's being either ON or OFF. A large degree of independence from variations in the parameters of the system, such as contact resistances and power supply voltage, has been reached. This conclusion is one of the most important reasons for basing equipment more and more on digital techniques, in which the information is represented by YES/NO decisions.

The accuracy of the solution in Figure 0.3 appears to be less than that in Figure 0.1. Small rotations are not visible in the system in Figure 0.3, unless they happen to be in the transition area between two segments. Figure 0.4.a shows how by doubling the number of segments the uncertainty in the position can be reduced and thus the accuracy increased.

With a sufficiently large number of segments the accuracy of the solution in Figure 0.3 can be better than that in Figure 0.1. This is caused by the great influence of battery supply voltage variations and contact resistances in the setup in Figure 0.1; these result in an uncertainty about the exact position that corresponds to a certain meter deflection. These influences are not considered interferences in the setup in Figure 0.3.

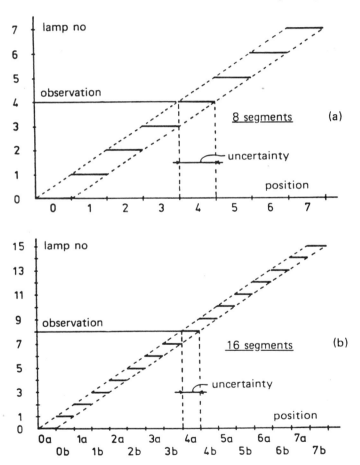

Figure 0.4 Influence of increasing the number of segments.

The encoding of information

In Figure 0.3 the position of the rotating spindle has been encoded in an *eight-bit code*:

- position 0 ↔ 10000000
- position 1 ↔ 01000000
  .
  .
  .
- position 7 ↔ 00000001

A 1 noted at a certain position in a code word means that the corresponding lamp must switch ON. This kind of encoding is not efficient. Figures 0.5/6 show that the position may be registered in a *three-bit code* with another kind of *code shaft*. Of the two types of code shaft, that of Figure 0.6 is to be preferred, for the following reason:

- Suppose that the brushes in Figure 0.5 are in position 7, i.e. all three are grounded. When the spindle turns to position 0 (not grounded) a number of wrong positions may be indicated, depending on the exact alignment of the brushes. Suppose that the brushes on the outermost and innermost rings are not in their correct positions so that position 7 [111] is changed into [011] and via [010] finally into position 0 [000]. The 1 denotes that a brush is grounded, and the 0 that it is not.

During the transition from position 7 to position 0 it appears that positions 3 [011] and 2 [010] occur.

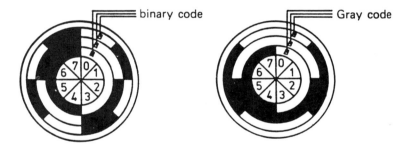

Figure 0.5 Code shaft in the binary code.   Figure 0.6 Code shaft in the Gray code.

The code shaft in Figure 0.6 is not affected by these transient phenomena, because of the *progressive code* applied. This code is known as the reflected binary code, also called the Gray code. At every transition the signal changes at only one position. One has to choose, then, between two neighbouring positions, which is exactly where the spindle is found.

The above example is not meant to solve the problem of how to determine the position of a spindle definitely. There are better methods, based on capacity variations, for example. We may, however, draw some important conclusions:

- For many systems it is advantageous to change from continuous registration methods to methods using a discrete range of values, because of reliability, reproducibility and noise immunity.

- The conversion to a system using discrete registration methods is in itself no guarantee of reliability and noise immunity. This is shown by the comparison of the code shafts.

- The accuracy of discrete systems need not be less than that of systems with continuous registration methods.

Figure 0.7 summarises the possibilities for encoding signal values. Figure 0.7.a shows how an analog signal can vary continuously in time. This signal is quantised by sampling it. The signal is considered to be constant until the next sampling moment. This results in the curve of Figure 0.7.b, which approaches the analog signal as well as possible. The approximation is improved when:

- the number of levels into which the sampled value can be divided is larger;
- the signal is sampled more frequently.

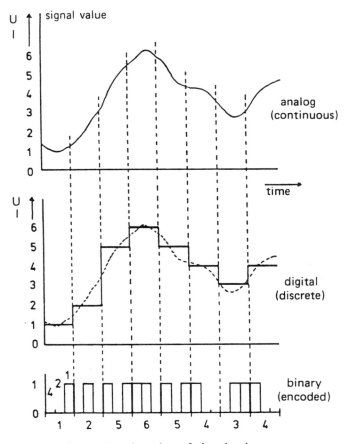

Figure 0.7 The encoding and registration of signal values.

There is a difference between the sampled signal and its source. This is called *quantisation noise*. Figure 0.7.c indicates how each discrete signal level can be represented in a 3-bit code, the elements of which have the weights of 4, 2 and 1.

The field of digital techniques is concerned with the design and construction of digitally operating equipment. The applications may vary from simple relay circuits to digital computers and microprocessors. To obtain a good result, it is essential to select the proper system architecture. This requires:

- know-how about the properties of the available components (what is possible);
- insight into the factors that may adversely influence the correct functioning of the circuit or system (interference factors);
- design methods which take into account the noise sources in advance (system implementation).

Our treatment of the components will be restricted to those properties that are of importance in the logic design of the circuit, in short, the implementation. The reader of this book will look in vain for an explanation of, for example, the conductivity mechanism in semiconductors and other subjects. The designer of digital equipment does not use this knowledge directly during the design process.

Binary or multivalent components?

Almost without exception only binary components are applied in digital techniques:

- a contact is open or closed;
- a relay is activated or not;
- a voltage is high or low;
- a lamp is on or off;
- a transistor conducts or it doesn't, etc.

The great reliability of digital circuits and the ease with which components can be replaced are the most important reasons for the almost exclusive use of binary components. Only in special applications are more than two signal levels used.

This book deals with the basic knowledge in the field of digital techniques. We treat not only the logical properties of the components, whose description is based on *switching algebra*, but also the physical and electronic properties of the components, insofar as they are of importance in the implementation and realisation of a digital circuit.

1
Number representations and codes

1.1 The decimal and binary number systems

Through the ages several *number representations* have arisen and have from time to time been replaced by others. An example is the Roman number system, which was not only in use in Roman times, but was also widely used in later centuries. We only have to look at the faces of antique clocks to be reminded of this. Nowadays this system is a mere curiosity, for it has been replaced by the decimal number system.

An example of an encoding method for letters and numbers is the Morse code, which was compiled in the early days of telegraphy. This code, too, has fallen into disuse due to the development of transmission technology.

The frequent use of the *binary number system* is of a more recent date. The rise of this system runs parallel to that of automation. So the choice of the number system is not only determined by history, but it is also influenced by technology, as we shall see in the case of computers.

The decimal number system

The *decimal number* 723 represents a number that is equal to 'seven times hundred' plus 'two times ten' plus 'three times one'. This we know because of the interpretation rules of the decimal number system. Should there be any confusion with other number representation systems, we represent the decimal number 723 as

$$723_{DEC}.$$

The rules for the decimal number system are:

- a number is represented by a series of digits;
- each digit belongs to $\{0, 1, 2, 3, 4, 5, 6, 7, 8, 9\}$;
- each digit contributes in accordance with its value, multiplied by its weight, which is a power of the *base* 10;
- the weight of any digit is determined by its *position* in the number representation, the right-hand digit having the weight of 10^0 (position 0), the digit next to it 10^1, etc.

Thus we may write the number 723_{DEC} as:

$$723_{DEC} = 7 \cdot 10^2 + 2 \cdot 10^1 + 3 \cdot 10^0.$$

A number can be represented in the decimal system in one way only.
Similar rules apply to other number representation systems.

The binary number system

In computers one frequently uses the binary number system. The reasons for this have already been mentioned in the introduction. The *binary number system* has base 2 and has the digits 0 and 1. The ith digit from the right gets a weight of 2^{i-1}. The binary number

$$N = a_{n-1}a_{n-2}...a_1a_0 \text{ BIN}$$

represents the number equal to

$$N = a_{n-1} \cdot 2^{n-1} + a_{n-2} \cdot 2^{n-2} + ... + a_1 \cdot 2^1 + a_0 \cdot 2^0.$$

Example

$$N = 100101_{BIN} = 1 \cdot 2^5 + 1 \cdot 2^2 + 1 \cdot 2^0 = 37_{DEC}.$$

$$N = 101.11_{BIN} = 1 \cdot 2^2 + 1 \cdot 2^0 + 1 \cdot 2^{-1} + 1 \cdot 2^{-2} = 5.75_{DEC}. \square$$

In the binary number system, as well, a number can only be represented in one way. Depending on the value of the number to be represented, we do this by adding more bits or taking them away. With n bits integers between

$$0 \leq N \leq 2^n - 1$$

can be represented in the binary number system. Table 1.1 lists the representations of the decimal numbers 0 through 15 in the binary number system.

1 bit	2 bits	3 bits	4 bits	Decimal
0	00	000	0000	0
1	01	001	0001	1
	10	010	0010	2
	11	011	0011	3
		100	0100	4
		101	0101	5
		110	0110	6
		111	0111	7
			1000	8
			1001	9
			1010	10
	Binary		1011	11
			1100	12
			1101	13
			1110	14
			1111	15

Table 1.1 Representation of numbers in the binary number system.

Conversion decimal to binary and binary to decimal

To convert decimal numbers to binary numbers and the reverse, there are several known algorithms. As an example of a decimal-to-binary conversion we follow the conversion of the number 47_{DEC} into its binary representation. As it holds that

$$2^6 > 47 > 2^5$$

at least six bits are needed for the binary representation. The *bits* (binary digits) have the following weights:

$$\begin{array}{cccccc} 32 & 16 & 8 & 4 & 2 & 1 \\ \downarrow & \downarrow & \downarrow & \downarrow & \downarrow & \downarrow \\ a_5 & a_4 & a_3 & a_2 & a_1 & a_0 \end{array}$$

From the number 47 we can subtract $2^5 = 32$, so that $a_5 = 1$. The remainder is $47 - 32 = 15$, from which $2^4 = 16$ can no longer be subtracted, so that $a_4 = 0$. We can, however, subtract 8 from 15, so that $a_3 = 1$. The remainder is 7, to be written as $4 + 2 + 1$. So $a_2 = a_1 = a_0 = 1$. The binary representation of 47_{DEC} is therefore

$$47_{DEC} = 101111_{BIN}.$$

For manual calculations this method works satisfactorily. Another algorithm, based on repeated division by 2, also gives a correct result.

$$\left. \begin{array}{l} 47 : 2 = 23 + \text{rest } 1 \rightarrow a_0 = 1 \\ 23 : 2 = 11 + \text{rest } 1 \rightarrow a_1 = 1 \\ 11 : 2 = 5 + \text{rest } 1 \rightarrow a_2 = 1 \\ 5 : 2 = 2 + \text{rest } 1 \rightarrow a_3 = 1 \\ 2 : 2 = 1 + \text{rest } 0 \rightarrow a_4 = 0 \\ 1 : 2 = 0 + \text{rest } 1 \rightarrow a_5 = 1 \end{array} \right\} 47_{DEC} = 101111_{BIN}.$$

The conversion procedure of repeatedly dividing by the base of the target number system (2 in the above example) can also be applied to convert the decimal number system to a number system with base g. We particularly have in mind the octal and hexadecimal number systems to be described later.

Binary numbers can also be converted to *decimal numbers* by adding the weights of those bits of the binary number representation that have the value of 1. For example

$$1011001_{BIN}$$

can be converted to the decimal representation:

$$1011001N = 1 \cdot 2^6 + 1 \cdot 2^4 + 1 \cdot 2^3 + 1 \cdot 2^0$$
$$= 64 + 16 + 8 + 1$$
$$= 89_{DEC}.$$

The BCD representation

It is customary in digital equipment to represent each digit by a series of binary digits 0 and 1. Decimal numbers are, as a rule, represented by the BCD code (Binary Coded Decimal). Every decimal digit, from 0 through 9, is represented by a 4-bit combination.

Example

The decimal number 974 is represented in the BCD code as

$$\left. \begin{array}{l} 9 \rightarrow 1001 \\ 7 \rightarrow 0111 \\ 4 \rightarrow 0100 \end{array} \right\} \quad 974_{DEC} = 1001\ 0111\ 0100_{BCD}.\ \square$$

From this example it follows that the weights of the bits in the BCD code are:

$$\underbrace{8\ 4\ 2\ 1}_{\times 10^2}\ \underbrace{8\ 4\ 2\ 1}_{\times 10^1}\ \underbrace{8\ 4\ 2\ 1}_{\times 10^0}$$

$$N = a_{11}\ a_{10}\ a_9\ a_8\quad a_7\ a_6\ a_5\ a_4\quad a_3\ a_2\ a_1\ a_0\ \text{BCD}$$

or

$$N = a_{11}\ a_{10}\ a_9\ a_8\quad a_7\ a_6\ a_5\ a_4\quad a_3\ a_2\ a_1\ a_0\ \text{BCD}$$

with weights for the BCD code:
- 100, 200, 400, 800
- 10, 20, 40, 80
- 1, 2, 4, 8

For larger numbers this representation is expanded accordingly.

The BCD code can quite easily be interpreted by the user. The same cannot be said of the binary code when the number of bits increases. The price to be paid is that the BCD code needs more bits than the corresponding binary representation. The number 974_{DEC} can be represented by ten bits in the binary code, while the corresponding BCD representation takes twelve.

Note
The subscripts DEC, BIN or BCD indicating which number representation system is being applied are usually omitted when it is clear from the context which number system is being used. When different number systems are used in one text it is advisable to apply the subscripts DEC, BIN or BCD systematically. □

1.2 The octal and hexadecimal number systems

As stated earlier, computers and other types of digital equipment operate internally with the binary number system. This is the case, for example, for the addressing of memory locations. Any user wanting to indicate memory addresses directly has to do this through a binary encoded address. The address is then represented by a series of zeros and ones, also called a *bit string*. Manually typing in a bit string, especially when it is longer than four or five bits may result in many errors. To avoid these mistakes, it is preferable to use the octal or hexadecimal number systems. These number systems have bases that are powers of 2.

The octal number system

The *octal number system* has base 8 and uses the digits 0 through 7. For the rest, rules similar to those of the decimal system apply. Thus 7516_{OCT} represents a number that is equal to

$$\begin{aligned}7516_{OCT} &= 7 \cdot 8^3 + 5 \cdot 8^2 + 1 \cdot 8^1 + 6 \cdot 8^0 \\ &= 7 \cdot 512 + 5 \cdot 64 + 1 \cdot 8 + 6 \cdot 1 \\ &= 3918_{DEC}.\end{aligned}$$

The number 7516_{OCT} may also be represented by a bit string:

7516_{OCT} via 111 101 001 110 to 111101001110_{BIN}.

Converting a *bit string* to the corresponding *octal representation* is easy. The bit string is partitioned into 3-bit groups, starting from the binary point. For each group the decimal digit that corresponds to that 3-bit binary code is noted. For example:

bit string 10110110101 → | 010 | 110 | 110 | 101 |
                    ↓   ↓   ↓   ↓
                    2   6   6   5   = 2665_{OCT}.

From the above it is apparent that the octal number system is not so much to be used to calculate with, but to handle bit strings. This is also true for the number system described below.

The hexadecimal number system

Computers usually have a word length that is a multiple of four bits (8-16-32, for example). The octal number system, with groups of three bits, does not fit well into this scheme. The *hexadecimal number system* does not have this disadvantage. It has 16 as its base and therefore has 16 digits, 6 of which are represented by the letters A through F for lack of sufficient decimal symbols.

decimal   0   8 9 10 11 12 13 14 15
    ↓   ↓ -----↓ ↓ ↓ ↓ ↓ ↓ ↓ ↓
hexadecimal 0   8 9 A B C D E F

The advantage here as well is that a bit string can be converted directly into the hexadecimal representation and vice versa. A hexadecimal input of bit strings is encountered frequently in practice. For example:

bit string   10110110101 → |0101|1011|0101|
                                  ↓   ↓   ↓
                                  5   B   5   = $5B5_{HEX}$.

Figure 1.1 shows a keyboard designed for the hexadecimal input of either numbers or bit strings.

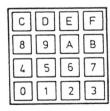

Figure 1.1 Hexadecimal keyboard.

Binary numbers can easily be converted into *hexadecimal numbers*. Even the weights of the bits correspond when the hexadecimal digits are written as a 4-bit binary number. Decimal may be converted to hexadecimal by dividing by the base of the target number system.

Example
Convert $48{,}692_{DEC}$ into the hexadecimal representation by repeatedly dividing by 16.

$$\left.\begin{array}{l} 48{,}692 = 16 \cdot 3{,}043 + 4 \\ 3{,}043 = 16 \cdot 190 + 3 \\ 190 = 16 \cdot 11 + 14\ (E) \\ 11 = 16 \cdot 0 + 11\ (B) \end{array}\right\} \quad 48{,}692_{DEC} = BE34_{HEX}.\ \square$$

Explanatory note
After dividing once by 16 a total of 3,043 multiples of 16 have been taken away from 48,692. The remainder of 4 is the least significant digit in the hexadecimal number representation. After the next division, 3,043 by 16, 190 multiples of 16^2 have been taken away. Three multiples of 16 are left. This brings us to the next digit, 3.

Then 190 is divided by 16, which results in 11 multiples of 16^3 with 14 remaining multiples of 16^2. At this point we have found the third and fourth digits of the hexadecimal representation, namely E and B. $\square$

Let us now try it the other way round, converting $BE34_{HEX}$ into decimal. This can be done either by the weights of the digits, or by repeated multiplication. The result is the same.

$$\begin{array}{l} BE34_{HEX} = ((11 \cdot 16 + 14) \cdot 16 + 3) \cdot 16 + 4_{DEC} \\ \qquad\qquad\ \ \downarrow\qquad\ \ \downarrow\qquad\ \downarrow\qquad\ \downarrow \\ \qquad\qquad\ \ B\qquad\ \ E\qquad\ 3\qquad\ 4 \\ = ((176 + 14) \cdot 16 + 3) \cdot 16 + 4_{DEC} \\ = (190 \cdot 16 + 3) \cdot 16 + 4_{DEC} \\ = 3{,}043 \cdot 16 + 4_{DEC} \\ = 48{,}692_{DEC}. \end{array}$$

In general

$$a_3 a_2 a_1 a_0{}_{HEX} = ((a_3 \cdot 16 + a_2) \cdot 16 + a_1) \cdot 16 + a_0{}_{DEC}.$$

Note
In analogy to the BCD representation of decimal numbers, one could conceivably introduce 'BCO' and 'BCH' representation to stand for the binary 3-bit and 4-bit encodings of the digits of respectively octal and hexadecimal numbers, even though it is not common practice to do so. $\square$

Table 1.2 shows the representation of the decimal numbers 1 through 20 in the number systems introduced above. The weights of the digits are specified at the top of every column.

1.3 Codes for alphanumeric character sets

An information processing system must be able to process texts, commands, calls and signals along with numbers. There are several internationally recognised standard *alphanumeric character sets* for this purpose. We mention only those most frequently in use.

DEC	BIN	OCT	HEX	BCD
$10^1, 10^0$	$2^4, 2^3, 2^2, 2^1, 2^0$	$8^1, 8^0$	$16^1, 16^0$	$8,4,2,1\ 8,4,2,1$
00	00000	00	00	0000 0000
01	00001	01	01	0000 0001
02	00010	02	02	0000 0010
03	00011	03	03	0000 0011
04	00100	04	04	0000 0100
05	00101	05	05	0000 0101
06	00110	06	06	0000 0110
07	00111	07	07	0000 0111
08	01000	10	08	0000 1000
09	01001	11	09	0000 1001
10	01010	12	0A	0001 0000
11	01011	13	0B	0001 0001
12	01100	14	0C	0001 0010
13	01101	15	0D	0001 0011
14	01110	16	0E	0001 0100
15	01111	17	0F	0001 0101
16	10000	20	10	0001 0110
17	10001	21	11	0001 0111
18	10010	22	12	0001 1000
19	10011	23	13	0001 1001
20	10100	24	14	0010 0000

Table 1.2 Number representations.

The international telegraph alphabet CCITT nr. 2

This character encoding, which is shown in Table 1.3, is used in telex and telegraph connections. This alphabet comprises not only letters and digits but also some commands and signals. In the customary *five-bit code* only 32 combinations are available. This number is too small to encode all letters and digits. Therefore the code is provided with a so-called *letter/figure shift*. After combination nr. 29 (11111) all subsequent 5-bit combinations are interpreted as *letters*. Once combination nr. 30 (11011) has been received all subsequent combinations represent digits and control signs. This nearly doubles the available character set.

The CCITT nr. 3 alphabet (TOR or Teletype-on-Radio) is also shown in Table 1.3. This code is a *three-out-of-seven code*. This alphabet comprises 35 combinations, three more than in CCITT nr. 2. Due to the three-out-of-seven encoding this code has *error detecting properties*. If the received character does not have exactly three ones, something has gone wrong during transmission. An error has occurred. The opposite is not always true, though!

	CCITT nr. 2	Meaning		CCITT nr. 3
nr.	Code combinations	Letter case	Figure case	Code combinations
1	1 1 0 0 0	A	–	0 0 1 1 0 1 0
2	1 0 0 1 1	B	?	0 0 1 1 0 0 1
3	0 1 1 1 0	C	:	1 0 0 1 1 0 0
4	1 0 0 1 0	D	Who are you?	0 0 1 1 1 0 0
5	1 0 0 0 0	E	3	0 1 1 1 0 0 0
6	1 0 1 1 0	F	free	0 0 1 0 0 1 1
7	0 1 0 1 1	G	free	1 1 0 0 0 0 1
8	0 0 1 0 1	H	free	1 0 1 0 0 1 0
9	0 1 1 0 0	I	8	1 1 1 0 0 0 0
10	1 1 0 1 0	J	bell	0 1 0 0 0 1 1
11	1 1 1 1 0	K	(	0 0 0 1 0 1 1
12	0 1 0 0 1	L	)	1 1 0 0 0 1 0
13	0 0 1 1 1	M	.	1 0 1 0 0 0 1
14	0 0 1 1 0	N	,	1 0 1 0 1 0 0
15	0 0 0 1 1	O	9	1 0 0 0 1 1 0
16	0 1 1 0 1	P	0	1 0 0 1 0 1 0
17	1 1 1 0 1	Q	1	0 0 0 1 1 0 1
18	0 1 0 1 0	R	4	1 1 0 0 1 0 0
19	1 0 1 0 0	S	'	0 1 0 1 0 1 0
20	0 0 0 0 1	T	5	1 0 0 0 1 0 1
21	1 1 1 0 0	U	7	0 1 1 0 0 1 0
22	0 1 1 1 1	V	=	1 0 0 1 0 0 1
23	1 1 0 0 1	W	2	0 1 0 0 1 0 1
24	1 0 1 1 1	X	/	0 0 1 0 1 1 0
25	1 0 1 0 1	Y	6	0 0 1 0 1 0 1
26	1 0 0 0 1	Z	+	0 1 1 0 0 0 1
27	0 0 0 1 0	Carriage return		1 0 0 0 0 1 1
28	0 1 0 0 0	Line feed		1 0 1 1 0 0 0
29	1 1 1 1 1	Letter shift		0 0 0 1 1 1 0
30	1 1 0 1 1	Figure shift		0 1 0 0 1 1 0
31	0 0 1 0 0	Space		1 1 0 1 0 0 0
32	0 0 0 0 0	Not normally used		0 0 0 0 1 1 1
33		Signal α		0 1 0 1 0 0 1
34		Signal β		0 1 0 1 1 0 0
35		Signal repetition		0 1 1 0 1 0 0

Table 1.3 The telegraph and the TOR alphabets.

The ASCII code

This code, the *American Standard Code for Information Interchange*, is often used in computers and peripherals and is, in practice, the most frequently applied standard in data communication networks. The code represents alphanumeric data, and reserves a large number of combinations for control commands and signals. See Table 1.4.

ASCII		commands		characters					
	b_6	0	0	0	0	1	1	1	1
	b_5	0	0	1	1	0	0	1	1
	b_4	0	1	0	1	0	1	0	1
$b_3\ b_2\ b_1\ b_0$									
0 0 0 0		NUL	DLE	SP	0	@	P	`	p
0 0 0 1		SOH	DC1	!	1	A	Q	a	q
0 0 1 0		STX	DC2	"	2	B	R	b	r
0 0 1 1		ETX	DC3	#	3	C	S	c	s
0 1 0 0		EOT	DC4	$	4	D	T	d	t
0 1 0 1		ENQ	NAK	%	5	E	U	e	u
0 1 1 0		ACK	SYN	&	6	F	V	f	v
0 1 1 1		BEL	ETB	'	7	G	W	g	w
1 0 0 0		BS	CAN	(	8	H	X	h	x
1 0 0 1		HT	EM	)	9	I	Y	i	y
1 0 1 0		LF	SUB	*	:	J	Z	j	z
1 0 1 1		VT	ESC	+	;	K	[	k	{
1 1 0 0		FF	FS	,	<	L	\	l	|
1 1 0 1		CR	GS	-	=	M	]	m	}
1 1 1 0		SO	RS	.	>	N	^	n	~
1 1 1 1		SI	US	/	?	O	—	o	DEL

Table 1.4 The ASCII code.

We will not go into the meaning of the commands here. This 7-bit code, $b_6 b_5 b_4 b_3 b_2 b_1 b_0$, contains $2^7 = 128$ code combinations. This is sufficient to encode all digits and letters as well as a number of signs and control commands.

The complete ASCII set of 128 characters is not always used. Sub-sets, consisting of 96 or 64 or 48 characters, are often applied in peripherals. The 48-character set, for example, contains the digit and capital letter columns of Table 1.4.

An eighth bit is usually added to the seven data bits. By virtue of this *parity bit* the sum of all bits in every 8-bit word is made even or odd, and consequently the code is endowed with certain *error detecting properties* (see Chapter 7).

The EBCDIC code

This code, the Extended BCD Interchange Code, is an 8-bit code. The code table is given in Table 1.5. The EBCDIC code is used in computer systems (IBM) to store data and programs. It contains several commands in addition to letters and digits. We note that not all of the 256 combinations are assigned. Some of them can be freely chosen by the user (which may result in compatibility problems).

EBCDIC	b_0	0	0	0	0	0	0	0	0	1	1	1	1	1	1	1	1
	b_1	0	0	0	0	1	1	1	1	0	0	0	0	1	1	1	1
	b_2	0	0	1	1	0	0	1	1	0	0	1	1	0	0	1	1
	b_3	0	1	0	1	0	1	0	1	0	1	0	1	0	1	0	1
$b_7 b_6 b_5 b_4$																	
0 0 0 0		NUL	DLE	DS		SP	&		_					{	}	\	0
0 0 0 1		SOH	DC1	SOS				/		a	j	~		A	J		1
0 0 1 0		STX	DC2	FS	SYN					b	k	s		B	K	S	2
0 0 1 1		ETX	TM	—						c	l	t		C	L	T	3
0 1 0 0		PF	RES	BYP	PN					d	m	u		D	M	U	4
0 1 0 1		HT	NL	LF	RS					e	n	v		E	N	V	5
0 1 1 0		LC	BS	ETB	UC					f	o	w		F	O	W	6
0 1 1 1		DEL	IL	ESC	EOT					g	p	x		G	P	X	7
1 0 0 0		GE	CAN							h	q	y		H	Q	Y	8
1 0 0 1		RLF	EM					`		i	r	z		I	R	Z	9
1 0 1 0		SMM	CC	SM		ç	!	¦	:								|
1 0 1 1		VT	CU1	CU2	CU3	.	$	,	#								
1 1 0 0		FF	IFS		DC4	<	*	%	@					ʃ		⌐	
1 1 0 1		CR	IGS	ENQ	NAK	(	)	_	'								
1 1 1 0		SO	IRS	ACK		+	;	>	=					↵			
1 1 1 1		SI	IUS	BEL	SUB	'	¬	?	"								EO

Table 1.5 The EBCDIC code.

Note

By adding an extra bit (*parity check*) or by making exclusive use of certain combinations from a larger set, *error detecting properties* can be built into a code. Take for example a *three-out-of-seven code*. It allows a number of errors to be detected, after which one can make the necessary inquiries. □

It is also possible to build in *error correcting properties* in a code by adding even more bits. When there are only a few errors it is not necessary to make further inquiries, as one can correct the errors directly.

1.4 Other codes

There are several other codes in current use. In computing, one encounters the *complement codes*; the one's complement and the two's complement. The *EXCESS-3 code* is used in BCD arithmetic. In systems with error detection the *m-out-of-n codes* are sometimes used. We have already introduced the 3-out-of-7 code, in which the CCITT nr. 3 alphabet is encoded. A single error in such a 7-bit code word can always be detected because the number of ones is no longer equal to three. However, not all errors involving two or more bits are recognised.

Progressive codes

A *progressive encoding* of a group of *consecutive* numbers, characters or events is characterised by the fact that code combinations differing in only one bit are assigned to each pair of consecutive elements in that group. An important representative of this class is the *reflected binary code* or *Gray code*. This code occurs when one reflects 2^k code combinations while adding a new bit. This bit is 0 for the existing combinations and 1 for the new combinations. The structure of this code is shown in Table 1.6.

Bit pattern	Combination
0 00 000 0000	0
1 01 001 0001	1
11 011 0011	2
10 010 0010	3
110 0110	4
111 0111	5
101 0101	6
100 0100	7
1100	8
1101	9
1111	10
1110	11
——— reflection line 1010	12
1011	13
1001	14
1000	15

Table 1.6 Structure of the reflected binary code.

Encoding shafts are an important application of progressive codes. (E.g. Figures 0.5 and 0.6.)

In general progressive codes are applied in those situations in which inconvenient *transient phenomena* may occur on the transition from one position/combination to another.

In going cyclically through a series of code combinations progressive encoding can only be done with an even cycle length. The cycle length does not necessarily have to be a power of 2. Take for example the combinations 0 through 4 followed by 11 through 15 in Table 1.6. This cycle of 10 combinations is also encoded progressively. A prerequisite is that the two groups must be chosen symmetrically with respect to the reflection line.

References

1. H.C.A. van Duuren, *Typendruktelegrafie over Radioverbindingen*, Het PTT Bedrijf, Vol. 1, n° 2, pp. 35–41.
2. F. de Fremery, *Communicatie in Informatieverwerkende Systemen*, Centrex, Eindhoven, 1975.
3. H.L. Garner, *Generalized Parity Checking*, IRE Tr. on Electronic Computers, Vol. EC-7, Sept. 1958, pp. 207–213
4. E.L. Gilbert, *How Good is Morse Code*, Information and Control, Vol. 14, 1969, pp. 559–565
5. P.E. Gosling and Q.L.M. Laarhoven, *Codes for Computers and Microprocessors*, Macmillan, London, 1980.

6. J.E. McNamara, *Technical Aspects of Data Communication*, Digital Equipment Corp., Bedford, Mass, 1978.
7. L.J. Zeckendorf, *Character Sets and Codes*, Informatie, Vol. 15, 1973, pp. 653–661.
8. H. Zemaneck, *Alphabets and Codes*, Electronische Rechenanlagen, Vol. 7, 1965, pp. 239–258.

2
Switching algebra

2.1 Digital circuits and symbolic logic

Every design of a circuit begins with a description of the desired external function. As a rule, the first, verbal, description is vague. The specifications have not been evaluated thoroughly. In a number of discussions between the customer and the designer a more complete picture of the customer's intentions gradually evolves. The foremost task of the designer is to formulate the user's wishes so as to make it possible to come up with a device or a program that meets the user's objectives.

This book treats the design of *digital circuits*. These circuits operate on discrete signals, usually at two levels. This means that in the final design stage everything must be described in two-valued variables. The value of such a variable can be represented directly by one of the two signal levels of the components.

Propositions

For the specification of digital circuits *propositions* are used. Propositions are *statements* that can only be 'TRUE' or 'FALSE', denoted as T or F. The designer tries to formulate his problem such that one or more propositions can describe it. By defining the relation between the propositions the function of the circuit to be designed is described formally.

Example

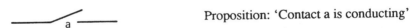
Proposition: 'Contact a is conducting'

Figure 2.1 The description of a contact.

The function of contact a in Figure 2.1 can be described by applying proposition A:

'Contact a is conducting'.

In the position indicated in Figure 2.1 proposition A is false. In a formula:

A = F.

If the contact is closed we indicate this as:

A = T. □

Everything cannot be expressed by propositions. Many *intuitive notions*, such as nearly, perhaps, etc. cannot be, or not completely, expressed this way. The same holds for the proposition

'It is raining'.

In cloudless skies or during torrential rains it is not difficult to assign a *truth value* to this proposition. However, it is often hard to tell whether it is raining or not in humid weather. In that case it would be advisable to formulate the proposition as

'I think it is raining'.

In this formulation the *subjective element* is incorporated.

With digital techniques such problems also arise. Propositions appear to be suitable for describing the *behaviour* of digital circuits in the *static state*. Their behaviour in the *dynamic state*, in which transient phenomena may occur, turns out to be difficult to describe, if at all.

Despite these restrictions, propositions are a very useful aid in describing the function of a circuit or during the design process. In the *IEC system for symbols of digital circuits* it has been agreed to formulate propositions as much as possible *in the affirmative*. For contacts, for example, the operation is defined as

'The contact is conducting'

instead of

'The contact does not conduct',

even though the latter is, formally speaking, possible. This rule has been agreed upon in order to avoid, as much as possible, confusion in the *interpretation* of formulas.

Truth tables

For the description of most problems several propositions are needed. Tables which specify the relation between the truth values of different propositions are called *truth tables*.

Example

The series interconnection of two contacts a and b conducts when both contact a and contact b are conducting. The propositions A, B and S which describe the function of the *series interconnection* are defined in Figure 2.2.

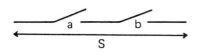

A: 'Contact a is conducting',

B: 'Contact b is conducting',

S: 'Connection S is conducting'.

Figure 2.2 Series interconnection of two contacts.

The relation between the truth values of propositions A, B and S is shown in Table 2.1. This table follows directly from the figure. □

A	B	S
F	F	F
F	T	F
T	F	F
T	T	T

Table 2.1 Truth table of the series interconnection.

The advantage of using propositions is that the relation between physical or other phenomena, about which statements are made by propositions, can be described in a realisation-independent way. This holds both for designing electronic digital circuits and, for example, for pneumatic logic. Decisions about the *realisation form*, such as representing T and F by voltage levels or by the conductivity of a network, can be postponed until a later stage.

Moreover the laws of *symbolic logic* can be applied to propositions. They can be used to simplify formulas, or make them 'fit' the available components.

2.2 Operations in symbolic logic

The concepts *proposition* and *truth value* stem from *symbolic logic*. An important publication in this field is

'An Investigation of the Laws of Thought'

by George Boole in 1854. The *switching algebra* to be introduced in Section 2.4 is therefore called *Boolean algebra*. A mathematician, however, understands more by a Boolean algebra than the simple collection of postulates and theorems used by a logic designer.

Another important publication is Shannon's 1938 article

'A Symbolic Analysis of Relay and Switching Circuits'.

It describes how the rules of symbolic logic can be applied in circuits with contacts, a milestone in the design of digital circuits!

In digital technique the logic values T and F are often represented by 1 and 0. In that case

$$T \leftrightarrow 1 \text{ and } F \leftrightarrow 0. \tag{2.1}$$

The 1 and 0 are used here as *truth values* or *logic values*. The use of 1 and 0 in digital technique is broader, however. They may also represent the numeric value of a bit in a number representation. This may seem confusing, but is not in general.

The logic operation AND

The function of the *series interconnection* of contacts in Figure 2.2 has been described by the propositions A, B and S, as given in the figure. This results in truth Table 2.1.

Table 2.2 also specifies for the series interconnection the relation between the truth values of propositions A and B, on the one hand, and of S, on the other. Table 2.2.a uses the truth values T and F. Table 2.2.b gives the same information but applies the logic values 1 and 0, which is more customary.

A	B	S = A AND B		A	B	S = A·B
F	F	F		0	0	0
F	T	F	$F \leftrightarrow 0$	0	1	0
T	F	F	$T \leftrightarrow 1$	1	0	0
T	T	T		1	1	1

Table 2.2 The logic operation AND.

The relation between the propositions A, B and S can be expressed in a formula by the logic operation AND:

$$S = A \text{ AND } B.$$

The *logic operation AND* formulates that S is only true if A and B are both true. Table 2.2 defines this logic operation. Several different symbols are encountered in the literature to denote the operation AND:

AND, $\wedge$, $\cdot$, *.

We have opted in this text to use the dot. The formula which expresses the relation between propositions A, B and S is then

$$S = A \text{ AND } B \quad \text{or} \quad S = A \cdot B, \tag{2.2}$$

or in short

$$S = AB.$$

The logic operation OR

Table 2.3 defines the *logic operation OR*. Here proposition S is true if A is true, if B is true or if both A and B are true.

A	B	S = A OR B		A	B	S = A + B
F	F	F		0	0	0
F	T	T	$F \leftrightarrow 0$	0	1	1
T	F	T	$T \leftrightarrow 1$	1	0	1
T	T	T		1	1	1

Table 2.3 The logic operation OR.

The operation OR is usually represented by one of the symbols

OR, V, +.

In this book we prefer the '+' sign:

$$S = A \text{ OR } B \quad \text{or} \quad S = A + B. \tag{2.3}$$

Example *Parallel interconnection of contacts*

In Figure 2.3 a parallel interconnection of two contacts a and b is drawn. The function of this circuit is described by the propositions A, B and S, as they were introduced for the series interconnection of contacts.

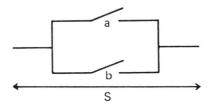

Figure 2.3 Parallel interconnection of contacts.

When deriving the relation between the propositions A, B and S for this parallel interconnection of contacts it turns out to be the same as in Table 2.3. The parallel interconnection of two contacts may therefore be considered to be a circuit that can realise the logic operation OR. □

The logic operation NOT

The *logic operation NOT* is used to indicate the negation of a statement or proposition. Some possible ways to represent it are:

$$\text{NOT A}, \overline{A}, \neg A, \sim A \text{ and } A'. \tag{2.4}$$

As a rule the operation NOT is denoted in this text as a bar over the proposition or formula. The operation NOT is defined in Table 2.4.

A	$\overline{A}$		A	$\overline{A}$
F	T	$F \leftrightarrow 0$	0	1
T	F	$T \leftrightarrow 1$	1	0

Table 2.4 The logic operation NOT.

From Table 2.4 it follows that if proposition A is true, proposition $\overline{A}$ (pronounce: not A) is false. When A is false, proposition $\overline{A}$ is true.

As an example of an application of the operation NOT we will define in the following example the *make contact* and the *break contact* of a push-button.

Example *Specification of a push-button*
One can describe the function of a *push-button* with the propositions:

D: 'The button is being pressed',

G: 'Its contact is conducting.'

When the relation between the truth values of D and G is as follows:

$$D = F \leftrightarrow G = F \text{ and } D = T \leftrightarrow G = T,$$

in other words

$$D = G,$$

then pressing the button results in a conducting contact. The contact is called a *make contact*.
If the relation turns out to be

$$D = F \leftrightarrow G = T \text{ and } D = T \leftrightarrow G = F,$$

in other words

$$D = \overline{G},$$

the contact is called a *break contact*. □

Note
When we represent T and F by 1 and 0 the relation for the make contact is

$$D = 0 \leftrightarrow G = 0 \quad \text{and} \quad D = 1 \leftrightarrow G = 1$$

and for the break contact

$$D = 0 \leftrightarrow G = 1 \quad \text{and} \quad D = 1 \leftrightarrow G = 0. \ \square$$

Example

Figure 2.4 shows a realisation of the logic operation NOT. On the push-button a break contact is used in the chain S_1, and a make contact in the chain S_2. The dashed line indicates the position of the two contacts when the button is pushed. The relation between the propositions S_1 and S_2 of this scheme is:

$$S_1 = \text{NOT } S_2 \text{ or } S_1 = \overline{S}_2,$$
$$\text{NOT } S_1 = S_2 \text{ or } \overline{S}_1 = S_2,$$

in which S_1 and S_2 stand for:

S_1: 'The chain S_1 is conducting';

S_2: 'The chain S_2 is conducting'. $\square$

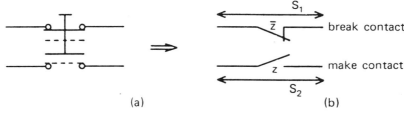

Figure 2.4 A realisation of the operation NOT.

The inclusive OR and the exclusive OR

There are two definitions of the logic operation OR which are in use, the so-called *inclusive OR* and the *exclusive OR* (see Tables 2.5.a and 2.5.b).

A	B	A + B		A	B	A ⊕ B	
F	F	F		F	F	F	
F	T	T	inclusive OR	F	T	T	exclusive OR
T	F	T		T	F	T	
T	T	T	(a)	T	T	F	(b)

Table 2.5 The inclusive and the exclusive OR.

The definition of Table 2.5.a is the most common. We will therefore always interpret OR as the inclusive OR operation in the following, unless explicitly stated otherwise. The exclusive OR is usually denoted by

EX-OR or ⊕.

In this text we shall indicate the exclusive OR by the sign ⊕.

Example

Propositions:

A: 'Switch a is in the normal position';

B: 'Switch b is in the normal position';

L: 'The lamp L is on'.

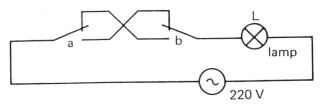

Figure 2.5 Hotel circuit.

In bedrooms, for example, the light can often be switched on and off in two places. See Figure 2.5 for the electrical diagram. When both switches are in the so-called *normal position* (the position drawn), the light is off. If one of the switches is put in the *working position* the light is on. When both switches are in the working position, the light is again off. The truth table, which specifies the relation between the propositions A, B and L, turns out to be the same as Table 2.5.b. The function of the circuit may be described by the formula

$$L = A \oplus B. \square$$

2.3 Composition rules for formulas

With the operations AND, OR and NOT expressions (formulas) may be formed from propositions. These expressions are in turn also propositions, with truth values T or F. For example

$$A + B,$$
$$A \cdot B,$$
$$\overline{A},$$

but also

$$A + B + C,$$
$$A + B + \overline{C}.$$

In *symbolic logic* it is formally defined which expressions are allowed and which are not. An example of the latter category is

$$A + + B.$$

There are rules for the interpretation of formulas. We will give these first. The formal laws of switching algebra are treated in Section 2.4.

Order of precedence of operations

When different logic operations are used in a formula, the order of precedence is:
- first do all negations;
- then all AND operations;
- then all OR operations.

This order can be changed by *means of parentheses*.

Example
Suppose $A = T$, $B = F$ and $C = F$. Then
$$A + B \cdot C = T + F \cdot F = T + F = T,$$
but
$$(A + B) \cdot C = (T + F) \cdot F = T \cdot F = F. \square$$

Before introducing the laws of switching algebra formally we give some examples of the description of some problems in a logic formula.

Example
From Table 2.5.b it follows that the proposition
$$A \oplus B$$
is true when

> A is true AND B is false
>
> OR
>
> A is false AND B is true.

In a formula:
$$A \oplus B = (A \text{ AND } (\text{NOT } B)) \text{ OR } ((\text{NOT } A) \text{ AND } B),$$

in which the parentheses have been used for clarity. After all, based on the precedence rules they are not necessary. The proposition $A \oplus B$ may now be written as

$$A \oplus B = A \cdot \overline{B} + \overline{A} \cdot B. \tag{2.5}$$

The *exclusive OR* operation can be expressed in the inclusive OR and the AND operations. $\square$

Example
The proposition calculus can also be used for non-technical applications. Let us say someone has to go shopping and has, for example, on his shopping list

> butter (t)
> sugar (u)
> medicines (v)
> candy (w)
> cookies (x)
> bread (y)
> meat (z).

Table 2.6 gives a picture of where the various articles can be bought.

		t	u	v	w	x	y	z	• = available
Grocer	A	•	•		•	•	•		
Chemist	B			•	•				
Dairy	C	•							
Baker	D				•	•			
Butcher	E							•	

Table 2.6 Availability of articles.

Let S be the proposition

$$S: \text{'Shopping done'};$$

then S is only true when all items have been bought. To buy butter (t) one has to go to the grocer or to the dairy. At least one of the propositions

$$A: \text{'Been to grocer'}$$

or

$$C: \text{'Been to dairy'}$$

must then be true (see Table 2.6). In terms of a logic formula:

$$S_{\text{butter}} = A + C.$$

Similarly, sugar has to be bought. According to Table 2.6

$$S_{\text{sugar}} = A.$$

Then

$$S_{\text{butter AND sugar}} = (A + C) \cdot A.$$

All items can be bought when

$$S_{\text{butter AND sugar AND ... AND meat}} \tag{2.6}$$
$$= (A + C) \cdot A \cdot B \cdot (A + B) \cdot (A + D) \cdot (A + D) \cdot (A + E)$$

is true.

After further consideration of Table 2.6 it turns out that it is sufficient to go to the grocer and to the chemist. Indeed, when filling in $A = T$ and $B = T$ in (2.6) it turns out that $S = T$ also. The articles on the shopping list can be bought when

$$S = A \cdot B \tag{2.7}$$

is true. □

The laws by which S in the form of (2.6) can be transformed into the form of (2.7), in other words the laws by which logic formulas can be processed, are introduced in the next section.

2.4 Switching algebra

Binary and logic variables.

The *truth values* of a proposition are T and F. A variable with range T and F is called a *logic variable*. A proposition can therefore be *interpreted* as a logic variable or *Boolean variable*.

A logic variable which has the value T or F is a special case of a *binary variable*. A binary variable is a variable which can have exactly two values. For a binary adder section each input and output has the numeric value 1 or 0. If each input and output is assigned a variable which has the value of this input or output (1 or 0), these variables are binary variables.

A circuit or system described with binary variables can be directly described by logic variables. As an intermediate step propositions must be introduced which define the relation between both descriptions. Without an explicit specification by propositions the relation is, of course, lost.

In digital technique formulas are usually expressed in logic variables, which can have the values T or F. As an introduction to *switching algebra* we compare two formulas, both describing the logic operation OR. See Table 2.7. In it T and F are represented by the more customary 1 and 0. It always holds that T $\Leftrightarrow$ 1.

y	z	S
0	0	0
0	1	1
1	0	1
1	1	1

$S = y + z$
$S = \bar{y}z + y\bar{z} + yz$

Table 2.7 Truth table of the logic operation OR.

In addition to the table there are two formulas which both describe how the logic variable S depends on the value of the logic variables y and z. No significance should be attached to the fact that S is written as a capital and y and z as lower-case letters (in digital technique S usually corresponds to the output of the circuit and y and z to the inputs, which explains the difference).

The formula

$$S = y + z$$

corresponds to the way in which the operation OR has been introduced. The other form of the formula,

$$S = \bar{y} \cdot z + y \cdot \bar{z} + y \cdot z,$$

resembles the way in which the truth table is interpreted, namely row by row. Since both formulas correspond to the same truth table it must be possible to convert one into the other. We will trace the laws which make such a conversion possible.

$$
\begin{aligned}
S &= \bar{y}\cdot z + y\cdot \bar{z} + y\cdot z \\
&= \bar{y}\cdot z + y\cdot z + y\cdot \bar{z} \\
&= \bar{y}\cdot z + y\cdot z + y\cdot z + y\cdot \bar{z} \\
&= (\bar{y} + y)\cdot z + y\cdot (z + \bar{z}) \\
&= 1\cdot z + y\cdot 1 \\
&= y + z
\end{aligned}
$$

provided that:
a permutation of the order is allowed,
$y\cdot z = y\cdot z + y\cdot z$,
e.g. $\bar{y}\cdot z + y\cdot z = (\bar{y} + y)\cdot z$,
$\bar{y} + y = 1$ and $z + \bar{z} = 1$,
$1\cdot z = z$ and $y\cdot 1 = y$.

After every line in the above deduction it is specified under which condition(s) that step is allowed.

When we verify the above-mentioned conditions for contact circuits it turns out that they are always met. The same proves to be true for an electronic or mechanical realisation of logic operations and functions.

The set of rules under which certain formulas can be converted into other (usually simpler) formulas is called switching or Boolean algebra. The set given below appears to be convenient in practice.

Switching algebra

Order of operations

In switching algebra a formula is interpreted in the following order (see also Section 2.2):

– first do all negations;

– then all logic operations AND, indicated by ·;

– finally the operations OR, indicated by +.

This order can be changed by parentheses, just as in ordinary algebra. All logic formulas must be interpreted in accordance with these rules.

Rules for constants

For the constants 0 and 1 the following rules apply:

$$
\begin{array}{lll}
0 + 0 = 0 & 0\cdot 0 = 0 & \\
0 + 1 = 1 & 0\cdot 1 = 0 & \bar{0} = 1 \\
1 + 0 = 1 & 1\cdot 0 = 0 & \bar{1} = 0 \\
1 + 1 = 1 & 1\cdot 1 = 1 &
\end{array}
$$

The 0 and 1 are the only constants in switching algebra. They represent the logic values 'FALSE and 'TRUE'. The symbols + and · stand for OR and AND.

In formulas in which + and · appear, less space is maintained around the ·, which better illustrates the hierarchy of these operations.

Rules for logic variables

1. *The idempotence laws*

$$z + z = z \qquad (1.a)$$

$$z\cdot z = z \qquad (1.b)$$

These laws say that identical terms may be repeated in a sum or product. For contact circuits these laws mean that the same contact chain may be repeated in series and parallel connections, but that this is, in fact, superfluous. For switching algebra these laws mean that coefficients and powers of variables other than 0 and 1 are not used.

2. *The associative laws*

$$(x + y) + z = x + (y + z) \tag{2.a}$$

$$(x \cdot y) \cdot z = x \cdot (y \cdot z) \tag{2.b}$$

The associative laws say that for the interpretation of a logic formula the order of identical logic operations is not important.
Neither is the order of the variables in a sum or product. This is a result of the commutative laws:

3. *The commutative laws*

$$y + z = z + y \tag{3.a}$$

$$y \cdot z = z \cdot y \tag{3.b}$$

According to these laws the order of the terms in a sum or the factors in a product may be arbitrarily interchanged.

4. *The distributive laws*

$$x + y \cdot z = (x + y) \cdot (x + z) \tag{4.a}$$

$$x \cdot (y + z) = x \cdot y + x \cdot z \tag{4.b}$$

These laws show how parentheses can be put in or taken out of a formula. It might be interesting to take a look at the contact circuits that correspond to these expressions. Figure 2.6 shows the circuits corresponding to law 4.a.

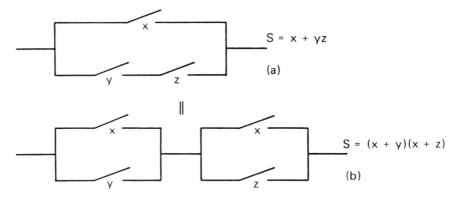

Figure 2.6 The distributive law in contact circuits.

The circuits are logically equivalent. The first circuit contains three contact pairs and is preferable from the point of view of the material used. The distributive laws can be used as an aid in simplifying circuits.

5. *The modulus laws*

$$0 + z = z \qquad (5.a)$$
$$1 \cdot z = z \qquad (5.b)$$
$$1 + z = 1 \qquad (5.c)$$
$$0 \cdot z = 0 \qquad (5.d)$$

The modulus laws describe how formulas with constants and logic variables must be interpreted.

6. *The negation laws*

$$\bar{z} + z = 1 \qquad (6.a)$$
$$\bar{z} \cdot z = 0 \qquad (6.b)$$
$$\overline{(\bar{z})} = z \qquad (6.c)$$
$$\left. \begin{array}{l} \overline{y + z} = \bar{y} \cdot \bar{z} \\ \overline{y \cdot z} = \bar{y} + \bar{z} \end{array} \right\} \text{DeMorgan laws} \qquad \begin{array}{l} (6.d) \\ (6.e) \end{array}$$

The negation laws 6.a through 6.e describe the rules by which the expressions containing negations must be interpreted. Laws 6.d and 6.e are called *DeMorgan laws*. Like Boole, DeMorgan is one of the founders of symbolic logic.

7. *The absorption laws*

$$z + y \cdot z = z \qquad (7.a)$$
$$z \cdot (y + z) = z \qquad (7.b)$$
$$y + \bar{y} \cdot z = y + z \qquad (7.c)$$
$$y \cdot (\bar{y} + z) = y \cdot z \qquad (7.d)$$
$$x \cdot y + \bar{x} \cdot z + y \cdot z = x \cdot y + \bar{x} \cdot z \qquad (7.e)$$
$$(x + y) \cdot (\bar{x} + z) \cdot (y + z) = (x + y) \cdot (\bar{x} + z) \qquad (7.f)$$

The absorption laws are reduction laws. These laws indicate when a term in a formula or branch in a circuit is redundant. Sometimes, however, these laws can be turned round. By adding terms it is sometimes possible to eliminate the negative effects of transient phenomena (see Section 4.4).

Some laws in the set of laws mentioned under 1 through 7 are redundant, they can be deduced from others. The set as given turns out to be convenient in practice, however. The absorption laws can save a lot of work, even though they are in fact superfluous for the definition of switching algebra.

Note

Instead of negation one also runs across the term inversion of variables and formulas. Nevertheless, negation and inversion are not identical, as we will discuss in a subsequent chapter. □

How to prove the laws?

Formal proof of the correctness of the laws described above can be given by a truth table. All possible combinations of the truth values of the logic variables are written to the left of the vertical line. On the right-hand side one puts in two columns the truth values of the logic expressions to the left and to the right of the equality sign in the law. When these two columns are equal the law has been validated.

Example
Prove DeMorgan laws. □

y	z	$\overline{y+z}$	$\overline{y}\cdot\overline{z}$
0	0	1	1
0	1	0	0
1	0	0	0
1	1	0	0

Table 2.8

y	z	$\overline{y\cdot z}$	$\overline{y}+\overline{z}$
0	0	1	1
0	1	1	1
1	0	1	1
1	1	0	0

Table 2.9

The proofs of the absorption laws can be given by applying the laws given previously.

Example
Proof of law 7.a:

$z + y\cdot z = 1\cdot z + y\cdot z$ (via 5.b)
$ = (1 + y)\cdot z$ (via 3.b and 4.b)
$ = 1\cdot z$ (via 5.c)
$ = z$ (via 5.b).

Proof of law 7.d:

$y\cdot(\overline{y} + z) = y\cdot\overline{y} + y\cdot z$ (via 4.b)
$\phantom{y\cdot(\overline{y} + z)} = 0 + y\cdot z$ (via 6.b)
$\phantom{y\cdot(\overline{y} + z)} = y\cdot z$ (via 5.a). □

Extension of the laws

A proposition may be:

– a simple proposition;

– a composite proposition, consisting of one or more propositions and the connectives AND, OR and NOT (composed according to the rules described in logic).

The same holds for logic variables. Every allowable expression in logic variables is, likewise, a logic variable. This property makes it possible to extend the interpretation of the laws.

Example *Extension of DeMorgan:* $\overline{y+z} = \overline{y}\cdot\overline{z}$

$\overline{x + y + z} = \overline{(x + y) + z}$
$\phantom{\overline{x + y + z}} = \overline{(x + y)}\cdot\overline{z}$
$\phantom{\overline{x + y + z}} = (\overline{x}\cdot\overline{y})\cdot\overline{z}$
$\phantom{\overline{x + y + z}} = \overline{x}\cdot\overline{y}\cdot\overline{z}.$

Similarly:
$$\overline{x \cdot y \cdot z} = \overline{x} + \overline{y} + \overline{z}. \square$$

Example *Application of the law*: $y + \overline{y} \cdot z = y + z$
From this law it follows that
$$\begin{aligned} a + b + \overline{a} \cdot \overline{b} \cdot c &= a + b + (\overline{a} \cdot \overline{b}) \cdot c \\ &= (a + b) + \overline{(a + b)} \cdot c \\ &= (a + b) + c \\ &= a + b + c. \square \end{aligned}$$

Example $x \cdot (y + z) = x \cdot y + x \cdot z$ (distributive law)
We ask ourselves what the following is:
$$a \cdot (b_1 + b_2 + \ldots + b_n).$$

This can be worked out as:
$$\begin{aligned} a \cdot (b_1 + b_2 + b_3) &= a \cdot ((b_1 + b_2) + b_3) \\ &= a \cdot (b_1 + b_2) + a \cdot b_3 \\ &= a \cdot b_1 + a \cdot b_2 + a \cdot b_3. \end{aligned}$$

$$\begin{aligned} a \cdot (b_1 + b_2 + b_3 + b_4) &= a \cdot ((b_1 + b_2 + b_3) + b_4) \\ &= a \cdot (b_1 + b_2 + b_3) + a \cdot b_4 \\ &= a \cdot b_1 + a \cdot b_2 + a \cdot b_3 + a \cdot b_4. \end{aligned}$$

etc. Finally,
$$a \cdot (b_1 + b_2 + \ldots + b_n)$$
can be elaborated with *complete induction* to n, as:
$$a \cdot (b_1 + b_2 + \ldots + b_n) = a \cdot b_1 + a \cdot b_2 + \ldots + a \cdot b_n. \square$$

2.5 The minterm form of functions

The right-hand part of Table 2.10 specifies the general form of a truth table of a function S in three variables x through z.

minterm	x	y	z	S
$m_0 = \overline{x}\overline{y}\overline{z}$	0	0	0	f_0
$m_1 = \overline{x}\overline{y}z$	0	0	1	f_1
$m_2 = \overline{x}y\overline{z}$	0	1	0	f_2
$m_3 = \overline{x}yz$	0	1	1	f_3
$m_4 = x\overline{y}\overline{z}$	1	0	0	f_4
$m_5 = x\overline{y}z$	1	0	1	f_5
$m_6 = xy\overline{z}$	1	1	0	f_6
$m_7 = xyz$	1	1	1	f_7

Table 2.10 Truth table.

Every row in Table 2.10 can be *identified* with a product m_i of the variables x through z, with every variable in the product being used normally or negated (complemented). Such a product is called a *minterm*. If the weights

$$x \leftrightarrow 2^2, \qquad y \leftrightarrow 2^1, \qquad z \leftrightarrow 2^0$$

are assigned to the variables x through z, we immediately find the rank number i of the product m_i. Thus the number

$$i = 1 \cdot 2^2 + 1 \cdot 2^1 + 0 \cdot 2^0 = 6$$

corresponds to the combination $xy\bar{z} = 110$.

Table 2.10 leads to a *standard form* of a function in three variables:

$$\boxed{\begin{aligned} S &= m_0 \cdot f_0 + m_1 \cdot f_1 + m_2 \cdot f_2 + \ldots + m_6 \cdot f_6 + m_7 \cdot f_7 \\ &= \bar{x}\bar{y}\bar{z} \cdot f_0 + \bar{x}\bar{y}z \cdot f_1 + \bar{x}y\bar{z} \cdot f_2 + \ldots + xy\bar{z} \cdot f_6 + xyz \cdot f_7 \end{aligned}}$$
(2.8.a)
(2.8.b)

This sum-of-products form of a function in three variables is called the *minterm form*. The minterm form is directly read from the truth table. (For another number of variables a minterm form can be found in the same way.)

Properties

In the minterm form one product of the variables corresponds to each function value f_i. This product has the value 1 for exactly one input combination of the truth table. For all other input combinations it has the value 0.

Such a product, which can be interpreted as the *address* of the corresponding function value in the truth table, is called a *minterm*. A minterm is a product in all input variables, either normal or negated. It is evident that the following properties apply:

Property 1

$$m_i \cdot m_j = 0 \text{ for } i \neq j. \tag{2.9}$$

Property 2

$$m_i = 1 \text{ for only one input combination.} \tag{2.10}$$

Property 3

$$\sum_{i=0}^{7} m_i = 1. \qquad \text{(In general } \sum_{i=0}^{2^n-1} m_i = 1.\text{)} \tag{2.11}$$

(It has been indicated above how the rank number i of a minterm m_i is determined.)

Example

Table 2.11 specifies a circuit. Its function is to determine whether for three binary signals x, y and z with weights 4, 2 and 1, the value of xyz_{BIN} is greater than or equal to 3.

The corresponding minterm form of the function S in Table 2.11 is:

$$\begin{aligned} S &= \bar{x}\bar{y}\bar{z} \cdot 0 + \bar{x}\bar{y}z \cdot 0 + \bar{x}y\bar{z} \cdot 0 + \bar{x}yz \cdot 1 + x\bar{y}\bar{z} \cdot 1 + x\bar{y}z \cdot 1 + xy\bar{z} \cdot 1 + xyz \cdot 1 \\ &= \bar{x}yz + x\bar{y}\bar{z} + x\bar{y}z + xy\bar{z} + xyz. \end{aligned}$$

x	y	z	S
0	0	0	0
0	0	1	0
0	1	0	0
0	1	1	1
1	0	0	1
1	0	1	1
1	1	0	1
1	1	1	1

Propositions

x: 'signal x has value 1';
y: 'signal y has value 1';
z: 'signal z has value 1';

S: 'The value of xyz_{BIN} is greater than or equal to 3'.

Table 2.11 Truth table.

The minimal sum form is:

$$S = x + yz.$$

The minterm form generally differs from the minimal sum-of-products form. □

Venn diagram

In Figure 2.7.a a so-called Venn diagram for one variable z is shown. The area enclosed by the circle corresponds to z = 1, which can also be interpreted as the area in which the proposition Z is true. The area outside the circle is to be interpreted as z = 0 or as the area in which the proposition Z is false. The total area enclosed by the rectangle is called the *universe*, i.e. in it all possibilities are comprised, both z = 0 and z = 1.

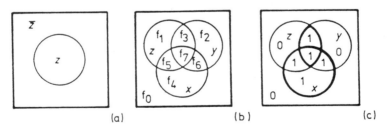

Figure 2.7 Venn diagrams.

Fig 2.7.b shows a Venn diagram for a randomly chosen function in three variables. In the diagram eight different areas are to be seen, which correspond to the eight minterms of a function in three variables. The corresponding function value has been placed in each area.

Venn diagrams can be used to simplify functions. In Figure 2.7.c the function values are annotated in accordance with the specification of the circuit given in Table 2.11. We can see immediately that the area where the function has the value of 1, consists of the union of the areas x and yz. The minimal form of the function as a sum-of-products can easily be recognised:

$$S = x + yz.$$

For more than four variables Venn diagrams soon become difficult to handle. In the next chapter a more stylised form of the Venn diagram is introduced, used especially for the simplification of functions. The Venn diagram has been brought in here simply to clarify the meaning of the word min(imal)term.

2.6 The maxterm form of functions

The minterm form of a function is used frequently because of its direct correspondence with the truth table. Sometimes another standard form (a product form) of a function is used, the so-called *maxterm form*. The maxterm form of a function in three variables is:

$$S = (x + y + z + f_0) \cdot (x + y + \bar{z} + f_1) \cdot (\ldots) \cdot (\bar{x} + \bar{y} + \bar{z} + f_7). \tag{2.12}$$

It is related to the minterm form as follows:
The negated form S of S is found by negating all function values. Then S can be restored by negating the expression. In a formula:

$$\bar{S} = m_0 \cdot \bar{f_0} + m_1 \cdot \bar{f_1} + \ldots + m_6 \cdot \bar{f_6} + m_7 \cdot \bar{f_7}$$
$$S = \bar{\bar{S}} = \overline{(m_0 \cdot \bar{f_0} + m_1 \cdot \bar{f_1} + \ldots + m_6 \cdot \bar{f_6} + m_7 \cdot \bar{f_7})}$$
$$= (\bar{m_0} + f_0) \cdot (\bar{m_1} + f_1) \cdot (\ldots) \cdot (\bar{m_6} + f_6) \cdot (\bar{m_7} + f_7),$$

from which it follows that

$$\boxed{S = (M_0 + f_0) \cdot (M_1 + f_1) \cdot (\ldots) \cdot (M_6 + f_6) \cdot (M_7 + f_7).} \tag{2.13}$$

Thus the relation between a *minterm* m_i and a so-called *maxterm* M_i is:

$$m_i = \overline{M_i} \quad \text{or} \quad \bar{m_i} = M_i. \tag{2.14}$$

A maxterm M_i fulfils the same role in product form (2.13) as m_i in the sum form (2.8) of the function. For every combination of the binary variables the value of the function S is determined by exactly one function value f_i. Whereas in the sum form all the remaining terms become 0, in a product form all the remaining factors of the product must be 1. The term 'maxterm form' can now also be explained. Every maxterm (maximal term) corresponds to an area as large as possible in a Venn diagram, not being the universe. Both forms, the minterm form and the maxterm form, are sometimes called the *canonical forms* of a logic function.

Both forms are hardly ever realised directly in this form in a circuit. Instead, a simpler expression which requires fewer components is used. In the next chapter some simplification methods are introduced.

Example

Determine the minterm form, the maxterm form, the minimal sum form and the minimal product form of the following function:

$$S = \bar{x}\bar{y}\bar{z} + x\bar{y} + x\bar{z} + yz.$$

Solution:
The minterm form is found by expansion, i.e. by introducing missing variables:

$$S = \bar{x}\bar{y}\bar{z} + x\bar{y}(\bar{z} + z) + x(\bar{y} + y)z + (\bar{x} + x)yz$$
$$= \bar{x}\bar{y}\bar{z} + x\bar{y}\bar{z} + x\bar{y}z + x\bar{y}z + xy\bar{z} + \bar{x}yz + xyz$$
$$= \bar{x}\bar{y}\bar{z} + \bar{x}yz + x\bar{y}\bar{z} + x\bar{y}z + xy\bar{z} + xyz.$$

The other product terms are missing because the corresponding function values are 0. So the function values that are 1 are:

$$f_0, f_3, f_4, f_5, f_6, f_7.$$

The function values that are 0 are:

$$f_1, f_2.$$

The maxterm form follows directly from this:

$$S = (\bar{m}_0 + f_0) \cdot (\bar{m}_1 + f_1) \cdot (\ldots) \cdot (\bar{m}_7 + f_7)$$
$$= \bar{m}_1 \cdot \bar{m}_2$$
$$= (x + y + \bar{z})(x + \bar{y} + z).$$

The minimal sum form is deduced from the minterm form as follows:

$$S = \bar{x}\bar{y}\bar{z} + \bar{x}yz + x\bar{y}\bar{z} + x\bar{y}z + xy\bar{z} + xyz$$
$$= (\bar{x}\bar{y}\bar{z} + x\bar{y}\bar{z}) + (\bar{x}yz + xyz) + x(\bar{y}\bar{z} + \bar{y}z + y\bar{z} + yz)$$
$$= \bar{y}\bar{z} + yz + x.$$

The minimal product form of the given function is (coincidentally) equal to the maxterm form. □

Notation

The following formulas hold for the minterm form and maxterm form of a function:

$$\boxed{\text{minterm form: } S = \sum_{i=0}^{2^k-1} m_i f_i} \qquad (2.15)$$

$$\boxed{\text{maxterm form: } S = \prod_{i=0}^{2^k-1} (M_i + f_i)} \qquad (2.16)$$

If the function values f_i are known, then all product terms with function value $f_i = 0$ in the minterm form can be omitted. In the maxterm form all factors with $f_i = 1$ are replaced by 1.

For the function specified in Table 2.11 this leads to

$$\sum_{i=0}^{2^3-1} m_i f_i = m_3 + m_4 + m_5 + m_6 + m_7, \qquad (2.17.a)$$

and

$$\prod_{i=0}^{2^3-1} (M_i + f_i) = M_0 \cdot M_1 \cdot M_2. \qquad (2.18.a)$$

Both formulas are sometimes written in a shorthand as

$$S = \sum (3,4,5,6,7) \qquad (2.17.b)$$

and

$$S = \prod (0,1,2) \qquad (2.18.b)$$

2.7 Example of a specification

The specification of a given problem is not always easy. Many specifications are not complete, even though they appear to be so at first sight. The following problem is an illustration.

Example

A club has two groups of members, a group of three persons {A, B, C} and the other members. The board must, according to its bye-laws, consist of *exactly* four members, of which the group {A, B, C} has to supply *at least half* of the board members.
In the following formulas the variables A, B or C stand for the proposition

'Member A/B/C has been appointed to the board'

and O_i for

'i other members have been appointed to the board.'

Which of the next formulas specifies, correctly and as simply as possible, the conditions under which the board has been made up (S = 1) in accordance with the regulations?

1. $S = AB + AC + BC + ABC$.
2. $S = AB\overline{C} + A\overline{B}C + \overline{A}BC + ABC$.
3. $S = ABO_2 + ACO_2 + BCO_2 + ABCO_1$.
4. $S = AB\overline{C}O_2 + A\overline{B}CO_2 + \overline{A}BCO_2 + ABCO_1$.

For the specification of this problem five logic variables are used, the variables A through C and the variables O_1 and O_2. We will take a look at the four alternatives given.

Alternative 1 drops out immediately, as does alternative 2. In both cases nothing has been specified about the variables O_1 and/or O_2. Expansion of alternative 1 into

$$\begin{aligned} S &= AB + AC + .. \\ &= AB(\overline{C} + C)(\overline{O}_1 + O_1)(\overline{O}_2 + O_2) + AC + .. \\ &= AB\overline{C}\overline{O}_1\overline{O}_2 + ... \end{aligned}$$

shows that S = 1 for the combination $AB\overline{C}\overline{O}_1\overline{O}_2$. In other words, only A and B are on the board. This is in conflict with the prerequisite that the board must consist of exactly four members. For similar reasons alternative 2 also drops out.

Alternative 3 is out as well. Expansion

$$S = ABO_2 + ACO_2 + ...$$
$$= AB(\overline{C} + C)(\overline{O}_1 + O_1)O_2 + ACO_2 + ...$$
$$= AB\overline{C}\overline{O}_1O_2 + ... + ABC\overline{O}_1O_2 + ...$$

suggests, among other things, that $ABC\overline{O}_1O_2$ describes a correct composition. The board would then consist of five members.

Alternative 4. Expansion of this formula gives

$$S = AB\overline{C}O_2 + \overline{A}BCO_2 + \overline{A}BCO_2 + ABCO_1$$
$$= AB\overline{C}(\overline{O}_1 + O_1)O_2 + A\overline{B}C(\overline{O}_1 + O_1)O_2 + \overline{A}BC(\overline{O}_1 + O_1)O_2 + ABCO_1(\overline{O}_2 + O_2)$$
$$= AB\overline{C}\overline{O}_1O_2 + AB\overline{C}O_1O_2 + ... + ABCO_1O_2.$$

The term $AB\overline{C}O_1O_2$ describes a correct composition of the board in accordance with the prerequisites.

There is something special about the term $AB\overline{C}O_1O_2$. This term cannot be 1 because the propositions

O_1: 'One member of the remaining group has been appointed to the board'

and

O_2: 'Two members of the remaining group have been appointed to the board'

can never be true at the same time. O_1 and O_2 represent *dependent logic variables. Even though the term $AB\overline{C}O_1O_2$ belongs to the formula, the combination for which this term becomes true never occurs.*

Expansion of the remaining terms of alternative 4 leads to the same conclusion. It turns out that alternative 4, *although not correct at first sight*, describes the problem correctly. □

Note

Interdependence of logic variables is a recurring problem. It is not always easy to recognise this. Much depends on the way the propositions have been formulated. If propositions O_1 and O_2 above were formulated as

O_i: 'At least i members of the remaining group have been appointed to the board',

the dependences would be different from those of the previously given definition. □

References

1. T. Bartee, I.L. Lebow and I.S. Reed, *Theory and Design of Digital Machines*, McGraw-Hill, New York, 1962.
2. G. Birkhoff and S. Maclane, *A Survey of Modern Algebra*, 4th ed., Macmillan, New York, 1977.
3. G. Boole, *An Investigation of the laws of Thought*, London, 1854.
4. Ph. Dwinger, *Introduction to Boolean Algebras*, Physica Verlag, Würzburg, 1961.
5. J.B. Fraleigh, *A first Course in Abstract Algebra*, Addison-Wesley, Reading, Mass., 1973.
6. P.R. Halmos, *Lectures on Boolean Algebras*, Van Nostrand, 1963.
7. J.P. Hayes, *An Unified Switching Theory with Applications to VLSI Design*, Proc. IEEE, Oct. 1982, pp. 1140–1151.

8. F.E. Hohn, *Applied Boolean Algebra*, Macmillan, New York, 1966.
9. E. Mendelson, *Boolean Algebra and Switching Circuits*, McGraw-Hill, New York, 1970.
10. S.R. Petrick, *A Direct Determination of the Irredundant Forms of a Boolean Function from the Set of Prime Implicants*, Computation laboratory of the AFCRC, Bedford, Mass., TR-56-110, Apr. 1956.
11. C.E. Shannon, *A Symbolic Analysis of Relay and Switching Circuits*, Trans.Am. Inst. Electr. Eng., Vol. 57, 1938, pp. 713–723.
12. J.E. Whitesitt, *Boolean Algebra and Its Applications*, Addison-Wesley, Reading, Mass., 1961.

3
The reduction of logic functions

3.1 Karnaugh maps

From Chapter 2 we know that the reduction of logic functions by means of the laws of switching algebra is a tedious job. Some systematic methods are known from the literature, e.g. those of Karnaugh and of Quine and McCluskey. Veitch suggested a reduction method, later improved by Karnaugh, using a graphical representation of a logic function. This method is suitable for manual use. The algorithm described by Quine-McCluskey has been developed for programming on digital computers. We will begin by introducing the Karnaugh map.

Truth tables and Karnaugh maps

Logic functions can be specified in a truth table. The tables in Figures. 3.1.a through 3.1.d show the general form of the truth table for functions of up to four variables. The *minterm form* of a logic function corresponds directly to the *truth table*. Every row corresponds to one minterm m_i with corresponding function value f_i.

Next to each truth table is the Karnaugh map. This map specifies the same information as the table. For each minterm m_i the function value f_i is placed in the cell that corresponds to m_i. This is done as follows. The cells are identified by labelling the edges of the map. Every line is labelled with one of the function variables. For all cells, lying above or below or to the left or the right of such a line, the variable assigned to it is thought to have the logic value 1 (True). For cells that are not 'below' the line the variable is thought to have the logic value 0. Lines and labels are placed in such a way that every cell corresponds to one minterm. The *function value* f_i corresponding to this minterm is written in the cell.

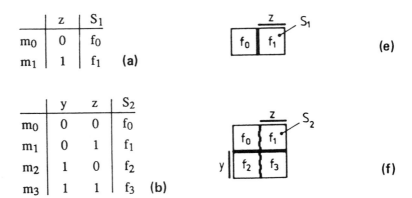

Figure 3.1 *(continued on next page).*

Figure 3.1 (continued).

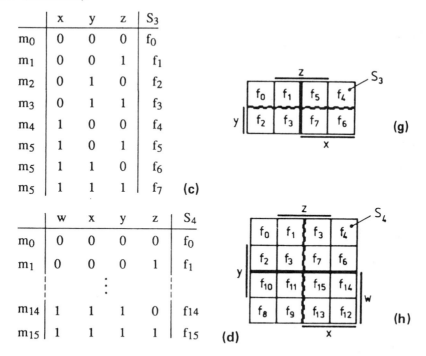

	x	y	z	S_3
m_0	0	0	0	f_0
m_1	0	0	1	f_1
m_2	0	1	0	f_2
m_3	0	1	1	f_3
m_4	1	0	0	f_4
m_5	1	0	1	f_5
m_5	1	1	0	f_6
m_5	1	1	1	f_7	(c)

	w	x	y	z	S_4
m_0	0	0	0	0	f_0
m_1	0	0	0	1	f_1
	⋮				
m_{14}	1	1	1	0	f_{14}
m_{15}	1	1	1	1	f_{15}	(d)

Figure 3.1 Truth tables and Karnaugh maps.

The minterm form of a function in two variables y and z, for example, is

$$S = m_0 f_0 + m_1 f_1 + m_2 f_2 + m_3 f_3$$
$$= \bar{y}\bar{z}f_0 + \bar{y}zf_1 + y\bar{z}f_2 + yzf_3.$$

The Karnaugh map for functions in two variables consists of four cells. The placement of the function value f_2 is as shown in Figure 3.2.

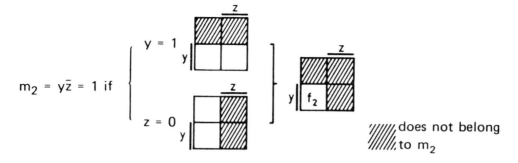

$m_2 = y\bar{z} = 1$ if $\begin{cases} y = 1 \\ z = 0 \end{cases}$

▨ does not belong to m_2

Figure 3.2 The placement of function values in a Karnaugh map.

The construction of Karnaugh maps

A Karnaugh map for functions in n variables is formed from the Karnaugh map of a function in n-1 variables by means of *reflection* around one of the edges. The bold lines in Figure 3.1 represent the reflection lines. Alternately, one reflects down and to the right. The already assigned labels are reflected as well. The new part of the map is labelled with the next variable. Constructed in this way a map for functions in n variables has exactly 2^n cells, enough to be able to note all 2^n function values.

When we observe how the function values have been placed in the Karnaugh map, there seems to be little regularity. Nevertheless there is. When, for example, in Figure 3.1.h the variables w through z have the weights 8, 4, 2 and 1 the *number* i of the function value f_i that corresponds to minterm m_i follows directly. This is based on the construction of Karnaugh maps by means of reflection.

3.2 Karnaugh maps and the reduction of logic functions

Karnaugh maps are a *visual tool* for the reduction of logic functions. To begin with, the concept 'minimal form' of a function has not been defined exactly. In general it is that form of a function that best suits the available components. It is the form that results in the most economic realisation of the circuit. Because the components have not been introduced, for the time being we will understand by the term *minimal form* of a function a *sum-of-products* with

- a minimal number of product terms;
- a minimal number of variables per product.

This form, as will be shown, does not always correspond to the minimal circuit.

The minimal form as *sum-of-products* can be easily read from the Karnaugh map. To get some insight we determine the simplest sum form of

$$S = m_1 + m_2 + m_3$$

of a function in two variables.

$$S = \bar{y}z + y\bar{z} + yz$$
$$= (\bar{y}z + yz) + (y\bar{z} + yz)$$
$$= (\bar{y} + y)z + y(\bar{z} + z)$$
$$= z + y$$
$$= y + z.$$

The most important laws used are

$$\bar{z} + z = 1$$

and

$$z + z = z.$$

The first law states that in the reduction of logic functions one can combine two product terms, which have all variables in common except one, and replace them by their common part.

The second law states that in a reduction process a term may be used more than once.

Figure 3.3 shows the function S

$$S = m_1 + m_2 + m_3 \Rightarrow f_0 = 0 \text{ and } f_1 = f_2 = f_3 = 1$$

in a Karnaugh map for two variables. In the construction of the map, cells of minterms which differ in one variable (and are thus logically adjacent) are also *adjacent* in the map. The reduction of the function S found previously can be recognised visually immediately.

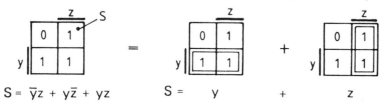

Figure 3.3 Reduction with the Karnaugh map.

Adjacent cells

The concept of *adjacent cells* must not be taken too literally in Karnaugh maps. Minterms m_i and m_j with function values $f_i = f_j = 1$, which correspond to cells which *lie symmetrically around one of the reflection axes*, are also logically adjacent. They can be combined in a product term in $n - 1$ variables (n is the number of variables of the map). See Figure 3.4 for some examples.

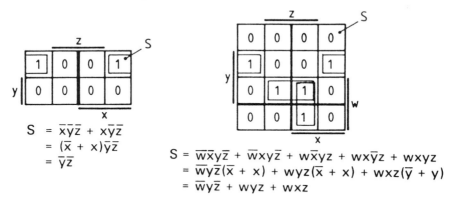

Figure 3.4 Adjacent cells in Karnaugh maps.

Four different minterms with $f_i = 1$ can be reduced to one product term in $n - 2$ variables if these minterms mutually differ in at most two variables. This condition is necessary, but also sufficient.

Example

$$S = wx\bar{y}\bar{z} + wx\bar{y}z + wxy\bar{z} + wxyz$$
$$= wx\bar{y}(\bar{z} + z) + wxy(\bar{z} + z)$$

$$= wx\bar{y} + wxy$$
$$= wx(\bar{y} + y)$$
$$= wx.$$

The Karnaugh maps of Figure 3.5 show three groups of four minterms, each with function value 1, which can be combined in product terms in $n - 2 = 2$ variables.

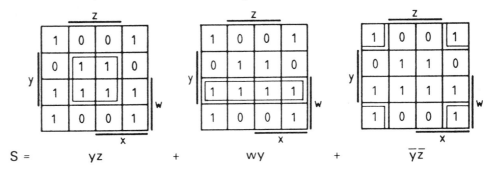

$$S = \quad yz \quad + \quad wy \quad + \quad \bar{y}\bar{z}$$

Figure 3.5 Karnaugh maps.

The minimal sum form (sum-of-products form) of the function specified in Figure 3.5 is

$$S = yz + wy + \bar{y}\bar{z}.$$

It is also possible to write this function as

$$S = yz + w\bar{z} + \bar{y}\bar{z}.$$

The corresponding product terms are indicated in the map of Figure 3.6. Apparently it is possible for functions with two or more 'minimal forms' to exist. □

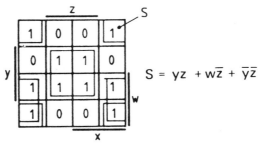

$$S = yz + w\bar{z} + \bar{y}\bar{z}$$

Figure 3.6 Karnaugh map.

Sets of 2^k minterms, which can be combined into one product term, are always situated *symmetrically* in Karnaugh maps with regard to the *reflection lines*. Such a group can be recognised easily. In general:

2^k different minterms can be combined into one product term in $n - k$ variables if and only if these 2^k minterms mutually differ in at most k variables. The remaining variables are similar in all minterms, i.e. normal or negated. When this condition is fulfilled the set of 2^k minterms is said to be logically adjacent.

Example

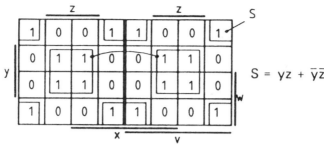

$S = yz + \bar{y}\bar{z}$

Figure 3.7 Karnaugh map.

3.3 Applications of Karnaugh maps

Karnaugh maps are a visual tool for the reduction of logic functions. The first example illustrates the reduction of a function. Besides this, there are some other applications of Karnaugh maps.

Example
Reduce the logic function

$$S = \bar{w}x + wx\bar{z} + wy\bar{z} + xy\bar{z} + \bar{w}\bar{x}yz + w\bar{x}yz$$

in four variables w, x, y and z.

It is possible to expand this function to the minterm form and then place the function values (0 or 1) in the Karnaugh map. With a little experience this (intermediate) step can be skipped and one can immediately place the function values corresponding to each product term in the map. For the term wx, for example, this is done as follows:

1. The term $\bar{w}x$ contains only minterms for which w = 0. The cells in the bottom half of the map in Figure 3.8.a are excluded.

2. Further, the term $\bar{w}x$ contains only minterms for which x = 1. The left-hand half of the map is excluded as well.

There is a part of the map that meets both requirements, viz. the four remaining cells in the upper right-hand corner of Figure 3.8.a. Similarly the ones corresponding to the remaining terms of the given function can be placed in Figure 3.8.a (compare Figure 3.2). Subsequently, the terms which correspond to the most reduced form of the given function are encircled in Figure 3.8.b. This form is:

$$S = \bar{w}x + x\bar{z} + \bar{w}yz + w\bar{x}y. \quad \square$$

Note
For the sake of completeness we remark that the order of the labels in a Karnaugh map can be chosen at random. With another order the position of the function values changes; this has no influence on the (minimal) form of the formula. $\square$

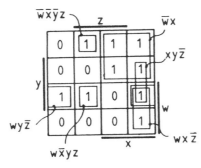
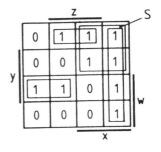

(a) Placement of function values.

(b) Cover of the minimal sum-of-products form.

Figure 3.8 Example of a reduction using the Karnaugh map.

Multi-output logic circuits

Often more than one logic function is defined on the same set of logic variables. During the realisation of a circuit several components can be used in different output functions. This may lead to a reduction in the number of components.

The functions S_1 and S_2 in Figure 3.9 are both defined on the same set of input variables $\{w, x, y, z\}$. The minimal sum-of-products forms of these functions are

$$S_1 = \overline{w}x + \overline{w}y + x\overline{y}\overline{z} + xyz + w\overline{x}y\overline{z}$$

$$S_2 = \overline{w}\overline{x}\overline{y} + y\overline{z} + wx\overline{z}.$$

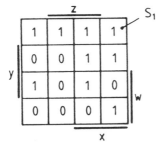
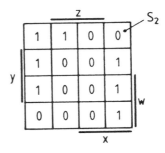

Figure 3.9 Specification of a circuit with two outputs S_1 and S_2.

Both functions can also be written as

$$S_1 = \underline{\overline{w}\overline{x}\overline{y}} + \underline{wx\overline{y}\overline{z}} + w\overline{x}y\overline{z} + \overline{w}x + xyz$$

$$S_2 = \underline{\overline{w}\overline{x}\overline{y}} + \underline{wx\overline{y}\overline{z}} + y\overline{z}.$$

Now the functions have some terms in common. These terms are underlined. The realisation of a circuit should therefore not automatically be based on the minimal form of the formula. The available components are an important factor in this selection process.

How to compare functions?

Two logic functions are identical if for each minterm the same function value has been specified. To test whether functions are identical one may expand them to the minterm form. After both sets have been sorted and compared the answer is known. If one compares the given functions in Karnaugh maps, it is no longer necessary to expand to the minterm form.

Example

S_1 and S_2 are given:

$$S_1 = \bar{w}\bar{x} + y\bar{z} + \bar{y}z$$

$$S_2 = w\bar{y}z + \bar{w}\bar{x}z + \bar{w}\bar{x}\bar{y}\bar{z} + xy\bar{z} + x\bar{y}z + \bar{x}y\bar{z}.$$

The Karnaugh maps are shown in Figure 3.10. The functions turn out to be identical. □

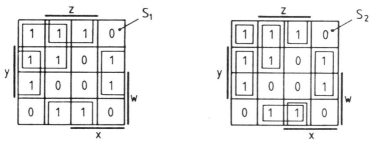

Figure 3.10 Karnaugh maps.

Logic operations on functions

In many applications the sum of, the product of or the difference between two functions has to be determined. Doing this with the laws of Boolean algebra can be a lot of work. Karnaugh maps make it easy.

Example

Two functions S_1 and S_2 in three variables are given. Figure 3.11 illustrates how their sum is taken. The other operations conform to this example. □

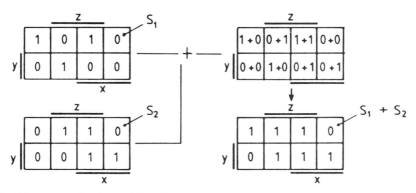

Figure 3.11 The determination of $S_1 + S_2$.

Karnaugh maps must be labelled identically when they are used for logic operations on functions, meaning they must have the same number of variables, and be labelled in the same order.

Completeness of specifications

For many problems the function of the circuit is specified in such terms as

'if condition 1 is true action 1 has to be executed',

'if condition 2 is true action 2 has to be executed',

etc. It must then be checked whether such a specification is *complete* and/or *unambiguous*. Karnaugh maps may simplify this verification.

Example
When it has been specified that

condition $\overline{w}\overline{x}$ true → action 1,

condition $\overline{w}yz + x\overline{y}$ true → action 2,

condition $wxy + w\overline{x}\overline{y}$ true → action 3,

condition $w\overline{x}yz + \overline{w}xy\overline{z}$ true → action 4,

the placement of the conditions in a Karnaugh map shows that two conditions coincide partly (for $\overline{w}\overline{x}yz$) and that for $w\overline{x}y\overline{z}$ no operation has been specified. See Figure 3.12. □

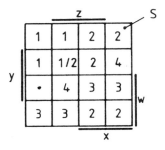

Figure 3.12 Karnaugh map.

3.4 The sum and product forms of functions

So far, we have only discussed how the minimal form of a logic function can be determined as a sum-of-products. This form of a logic function will henceforth be referred to as the *sum form*. For some applications the representation as a product-of-sums, called the *product form*, is simpler.

When the negation of the function S,

$$S = w\overline{y}z + wy\overline{z} + xyz + x\overline{y}\overline{z},$$

is determined by applying DeMorgan's negation laws, the negated function is obtained in the product form:

$$\overline{S} = \overline{(w\overline{y}z + wy\overline{z} + xyz + x\overline{y}\overline{z})}$$
$$= \overline{w\overline{y}z} \cdot \overline{wy\overline{z}} \cdot \overline{xyz} \cdot \overline{x\overline{y}\overline{z}}$$
$$= (\overline{w} + y + \overline{z})(\overline{w} + \overline{y} + z)(\overline{x} + \overline{y} + \overline{z})(\overline{x} + y + z).$$

This conclusion can be used to convert a function S, specified in the sum form, into the product form. The logic function S must then be placed in a Karnaugh map, in which one combines the zeros instead of the ones. This results in the minimal sum form of the function $\overline{S}$, while negation of $\overline{S}$ results in the minimal product form of the function S.

Example
Determine the minimal product form of the function

$$S = yz + wxy + w\overline{x}z + \overline{w}xz.$$

The minimal sum form of the function S follows from Figure 3.13:

$$\overline{S} = \overline{w}\overline{z} + \overline{x}\overline{z} + \overline{w}x\overline{y} + wx\overline{y}.$$

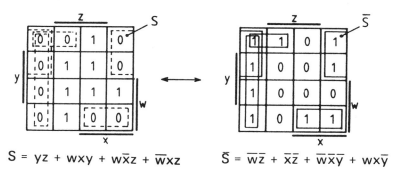

$$S = yz + wxy + w\overline{x}z + \overline{w}xz \qquad \overline{S} = \overline{w}\overline{z} + \overline{x}\overline{z} + \overline{w}x\overline{y} + wx\overline{y}$$

Figure 3.13 Conversion of the sum form into the product form.

The product form of the function S follows from negation:

$$S = (w + z)(x + z)(w + x + y)(\overline{w} + \overline{x} + y).$$

This product form has one letter less than the equivalent sum form. Whether this is important for the realisation of the circuit also depends on the available components. We will go into these aspects in more detail in Chapter 5. □

3.5 Incompletely specified functions

Often not all function values of a logic function have been specified. Specification is, for example, not necessary when certain input combinations do not occur. In decimal counters, for example, the content 9 must be detected in every section. Depending on the previous section, a carry must be forwarded to the next section if the content is 9. If so, this section gets, after the next count step, the content 0. (For details about the structure of the carry mechanism see Chapter 13.) Figure 3.14 specifies the carry mechanism for a BCD counter.

In a BCD counter section the carry has the value 0 (no transport) for the combinations wxyz = 0000 through wxyz = 1000. With wxyz = 1001 (content 9) the carry must be 1.

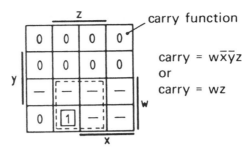

Figure 3.14 Carry mechanism of a 4-bit BCD counter section.

The other six input combinations are not used in the BCD code. For these combinations it is not necessary to specify function values. A choice of 0 or 1 is usually denoted by a dash, which means *don't care*.

In Figure 3.14 we see that, depending on how it is decided to choose the don't cares, the formula of the carry is

$$S = w\bar{x}\bar{y}z$$

(all don't cares are 0) or

$$S = wz$$

(three don't cares are chosen to be 1, three 0). The latter formulation is simpler. Don't cares can often be used to reduce logic functions.

Other examples of don't cares are found, for example, in the selection switches of radios and television sets. Here a mechanical or electrical lock sees to it that not all positions of the switches are possible.

Example

A don't care is sometimes hard to recognise. In robotics, for example, often a part of the system must react after receiving a specific row of input symbols. For example, a certain circuit must be triggered after receiving a row of three symbols, 010. (How this row of symbols is received is left out of consideration for the time being.)
If it is also given that in each row of three symbols only one 1 occurs the number of different input rows is reduced from eight to three. This is a source of don't care conditions. It is more difficult to see, under the given restriction of the number of possible input rows of symbols to 100, 010 and 001, that for detection it is important to see whether the second symbol is a 1. The first and third symbols are in this case not relevant for the decision procedure! □

Don't cares in a specification may have several sources:

– input combinations that do not occur, so that the corresponding output values can be specified at random;

– components or subsystems that are not necessary or are not used during certain process steps.

The first category specifies *don't happen conditions*, and the second *don't use conditions*.

Non-equivalent forms of logic functions

Freedom in a specification (in the form of don't cares) can be used in different ways. This results in different realisations, each with its own truth table. Of course, the differences are only related to the don't care conditions.

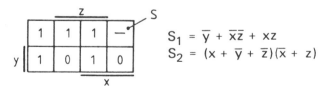

$$S_1 = \bar{y} + \bar{x}\bar{z} + xz$$
$$S_2 = (x + \bar{y} + \bar{z})(\bar{x} + z)$$

Figure 3.15 Karnaugh map.

In Figure 3.15, for example, one function value has been specified as a don't care. In determining the minimal sum-of-products form this don't care is chosen as 1. For the product-of-sums form 0 is preferred.

In fact, we touch here upon a recurring problem in the design verification process of circuits. When it is said that two circuits 'can do the same', this statement has to be checked with the *original specification*. Circuits can be compatible with regard to their *external behaviour*, as far as specified, and still have logic formulas which are *non-equivalent*.

3.6 A systematic procedure to find a cover in Karnaugh maps

In Karnaugh maps one tends to start to encircle areas as large as possible. This is not correct.

Example

The pattern of zeros and ones in Figure 3.16 is known as 'the sails of a windmill'. When one starts to encircle the largest term yz and subsequently the terms of $\bar{w}xz$, $\bar{w}\bar{x}y$, $w\bar{x}z$ and wxy it turns out that the minterms of the term yz are contained twice. The term yz is superfluous. □

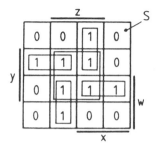

Figure 3.16 Karnaugh map.

At this point we define some concepts.

Minterm *(minimal product term)*
A *minterm* is a product term in which all variables occur once. When a minterm belongs to the function, the function value 1 is placed in the corresponding cell of the Karnaugh map.

Implicant *(product term)*
An *implicant* is a product of one or more variables that belongs to the function. In every corresponding cell the function value is 1.

Prime implicant *(prime term)*
A *prime implicant* is an implicant of the function which cannot be combined with other implicants into a product term in fewer variables.

Essential prime implicant
An *essential prime implicant* is a prime implicant that contains at least one minterm that is not covered by any other prime implicant. An essential prime implicant belongs *necessarily* to the minimal sum-of-products form of the function.

These definitions are valid for completely specified functions. For incompletely specified functions one usually takes the *don't cares* into consideration as well. They can increase the number of minterms covered by a prime implicant. Of course, every prime implicant then contains at least one minterm with a function value of 1. Don't cares that do not contribute to prime implicants are chosen as 0.

The *set of prime implicants* forms the basis for the determination of the minimal sum-of-products form. Each essential prime implicant belongs to it. If not all 1s in the map are covered by the essential prime implicants, a selection must be made from the remaining set of prime implicants.

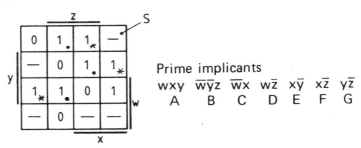

Figure 3.17 Karnaugh map with the set of prime implicants.

Example *Prime implicants*
The set of prime implicants of Figure 3.17 consists of the product terms A through G. Three 1s in the map are covered by only one prime implicant. These are indicated by a •. So the prime implicants A, B and C are essential prime implicants. They belong necessarily to the minimal sum form of the formula.

Because A, B and C are essential prime implicants the 1s indicated by * are also covered. One 1 has not yet been covered, the minterm $wxy\bar{z}$. For this 1 it is possible to choose from the prime implicants $D = w\bar{z}$, $F = x\bar{z}$ or $G = y\bar{z}$. These implicants have the same number of variables, so that three minimal sum-of-products forms are found:

$S = A + B + C + D$

$S = A + B + C + F$

$S = A + B + C + G$. □

Roughly speaking the following order must be taken into account when using Karnaugh maps.

- First look for those ones (1s) which cannot be combined with other ones (1s) or which can only be combined in one way. (These are the essential prime implicants.)
- For the remaining ones: take that prime implicant that is as large as possible and which also contains as many as possible of the ones (1s) still to be covered.

Often several solutions are possible. How these can be systematically generated is discussed in Section 3.7.

The set of prime implicants can be determined visually. A systematic method for this is dealt with in Section 3.8.

3.7 The formulation of covering problems

The problem discussed in the previous section is a special case of a category of covering problems that one encounters in logic circuit design. The problem is usually formulated as follows.

A set V is given:

$$V = \{v_1, v_2, ..., v_{n-1}, v_n\}.$$

A set W of subsets of V is also given:

$$W = \{w_1, w_2, ..., w_{k-1}, w_k\}.$$

Its elements are *non-disjunct subsets* of V. Depending on the application several covers of V must be determined from W.

- Determine a minimal subset of W, which at least contains all elements of V once. This subset of W is called a *minimal cover* of V.
- Each element of W has a price in accordance with a certain *cost function*. Determine from W a cover of V with the *lowest total costs*.

When using Karnaugh maps, V is the set of minterms m_i for which $f_i = 1$ and W is the set of prime implicants. The cost function is the number of variables per prime implicant.

How the elements of V are contained in the subsets of W is often specified in a *prime implicant table*, of which Table 3.1 is an example. This table shows how the minterms belonging to the function are covered by the prime implicants.

From Table 3.1 it follows that:

- minterm m_1 is only covered by prime implicant B. B is necessarily part of the selection.
- minterm m_5 can be covered by B, by C or by E.

- minterm m_6 can be covered by C, by F or by G.

⋮

- minterm m_{14} can be covered by D, by F or by G.

minterms		m_1	m_5	m_6	m_7	m_{10}	m_{11}	m_{14}
prime impl.		$\bar{w}\bar{x}\bar{y}z$	$\bar{w}x\bar{y}z$	$\bar{w}xy\bar{z}$	$\bar{w}xyz$	$w\bar{x}y\bar{z}$	$w\bar{x}yz$	$wxy\bar{z}$
A	$w\bar{x}y$	.	.	.	.	×	×	.
B	$\bar{w}\bar{y}z$	×	×	.	.	.	.	.
C	$\bar{w}x$	.	×	×	×	.	.	.
D	$w\bar{z}$	.	.	.	.	×	.	×
E	$x\bar{y}$	.	×	.	.	.	.	.
F	$x\bar{z}$	.	.	×	.	.	.	×
G	$y\bar{z}$	.	.	×	.	×	.	×

Table 3.1 Prime implicant table.

We can express these conditions in a logic function:

$$S = B \cdot (B + C + E) \cdot (C + F + G) \cdot C \cdot (A + D + G) \cdot A \cdot (D + F + G).$$

The logic variables A through G have the value 1 (True) if the corresponding prime implicant belongs to the selected cover. When $S = 1$ for certain values of A through G, a cover has been found. This is an efficient way to describe the covering problem.

The laws of switching algebra apply to the logic function defined. In particular, the application of the law $z(y + z) = z$ allows several factors in the formula to be omitted.

$$S = B \cdot (B + C + E) \cdot (C + F + G) \cdot C \cdot (A + D + G) \cdot A \cdot (D + F + G)$$

$$= B \cdot C \cdot A \cdot (D + F + G)$$

$$= A \cdot B \cdot C \cdot D + A \cdot B \cdot C \cdot F + A \cdot B \cdot C \cdot G.$$

This form cannot be reduced. Sum S is 1 if at least one of the terms is 1. A minimal cover has been found when

$A = B = C = D = 1 \rightarrow$ the cover consists of the prime implicants $w\bar{x}y$, $\bar{w}\bar{y}z$, $\bar{w}x$ and $w\bar{z}$;

$A = B = C = F = 1 \rightarrow$ the cover consists of the prime implicants $w\bar{x}y$, $\bar{w}\bar{y}z$, $\bar{w}x$ and $x\bar{z}$;

$A = B = C = G = 1 \rightarrow$ the cover consists of the prime implicants $w\bar{x}y$, $\bar{w}\bar{y}z$, $\bar{w}x$ and $y\bar{z}$.

The idea of formulating covering problems by logic functions stems from Petrick. Such functions are therefore called *Petrick functions*.

Strictly speaking, we must formulate the Petrick function for the entire prime implicant table. Yet a limited Petrick function will suffice. We can see directly that some prime implicants have to belong to the cover. These are the *essential prime implicants*.

In the table the *essential prime elements* of W are easy to recognise. If there is only one cross in a column we are dealing with an essential prime element.

An essential prime implicant means that all of its minterms are automatically included. Table 3.2 illustrates the consequences of the essential prime implicants.

	m_1	m_5	m_6	m_7	m_{10}	m_{11}	m_{14}
A	.	.	.	.	×	⊗	.
B	⊗	×	.	.	.	.	.
C	.	×	×	⊗	.	.	.
D	.	.	.	.	×	.	×
E	.	×	.	.	.	.	.
F	.	.	×	.	.	.	×
G	.	.	×	.	×	.	×

⊗ necessary
× consequence of a choice

Table 3.2 Prime implicant table with implications of the essential prime implicants.

A Petrick function which includes those elements of V not yet covered must be specified next. For Table 3.2 this function is

$$S = A \cdot B \cdot C \cdot (D + F + G).$$

It is therefore useful to solve a covering problem in two steps:

– first find all essential prime implicants;

– then try to cover the remaining minterms as favourably as possible.

Another aspect of the covering problem is that some rows of a prime implicant table are *dominant* over others. An example is given in Table 3.3. In this table the row after w_i is dominant over the row after w_j.

	v_1	v_2	v_3	v_4	v_5	v_6
wi	×	×	.	×	×	×
wj	.	×	.	×	.	.

Table 3.3 Prime implicant table.

In the determination of a minimal cover the row w_j can be left out of consideration, at least when the value of the cost function of m_i is smaller than or equal to that of m_j. It is therefore advisable to check whether there are any dominant rows in a table (after the essential elements from W have been determined). Whether a row can be left out of consideration then depends on the cost function.

Example
The function

$$S_1 = 0, 6, 7, 8, 13, 14$$
$$S_d = 1, 3, 5$$

is given, in which the numbers of minterms for which $f_i = 1$ are stated after S_1 and the numbers of minterms for which f_i is not specified after S_d. Table 3.4 is the prime implicant table of this function. See also Figure 3.18.

	m_0	m_6	m_7	m_8	m_{13}	m_{14}
$\bar{x}\bar{y}\bar{z}$ A	(×)	.	.	⊗	.	.
$\bar{w}\bar{x}\bar{y}$ B	×	.	.	.	.	.
$\bar{w}z$ C	.	.	×	.	.	.
$\bar{w}xy$ D	.	×	×	.	.	.
$xy\bar{z}$ E	.	(×)	.	.	.	⊗
$x\bar{y}z$ F	.	.	.	.	⊗	.

Table 3.4 Prime implicant table.

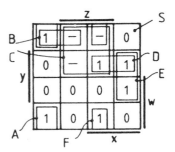

Figure 3.18 Karnaugh map.

In Table 3.4 row A is dominant over row B and row D is dominant over row C. Moreover, A, E and F are essential prime implicants. So when the realisation costs of the given prime implicants are equal, the rows B and C can be omitted. However, prime implicant C contains one variable less than prime implicant D. In gate circuits (see Chapter 5) a realisation of D requires more inputs than a realisation of C. This will sometimes lead to a reduction in costs. If in a realisation a term in two variables costs k_1 cost units and a term in three or four variables k_2 cost units, with $k_2 > k_1$, then row B may be left out of consideration, but not row C.

For table 3.4 the following Petrick function can be formed:

$$S = (A + B) \cdot (D + E) \cdot (C + D) \cdot A \cdot F \cdot E$$
$$= A \cdot E \cdot F \cdot (C + D)$$
$$= A \cdot E \cdot F \cdot C + A \cdot E \cdot F \cdot D.$$

Because the prime implicant C is simpler than D the first term is preferable. The cover then consists of the prime implicants:

$\bar{x}\bar{y}\bar{z}$, $xy\bar{z}$, $x\bar{y}z$ and $\bar{w}z$. □

The conclusion of this example is that realisation aspects, such as cost functions, can be included, in principle, in the covering algorithm.

3.8 The Quine-McCluskey algorithm

Karnaugh maps can be used for the reduction of logic functions of up to six variables. With more variables the problem must be handled in a more systematic way. The tabular method developed by Quine-McCluskey is suitable for computer calculations. Actually, there are no differences between a Karnaugh map or the Quine-McCluskey method.
The reduction of logic functions is done in two steps.

1. Determine the set of prime implicants.

2. Select a (minimal) subset of it, in which all minterms m_i with $f_i = 1$ are contained at least once.

As an example the set of prime implicants of the function specified in Figure 3.19 is determined. For this function the following is valid:

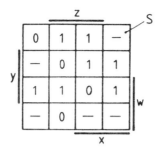

Figure 3.19 Karnaugh map.

$S_1 = 1, 5, 6, 7, 10, 11, 14;$

$S_d = 2, 4, 8, 12, 13.$

These sets of minterms form the base of Table 3.5, in which d denotes that the minterm originates from S_d, i.e. from the unspecified part of the function. The minterms in Table 3.5 are subsequently divided into three groups. Minterms from group I can be combined with minterms from group II only and minterms from group III with those from group II only. The criterion for this division is the number of negated variables in a (min)term. For product terms a *ternary* notation is introduced:

a. When a variable is used normally in a product this is indicated by 1.

b. When a variable is used negated in a product this is indicated by 0.

c. When a variable is absent in a product this is indicated by a dash.

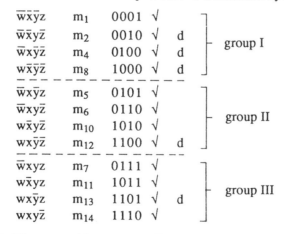

Table 3.5 Minterms with corresponding ternary notation.

Minterm m_1 from group I can be combined with minterm m_5 from group II into the implicant $\bar{w}\bar{y}z$, which is represented in the ternary notation as 0–01. In Table 3.5 the checks √ after m_1 and m_5 indicate that they are covered by a larger implicant. (The concept 'larger' is to be interpreted as containing more minterms.) Table 3.6 shows the result of combining all possible combinations of minterms from group I with those of group II and of minterms from group II with those from group III. It turns out that all the minterms from Table 3.5 are contained at least once in one of the implicants in Table 3.6. In this table the implicants are arranged according to the absence of variables. This decreases the amount of work in further attempts to combine implicants.

$\bar{x}yz$	m_2 +	m_{10}	–010 √		
$\bar{x}y\bar{z}$	m_4 +	m_{12}	–100 √	d	I
$\bar{x}\bar{y}z$	m_5 +	m_{13}	–101 √		
$\bar{x}y\bar{z}$	m_6 +	m_{14}	–110 √		
$\bar{w}\bar{y}z$	m_1 +	m_5	0–01		
$\bar{w}y\bar{z}$	m_2 +	m_6	0–10 √		II
$w\bar{y}\bar{z}$	m_8 +	m_{12}	1–00 √	d	
$wy\bar{z}$	m_{10} +	m_{14}	1–10 √		
$\bar{w}x\bar{z}$	m_4 +	m_6	01–0 √		
$\bar{w}xz$	m_5 +	m_7	01–1 √		III
$w\bar{x}\bar{z}$	m_8 +	m_{10}	10–0 √		
$wx\bar{z}$	m_{12} +	m_{14}	11–0 √		
$\bar{w}x\bar{y}$	m_4 +	m_5	010– √		
$\bar{w}xy$	m_6 +	m_7	011– √		IV
$w\bar{x}y$	m_{10} +	m_{11}	101–		
$wx\bar{y}$	m_{12} +	m_{13}	110– √	d	

Table 3.6 Table with implicants.

Subsequently, an attempt is made to combine the implicants from Table 3.6. It will be understood that this is only possible within groups I, II, III and IV in Table 3.6, i.e. two implicants can only be combined when the missing variable is the same in both. Table 3.7 shows the result. Notice that each of the implicants in Table 3.7 is generated twice!

It turns out that none of the implicants from Table 3.7 can be combined with any of the others. All implicants from this table prove to be prime implicants. The process described above will therefore be stopped.

$--y\bar{z}$	$m_2 + m_6 + m_{10} + m_{14}$	––10
$\bar{w}x--$	$m_4 + m_5 + m_6 + m_7$	01––
$-x\bar{y}-$	$m_4 + m_5 + m_{12} + m_{13}$	–10–
$-x-\bar{z}$	$m_4 + m_6 + m_{12} + m_{14}$	–1–0
$w--\bar{z}$	$m_8 + m_{10} + m_{12} + m_{14}$	1––0

Table 3.7 Table with implicants.

Summary

The generation of prime implicants is done as follows.

1. Start with the set of minterms for which the function value $f_i = 1$ or for which f_i is not specified. Mark the latter with a d. Make a table of all implicants formed by combining two minterms. Put a d by an implicant when the function value of *both* minterms is not specified.

2. Combine the implicants containing two minterms as much as possible into implicants containing four minterms. Mark these implicants with a d when *all* minterms covered by it have a don't care function value.

3. Repeat the process k times until no more implicants (each containing 2^k minterms) can be combined with others into new implicants.

During steps 1 through 3 all implicants that are contained at least once in a larger implicant are marked with a √. All implicants that are not marked with √ cannot be combined with others. They are prime implicants.

In the above example the following prime implicants are found:

$\bar{w}\bar{y}z$ and $w\bar{x}y$ (from Table 3.6)

and

$\bar{w}x$, $w\bar{z}$, $x\bar{y}$, $x\bar{z}$ and $y\bar{z}$.   (from Table 3.7).

When prime implicants which contain only minterms with unspecified function values are generated, they can be omitted.

Subsequently, a (minimal) subset must be selected from the set of prime implicants, such that the function is covered. This problem has already been discussed in the previous section.

The above method for generating prime implicants roughly describes the method for reducing logic functions developed by Quine-McCluskey. We have not aimed at completeness. There is, for example, the question of whether it is necessary to expand a function into the minterm form first before the prime implicants can be determined systematically. See, for example, Tison.

An adapted version of the Quine-McCluskey method can be developed for circuits with more than one output.

References

1. T.C. Bartee, *Computer Design of Multiple-Output logical Networks*, IRE Trans. Electron. Comput., Vol. EC-10, 1961, pp. 21–30.
2. R.K. Brayton, G.D. Hachtel, C.T. McMullen and A.L. Sangiovanni-Vincentelli, *Logic Minimization Algorithms for VLSI Synthesis*, Kluwer Academic, Boston, 1984.
3. S.J. Hong, R.G. Cain and D.L. Ostapko, *MINI: A Heuristic Approach for Logic Minimization*, IBM J. Res. Dev., Sept. 1974, pp. 443–458.
4. M. Karnaugh, *The Map Method for Synthesis of Combinational Logic Circuits*, Trans. AIEE, Vol. 72, part I, 1953, pp. 593–598.
5. E.J. McCluskey, *Minimization of Boolean Functions*, Bell Syst. Tech. J., Vol. 35, no. 5, 1956, pp. 1417–1444.
6. E.J. McCluskey and H. Schorr, *Essential Multiple-Output Prime Implicants, Mathematical Theory of Automata*, Proc. Polytechnic Inst. Brooklyn Symp., Vol. 12, April 1962, pp. 437–457.
7. S.R. Petrick, *A Direct Determination of the Irredundant Forms of a Boolean Function from the Set of Prime Implicants*, Computation Laboratory of the AFCRC, Bedford, Mass., TR-56-110, April 1956.
8. W.V. Quine, *The Problem of Simplifying Truth Functions*, Am. Math. Monthly, Vol. 59, no. 8, 1952, pp. 521–531.

9. D.C. Schmidt and L.E. Druffel, *An Extension of the Clause Table Approach to Multiple-Output Combinational Switching Networks*, IEEE Trans. on Computers, Vol. C-23, no. 4, April 1974, pp. 338–346.
10. J.R. Slagle, Chin-liang Chang and R.C. Lee, *A New Algorithm for Generating Prime Implicants*, IEEE Trans. on Computers, Vol. C-19, no. 4, April 1970, pp. 304–310.
11. P. Tison, *Generalization of Consensus Theory and Application to the Minimization of Boolean Functions*, IEEE Trans. on Electronic Computers, Vol. EC-16, no. 4, Aug. 1967, pp. 446–456.
12. E.W. Veitch, *A Chart Method for Simplifying Truth Functions*, Proc. ACM, Pittsburgh, May 1952, pp. 127–133.
13. H.A. Vink, B. van den Dolder and J. Al, *Reduction of CC-tables Using Multiple Implication*, IEEE Trans. on Computers, Vol. C-27, no. 10, Oct. 1978, pp. 961–966.

4
Branch type combinational circuits

4.1 Contacts and relays

Digital circuits are categorised into combinational and sequential circuits. While combinational circuits have no memory, sequential circuits do. In a *combinational circuit* the present input combination determines the output value(s), at least in the static state when possible transient phenomena have faded away. According to their internal structure combinational circuits can be classified into two groups:

– circuits of the branch type;

– circuits of the gate type.

In circuits of the *branch type* the information is transferred by means of the conductivity of a network between the inputs and outputs. Figure 4.1 explains this. The condition under which a relay, for example, must be activated or a lamp must be switched on is formulated by a logic formula. When it has the value of T (True) the network which realises the formula will, under certain conditions, conduct.

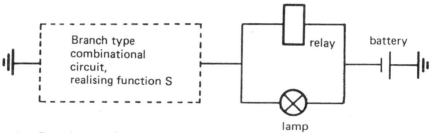

Figure 4.1 Branch type circuit.

In a gate type circuit the information is represented by a 'High' or a 'Low' voltage level at the inputs and outputs of the circuit. We will discuss this type of circuit in Chapter 5.

Until about 1950-1960 logic circuits were realised by *electromagnetic relays, relay contacts and switches*. The electromechanical components were then replaced by electronic devices, first by vacuum tubes and later by transistors. Since 1970 almost nothing but semiconductor integrated circuits are in use, although relays and contacts are still found in specific applications, especially those in which high currents and voltages have to be switched.

Relays

Figure 4.2 shows the structure of an *electromagnetic relay*. When the coil is energised a magnetic field is created in the electromagnet. This electromagnetic field displaces the armature, which opens or closes the contacts. When the current through the coil is broken the relay is deactivated, causing the contacts to spring back.

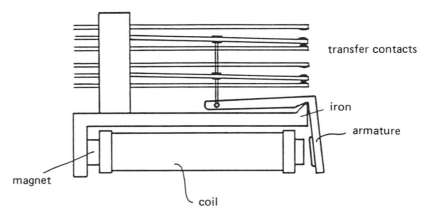

Figure 4.2 Relay with two transfer contacts.

Modern relays are often constructed as *reed relays* (Figure 4.3). The contact is mounted in a sealed glass tube, which is filled with an inert gas. The contacts can then no longer be corroded through oxidation by air. A coil is wound around the tube. Activation of the coil produces a magnetic field, which makes the contact springs attract each other, which closes or breaks the contact. Reed relays are maintenance-free and are capable of making a large number of connections or circuit-breaks.

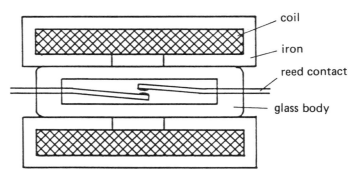

Figure 4.3 Reed relay.

Figure 4.4 is a by no means exhaustive list of some of the symbols used in drawing relay circuits.

Application

The next example describes a relay application. The operating instructions of several types of equipment require the hands of the operator to be out of the way when engines, hydraulic presses, knives, etc. are in motion. This is accomplished by mounting two push-buttons 'Left' and 'Right', which have to be pushed simultaneously, as shown in Figure 4.5.

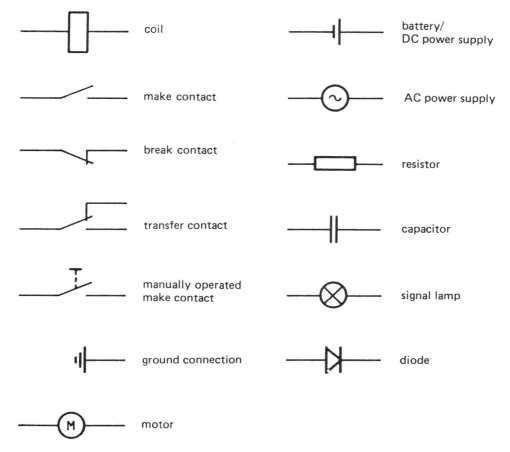

Figure 4.4 Symbols for electromechanical circuits.

The principle of this protection circuit is shown in Figure 4.5.a. The setup of Figure 4.5.b is found more often. The relay Z *separates* the operator circuit (low voltages) from the power circuit of the motor (high voltages 220/380V).

In this application the protection is effected by the *interconnection* of two contacts L and R in series. This series interconnection realises the AND function between the propositions 'Left pushed' and 'Right pushed'. When both are true the series interconnection conducts. See also Chapter 2.

A *parallel interconnection* of contacts realises the OR function. With a series/parallel interconnection of contacts it is possible to realise a random logic function, describing operating conditions of relays, lamps and other parts of the equipment. Before we go into further detail, we have to come to some agreements.

Drawing rules

Relay contacts are shown in diagrams in the position corresponding to the *normal (non-energized) state* of the relay.

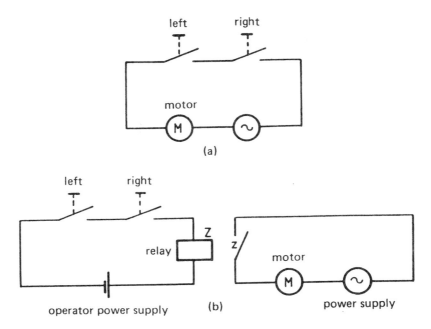

Figure 4.5 Protection circuit.

The push-button contacts are drawn in the position the contact is in when the push-button is *not being operated*. For (manually operated) switches with two stable positions, one has to come to an agreement about what the normal position is, and this is the position that is then drawn in a diagram.

In practice, this agreement may lead to the situation in which the positions of the contacts in a diagram are drawn in a configuration that does not occur during normal operation.

The assignment of variables

In a *normally open (make) contact* of a relay Z the logic variable z is used, and in a *normally closed (break) contact* $\bar{z}$ is used. In this way the correspondence between the active state of a relay Z and the state of the contacts is expressed the most clearly. Moreover, the lettering in the diagram corresponds to that in the logic formula, describing the function of the circuit. An exception is made for *transfer contacts*. In that case only z is used and $\bar{z}$ on the break side is usually left out. See Figure 4.6.

4.2 Designing in an AND-OR structure

The (external) function of a circuit being designed is usually specified by a logic function S, expressed in the *sum-of-products* form (AND-OR). The circuit corresponding to this formula basically consists of a number of series interconnections of contacts, connected in parallel. For each product in the formula the circuit has one branch. See Figure 4.7.
Figure 4.8 describes three alternative circuits, in which

$$S = \bar{x}\bar{y}\bar{z} + xyz$$

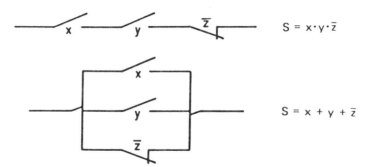

Figure 4.6 Assignment of variables.

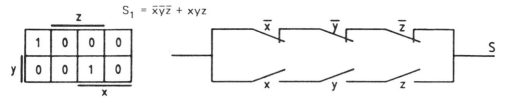

Figure 4.7 Realisation of $S = \bar{x}\bar{y}\bar{z} + xyz$.

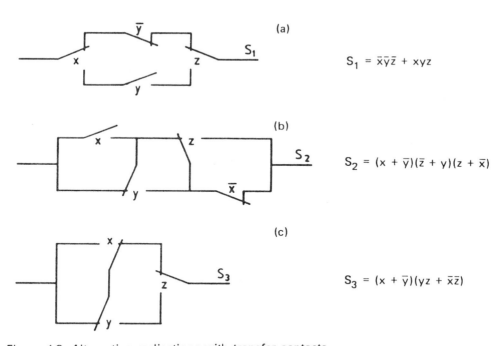

Figure 4.8 Alternative realisations with transfer contacts.

is also realised. The structure of the formula next to each circuit corresponds directly to the structure of the network. All circuits are logically equivalent.

In the circuits in Figures 4.8.a - 4.8.c *transfer contacts* are used. These are contacts with both a make and a break side, one of which is connected to the middle contact. The transfer contact has one contact spring less than separate make and break contacts.

We are confronted here with the fact that the minimal sum-of-products form of the function does not always correspond to the *structure* of the minimal circuit. The logic specification must be *mapped* and made to fit the available components. At this stage of the design process the *specific properties* of the components play an important role. This part of the design process requires a great deal of inventiveness and experience in manipulating logic functions.

Several methods of designing combinational contact circuits are known from the literature. Tables with 'minimal' solutions, for functions up to four variables, have also been published. In view of the relative insignificance of contact circuits in modern digital circuits we will not go into details.

The following *rules of thumb* may be a guide to the mapping of functions onto circuits with contacts. For a formula in the *sum form*, simple circuits are often found when as many as possible product terms start and/or end with the same variable. With such a structure one can work towards transfer contacts. Compare Figure 4.8.a with Figure 4.7. In the *product form* the factors have to be arranged so that factors with identical and/or negated variables are next to each other, see Figures 4.8.b and c. Sometimes two terms are multiplied in order to obtain a further reduction in the circuit. These rules usually result in circuits which approach optimality.

4.3 Designing through decomposition

Decomposition can be an aid in mapping logic formulas onto circuits with transfer contacts. The following theorem is then applicable.

Theorem
A logic function S

$$S = f(x_1, x_2, ..., x_n)$$

in the logic variables x_1 through x_n can be written as

$$S = f(x_1, x_2, ..., x_n)$$
$$= \bar{x}_1 \cdot f(0, x_2, ..., x_n) + x_1 \cdot f(1, x_2, ..., x_n). \tag{4.1}$$

The *subfunctions* or *residues* $f(0, x_2, ..., x_n)$ and $f(1, x_2, ..., x_n)$ originate from S by respectively substituting in S the values of 0 or 1 for x_1. These are functions of (at most) $n - 1$ variables.

Proof
For functions in three variables the proof, based on the *minterm form*, is as follows:

$$S = \bar{x}_1\bar{x}_2\bar{x}_3 f_0 + \bar{x}_1\bar{x}_2 x_3 f_1 + ... + x_1 x_2 x_3 f_7$$

$$= \bar{x}_1(\bar{x}_2\bar{x}_3 f_0 + \bar{x}_2 x_3 f_1 + x_2\bar{x}_3 f_2 + x_2 x_3 f_3) +$$
$$x_1(\bar{x}_2\bar{x}_3 f_4 + \bar{x}_2 x_3 f_5 + x_2\bar{x}_3 f_6 + x_2 x_3 f_7)$$
$$= \bar{x}_1 \cdot f(0,x_2,x_3) + x_1 \cdot f(1,x_2,x_3).$$

(Because of the commutative law decomposition can be performed to any variable x_i.)

Table 4.1 illustrates how subfunctions obtained by decomposition to x follow from the truth table of a function in three variables x, y and z.

x	y	z	S
0	0	0	f_0
0	0	1	f_1
0	1	0	f_2
0	1	1	f_3
1	0	0	f_4
1	0	1	f_5
1	1	0	f_6
1	1	1	f_7

$f(0,y,z) =$

y	z	S
0	0	f_0
0	1	f_1
1	0	f_2
1	1	f_3

$f(1,y,z) =$

y	z	S
0	0	f_4
0	1	f_5
1	0	f_6
1	1	f_7

Table 4.1 Decomposition of the truth table.

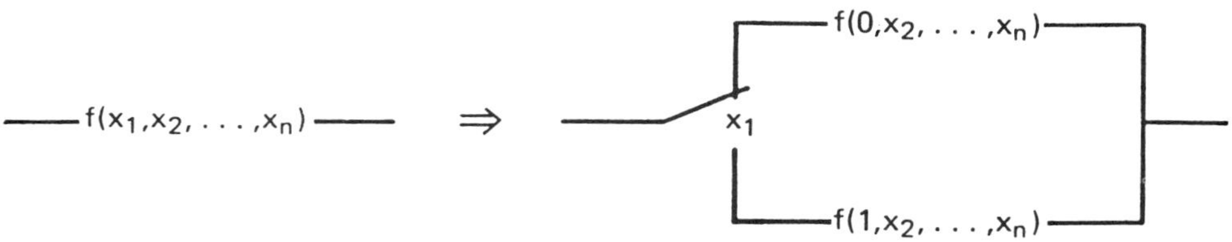

Figure 4.9 Decomposition to x with a transfer contact.

Figure 4.9 shows how decomposition is realised with a transfer contact. Some design methods for circuits with contacts are based on repeated decomposition. For functions in three variables this leads, through

$$S = f(x,y,z)$$
$$= \bar{x} \cdot f(0,y,z) + x \cdot f(1,y,z)$$
$$= \bar{x} \cdot \{\bar{y} \cdot f(0,0,z) + y \cdot f(0,1,z)\} + x \cdot \{\bar{y} \cdot f(1,0,z) + y \cdot f(1,1,z)\}$$

to

$$S = \bar{x}\bar{y}\bar{z} \cdot f(0,0,0) + \bar{x}\bar{y}z \cdot f(0,0,1) + \ldots + xyz \cdot f(1,1,1). \tag{4.2}$$

In (4.2) $f(0,0,0)$ through $f(1,1,1)$ are constants. They correspond to the function values f_0 through f_7:

$$f(0,0,0) = f_0 \qquad (4.3.\text{a})$$
$$f(0,0,1) = f_1 \qquad (4.3.\text{b})$$
$$\vdots$$
$$f(1,1,1) = f_7. \qquad (4.3.\text{h})$$

For the indices see Section 2.5.

Example

The realisation of the function S

$$S = \overline{w}\overline{x}z + \overline{w}\overline{y}z + wx\overline{z} + wy\overline{z} + \overline{x}\overline{y}z + xy\overline{z} \qquad (4.4)$$

can start by decomposing to the variable w. We then find

$$S = \overline{w}\cdot f(0,x,y,z) + w\cdot f(1,x,y,z)$$
$$= \overline{w}\cdot(\overline{x}z + \overline{y}z + \overline{x}\overline{y}z + xy\overline{z}) + w\cdot(x\overline{z} + y\overline{z} + \overline{x}\overline{y}z + xy\overline{z}).$$

Reduction of the subfunctions leads to

$$S = \overline{w}\cdot(\overline{x}z + \overline{y}z + xy\overline{z}) + w\cdot(x\overline{z} + y\overline{z} + \overline{x}\overline{y}z).$$

If the decomposition is continued to the variable x, we find:

$$S = \overline{w}\overline{x}\cdot f(0,0,y,z) + \overline{w}x\cdot f(0,1,y,z) + w\overline{x}\cdot f(1,0,y,z) + wx\cdot f(1,1,y,z)$$
$$= \overline{w}\overline{x}.z + \overline{w}x\cdot(\overline{y}z + y\overline{z}) + w\overline{x}\cdot(\overline{y}z + y\overline{z}) + wx\cdot\overline{z}.$$

Here the subfunctions $f(0,1,y,z)$ and $f(1,0,y,z)$ are identical. We combine the corresponding branches in the design:

$$S = \overline{w}\overline{x}\cdot z + (\overline{w}x + w\overline{x})\cdot(\overline{y}z + y\overline{z}) + wx\cdot\overline{z}.$$

Then only one branch with the subfunction

$$f = \overline{y}z + y\overline{z}$$

is left. It can easily be decomposed to y or z. Figure 4.10 shows the various steps of the decomposition of function (4.4) and its realisation with contacts.

The reader will have no problems in recognising the circuit of Figure 4.11 in Figure 4.10.c. □

In this method the result depends on the *order* of the variables to which the functions are decomposed. For functions in many variables it might be useful to start the decomposition with those variables for which as many as possible *identical subfunctions* (*residues*) are found.

Don't cares are another problem. The subfunctions do not have to be identical. It is sufficient for them to be compatible, i.e. they can be made identical by substituting don't cares. For functions with don't cares one must not automatically start with the *minimal* form of the formula (see also Scheinman and Roginsky).

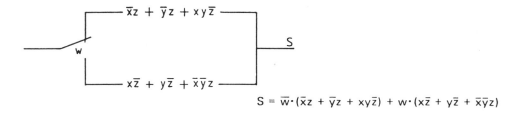

$$S = \overline{w} \cdot (\overline{x}z + \overline{y}z + xy\overline{z}) + w \cdot (x\overline{z} + y\overline{z} + \overline{x}\overline{y}z)$$

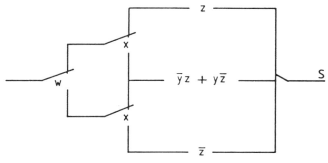

$$S = \overline{w}\overline{x} \cdot z + \overline{w}x \cdot (y\overline{z} + \overline{y}z) + w\overline{x} \cdot (y\overline{z} + \overline{y}z) + wx \cdot \overline{z}$$

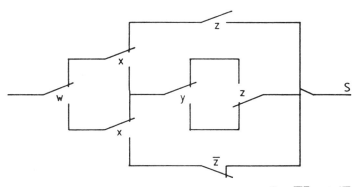

$$S = \overline{w}\overline{x} \cdot z + (\overline{w}x + w\overline{x}) \cdot (y\overline{z} + \overline{y}z) + wx \cdot \overline{z}$$

Figure 4.10 Realisation through decomposition to variables.

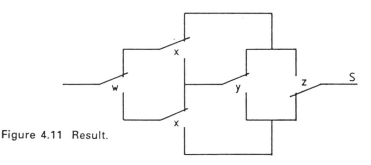

Figure 4.11 Result.

4.4 Practical aspects

During the realisation phase one must take into account all the properties of the components, both the logical and the physical. Contacts have positive and less positive properties. For example, they have a *low contact resistance*. In specific applications, such as the routing of speech signals in local exchanges in a telephone system, contacts (reed relays) are still being used. Also, the capability to switch high currents and voltages is based on the low contact resistance.

Contacts are *bidirectional*. They conduct in both directions equally well. For conducting a d.c. current or voltage in a branch type network this property is not always necessary. In fact, it can sometimes even be inconvenient, because *hidden loops* may be introduced.

Hidden loops

When contact circuits are designed through decomposition, branches are often combined because of identical or compatible subfunctions. This may introduce loops that cannot be recognised easily, hidden loops, so to speak. These loops go partly 'against the current'. See Figure 4.12.

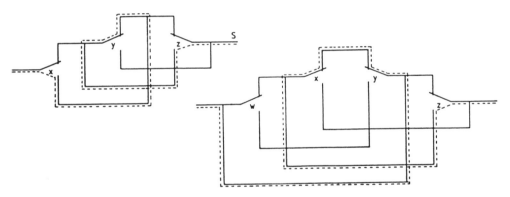

Figure 4.12 Circuits with hidden loops.

When one has introduced a hidden loop, its paths either belong to the logic function or not. In d.c. systems they can be blocked by *diodes*, if necessary. See Figure 4.13.

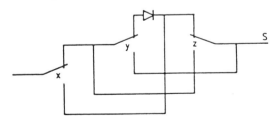

Figure 4.13 The elimination of loops.

Circuits with an AND-OR structure never have (hidden) loops. These only arise when paths are combined in order to obtain a reduction in the number of contacts. In this way, in Figure 4.12 a hidden loop has been introduced by realising the term $x\bar{z}$ via the contact $\bar{z}$ which was already present on the upper side. Loops are a specific problem in circuits with *bidirectional elements.* Some electronic circuits have a similar problem, although these circuits are usually of the gate type.

Transient phenomena

Propagation delay time is an important parameter. Relays and contacts switch *slowly*. Their reaction time is of the order of ms and higher. *Transient phenomena* therefore last a relatively long time and can negatively influence the reliability. Sometimes extra precautions are necessary to combat the consequences.

Example

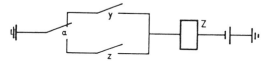

Figure 4.14 Circuit with a transient phenomenon.

In the circuit of Figure 4.14 a condition Y is tested when the transfer contact α is in the break side position. When Y = True the contact y is closed. Relay Z is then activated. The design objective of the circuit is for the relay to remain activated when the α-contact is switched. The contact z of relay Z is used to that end. When the α-contact in Figure 4.14 switches relatively slowly there is the possibility that the relay will be deactivated during the transition of α. (In practice, the probability of such deactivation is slight. The mechanical inertia of the relay bridges a certain switching time.) There are two approaches to this problem:

– adapt the components so that they better meet the logical model;

– keep the non-ideal character of the components in mind during the design process.

It is possible to construct a relay so that the make side of the transfer contact closes before the break side opens. Such a contact is called *make-before-break*. It switches as follows:

$$\begin{array}{ccc} \bar{\alpha} = 1 & \bar{\alpha} = 1 & \bar{\alpha} = 0 \\ \Rightarrow & \Rightarrow & \\ \alpha = 0 & \alpha = 1 & \alpha = 1. \end{array}$$

With a make-before-break contact the transient problem in Figure 4.14 has been eliminated. Practical details of this kind cannot be described in switching algebra. Then

$$\bar{\alpha} \cdot \alpha = 0$$

which excludes the possibility that $\bar{\alpha} = \alpha = 1$. □

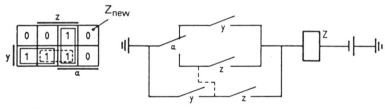

Figure 4.15 Circuit without transient phenomena.

Figure 4.15 shows a typical *digital solution* to the problem described above. The function of the circuit in Figure 4.14 is specified in Figure 4.15.a, in which the make chain αy and the hold chain αz are encircled. The switching of α can be bridged, according to Figure 4.15.a, with the dashed term yz. This term also belongs to the function of Z_{new} and is *independent* of α. With the additional path yz in Figure 4.15.b the transient phenomenon has been eliminated.

In this example we have seen that adding some *redundancy* to a circuit may allow the designer to eliminate transient phenomena. Sometimes this cannot be done 'digitally', so that mechanical means, such as make-before-break contacts and delayed activation and/or deactivation of relays, will have to be used.

Bouncing of contacts

Switching contacts often *bounce* back after having arrived on the other side of the (transfer) contact. Figure 4.16 illustrates this. Each of the points A and B is connected to the 5 volt power supply by a resistor R, or is grounded via the middle spring of the transfer contact. In this way, the position of a switch can be converted into a 'High' or a 'Low' voltage level. The timing diagrams shown are representative of the phenomena encountered in practice.

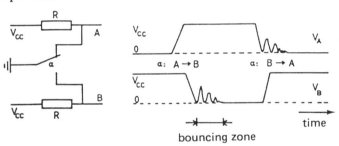

Figure 4.16 Bouncing of contacts.

In circuits with relays bouncing is hardly ever experienced as an inconvenience. This is because the *activating delay* of a relay is rather great and lasts longer than the bouncing zone in Figure 4.16. Because of its *mechanical inertia* a relay is able to bridge this.

The situation is totally different when push-buttons are used to command a much faster electronic circuit. On the contrary, such a circuit does interpret bouncing as a series of input pulses.

This problem can be circumvented by an anti-bounce circuit. This circuit will be discussed in Chapter 9.

The foregoing was intended merely as an introduction, rather than as an exhaustive treatment of transient phenomena and their elemination.

References

1. J. Appels and B. Geels, *Handboek der Relaisschakeltechniek*, Philips Technische Bibliotheek, Eindhoven, 1965.

2. M.A. Gavrilov, *Relaisschaltechnik*, Berlin, 1953.
3. M.L. Gayford, *Modern Relay Techniques*, Butterworth, London, 1969.
4. M.A. Harrison, *Introduction to Switching and Automata Theory*, McGraw-Hill, New York, 1965.
5. F.E. Hohn, *A Matrix Method for the Design of Relay Circuits*, IRE Trans. on CT., Vol. CT-2, 1955, pp. 154–161.
6. R. Holm, *Electrical Contacts*, Springer-Verlag, Berlin, 1967.
7. R.M.M. Oberman, *Disciplines in Combinational and Sequential Circuit Design*, McGraw-Hill, New York, 1970.
8. H.W. Ott, *Noise Reduction Techniques in Electronic Systems*, John Wiley & Sons, New York, 1976.
9. R.L. Peek and H.N. Wagar, *Switching Relay Design*, Van Nostrand, Princeton, New Yersey, 1955.
10. W.N. Roginsky, *Grundlagen der Structursynthese von Relaisschaltungen*, Oldenburg, München, 1962.
11. W.N. Roginsky, *A Graphical Method for the Synthesis of Multiterminal Networks*, Proc. International Symposium on the Theory of Switching, 1957.
12. A.H. Scheinman, *A Numerical-Graphical Method for Synthesising Switching Circuits*, Trans. AIEE, 1958.

5
The logic design of combinational logic circuits

5.1 Logic levels and truth values

Electronic logic circuits usually represent information by *voltage levels*. In the static state almost all logic circuits have two voltage levels 'High' (abbreviated to H or V_H) and 'Low' (L or V_L) at the inputs and outputs.

Figure 5.1 shows how the limits for High and Low levels for standard TTL circuits are defined. An input level

$$V_{in} < V_{IL(max)} = 0.8 \text{ V}$$

is interpreted as 'Low'. An input level

$$V_{in} > V_{IH(min)} = 2.0 \text{ V}$$

is considered 'High'. Input levels are allowed to vary within these ranges to a certain degree.

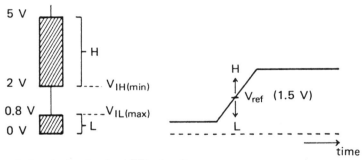

Figure 5.1 Logic levels in standard TTL circuits.

The output levels of standard TTL circuits are approximately:

$$V_{OL(max)} = 0.4 \text{ V} \quad \text{and} \quad V_{OH(min)} = 2.4 \text{ V}.$$

These values are well within the specified ranges. The margins are primarily needed to improve the *noise immunity* of the circuit. Chapter 6 goes into this subject more deeply. Each technology has its own specific limits for H and L levels. Sometimes these correspond to those of Figure 5.1. This is necessary when ICs realised in another technology are directly coupled with TTL ICs. Circuits that can be directly coupled with TTL ICs are called *TTL compatible*. (In practice other properties must also be taken into account before the designation 'compatible' applies.)

Threshold level

The *threshold* of an input lies between the levels $V_{IH(min)}$ and $V_{IL(max)}$. Input voltages above the threshold are interpreted as High, and below it as Low. The height of the threshold depends on the temperature, and other parameters. For standard TTL circuits the threshold voltage is between

$$1.2 \text{ V} < V_{threshold} < 1.6 \text{ V}.$$

In data books of components a *fixed level* is taken as *reference level* V_{ref} for the threshold, for example 1.5 V. All times concerning the description of the timing behaviour of a gate, e.g. the propagation delay time, are related to V_{ref}. This practice introduces a discrepancy between the circuit behaviour and its timing model. This difference has to be incorporated, as an uncertainty, in the specification of the parameters of the timing model. (See Chapter 6.)

The assignment of truth values

During the *specification phase of a system* the designer introduces propositions. In electronic logic circuits their truth values are usually represented by the voltages H and L.

Example

A machine consists of two modules, which together realise the desired function. The machine must sound an alarm when there is a breakdown in the operation. This alarm signal is deduced from the alarm signals of the separate modules. Table 5.1 specifies the function of the alarm circuit.

A1	A2	A
F	F	F
F	T	T
T	F	T
T	T	T

Proposition A1: 'Module 1 detects an alarm'
Proposition A2: 'Module 2 detects an alarm'

Proposition A: 'Machine is in the alarm state'

Table 5.1 Truth table.

If one wishes to realise this alarm circuit in an electronic circuit, operating with H and L logic levels, one first has to make a decision about the mapping of T and F on H and L. The two possible ways to do this are shown in Table 5.2.

Table 5.2.b differs from Table 5.2.c. *The convention of the mapping of T and F on H and L also determines the specification of the electronic circuit being designed.* The relation between a logic function and the circuit is only unambiguous when it is clearly specified how the truth values are assigned to H and L. See, for example, Table 5.3. Depending on the decision about how to interpret the H and L levels the circuit realises either the logic operation OR or the logic operation AND. □

Stated broadly, designing logic circuits consists of the following steps:

– make a logic specification, with the help of propositions and logic operations;

– map this onto a circuit that realises the desired behaviour.

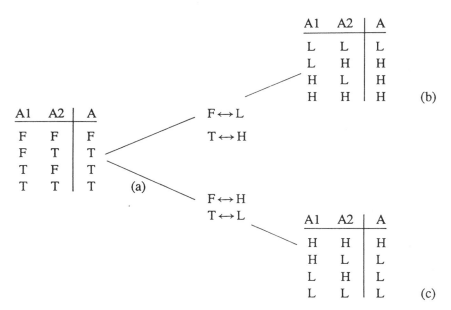

Table 5.2 Assignment of H/L logic levels.

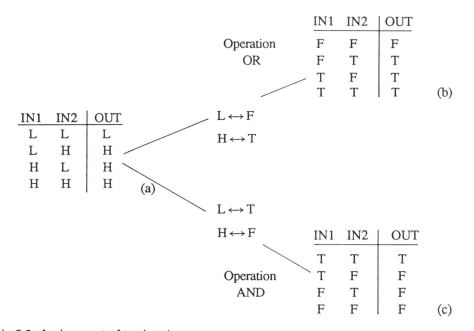

Table 5.3 Assignment of truth values.

In light of the above, a fixed convention for mapping truth values onto voltage levels is preferable. In the *design process* one usually opts for

$$L \leftrightarrow F \quad (= 0),$$
$$H \leftrightarrow T \quad (= 1).$$

This system is called the *positive logic convention*. However, the reverse, the *negative logic convention*, is also used:
$$L \leftrightarrow T \quad (= 1),$$
$$H \leftrightarrow F \quad (= 0).$$
In both cases the relation

$$\text{circuit} \leftrightarrow \text{logic function}$$

has been established unambiguously. One then usually talks in terms of *logic circuits*, even though, strictly speaking, this is actually not allowed. In practice, particularly in circuit maintenance and in documentation, it is advisable to work with the *mixed logic convention*. In that case, for every signal line it is agreed at which voltage level the corresponding proposition (signal name) is true. The reason for this can be traced back to the physical properties of the components. We will therefore postpone a further discussion about the assignment of truth values until later.

Logic function symbols

Table a in Figure 5.2 specifies, in voltage levels, the function of a given electronic circuit. Figure 5.2.b describes, in a timing diagram, how the output level OUT changes with varying input levels. With the convention

$$L \leftrightarrow F$$
$$H \leftrightarrow T$$

we assign a logic function to the circuit, which is described in Table c of Figure 5.2. This circuit then realises the logic operation AND, and is called an AND gate.

Logic functions are usually represented in drawings by a symbol. For the AND gate the symbol shown in Figure 5.2.d is used. There are circuits that can realise the other logic operations.

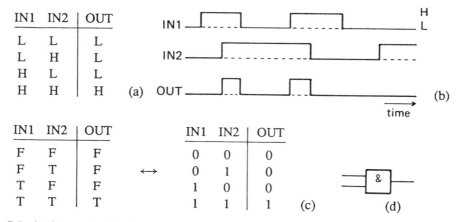

Figure 5.2 Assignment of truth values.

The remainder of this chapter is based on the assumption that there are components available in the various technologies that can realise any desired logic operation (for a given logic convention). From now on we will regard these components as a black box and we will trace how a specified logic function can be mapped onto these circuits. Physical parameters, including time, will be left out of consideration. This results in a design that can be directly *mapped* onto the available components. The components are represented by symbols, depicting their logic functions. Symbols for the most important components are shown in Figure 5.3.

Type/name	Logic function	Symbol
AND	$S = y.z$	 &
OR	$S = y + z$	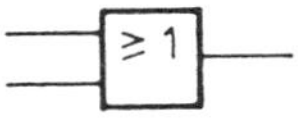 ≥1
NOT	$S = \bar{z}$	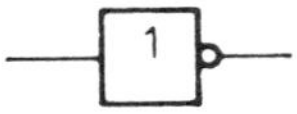 1
NAND	$S = \overline{y.z}$	 &
NOR	$S = \overline{y + z}$	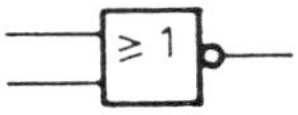 ≥1
EX-OR	$S = y \oplus z$	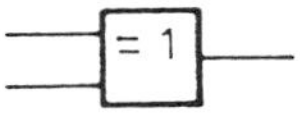 =1

Figure 5.3 Symbols representing logic functions according to the IEC system.

Until recently each logic function was denoted by a distinctly shaped symbol, (see Figure 5.4). Because these older symbols are still encountered quite often, we list them below:

Type/name	Symbol	Type/name	Symbol
AND	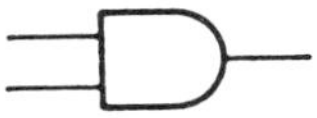	NAND	
OR		NOR	
NOT		EX-OR	

Figure 5.4 Distinctly shaped symbols.

5.2 Designing with ANDs, ORs and NOTs

The logic function of a circuit is usually specified in a truth table or by a logic formula, which must then be realised in hardware. *A formula cannot always be directly mapped onto hardware.* Sometimes a restriction of the number of inputs on the components forms a stumbling block. In other cases the IC manufacturer prefers other logic operations and functions than those used by the logic designer. The first example examines the realisation of logic formulas with components that perform the AND, the OR and the NOT operations. These components are called gates. The mapping must be based on the *precedence rules for the interpretation of logic formulas*.

Example

The current logic levels (L/H) of three signals x, y and z correspond to the truth values (0/1) of the variables/propositions of the same name. The logic function of the circuit to be realised is

$$S = \bar{x}\bar{y} + \bar{x}z + xy\bar{z}. \tag{5.1}$$

Formula (5.1) is mapped onto a circuit as follows. According to the precedence rules all negations have to be performed first. For (5.1) this means that first of all signals corresponding to x, y and z must be created, if they are not available. This can be done by three negators (logic function NOT).

Subsequently, in accordance with the precedence rules, the AND operations on the input signals have to be performed. The resulting three signals realise the product terms in (5.1). Next the OR operation has to be performed on these signals. This results in signal S in (5.1). See Figure 5.5. This structure can now be mapped directly onto the available components.

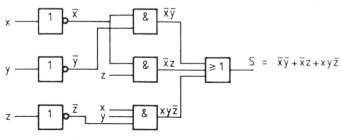

Figure 5.5 NOT-AND-OR design.

Figure 5.6.a describes the same design, in a more compact way. This method of drawing is supported in the IEC system for logic symbols. It allows a lot of space to be saved in drawings. The older distinctly shaped symbols hardly allow any compact drawing at all.

□

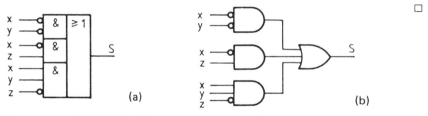

Figure 5.6 Compact method of drawing.

Example

Formula (5.2) is the product-of-sums form of (5.1). Note that the parentheses change the order of interpretation. This can also be seen in the circuit design.

$$S = (\bar{x} + y)(\bar{x} + z)(x + \bar{y} + z). \tag{5.2}$$

Figure 5.7 illustrates the circuit design. First is the negation level and then the AND and the OR levels follow. □

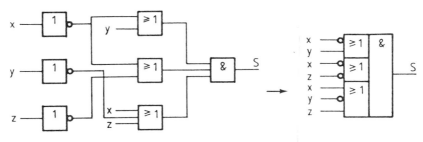

Figure 5.7 NOT-OR-AND design.

Restrictions

In practice, the number of inputs on a gate is restricted. The electrical properties of the transistors, for example, restrict the maximum allowable number of inputs in the design of an IC. When catalogue components are used there are even more restrictions, as one has to conform to the available components. Table 5.4 shows an example.

	number of inputs	gates per package
AND	2	4
	3	3
	4	2
OR	2	4
NOT	1	6

Table 5.4 AND/OR/NOT circuits per package (TTL, Texas Instr., 1985).

With the restrictions as shown in Table 5.4 designs of logic circuits must be specified in a form that can be mapped directly onto the available components. For AND and OR gates this is not difficult. Based on switching algebra we may, for example, write

$$S = (abcd) \cdot (efgh).$$

instead of

$$S = abcdefgh.$$

This form can be directly mapped onto the components of Table 5.4 (see Figure 5.8). For composing n-input OR gates similar rules apply.

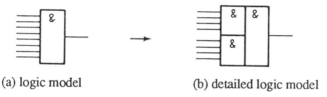

(a) logic model (b) detailed logic model

Figure 5.8 Transformation for 8-input AND function.

Open inputs

When standard components are applied, often not all inputs of a gate are used. In many ICs open inputs automatically pull to one of the logic levels. Nevertheless, one cannot leave these inputs open, because of the high *noise sensitivity of open inputs*. One can easily anticipate this problem.

With ANDs one can put a 1 on superfluous inputs. After all, we know from switching algebra that

$S = yz$ (2-input AND),
$ = 1 \cdot yz$ (3-input AND).

The alternative is to connect an input variable/signal twice:

$S = yz$ (2-input AND),
$ = y \cdot yz$ (3-input AND),
$ = yz \cdot z$ („).

For ORs something similar holds:

$S = y + z$ (2-input OR),
$ = 0 + y + z$ (3-input OR),
$ = y + y + z$ („),
$ = y + z + z$ („).

These solutions to the problem of unused inputs cause the outputs of controlling circuits to have a higher fanout load. Where this is a disadvantage, one must choose to apply a signal (H/L) that corresponds to 1 (AND) or 0 (OR), respectively.

Conclusion

Formulas, expressed in the operations AND, OR and NOT, can easily be mapped onto hardware. Extra conversions are often necessary to make a given formula fit the available components. For that reason several design levels must be distinguished, varying from a *theoretical logic model* to a *detailed or mappable logic model*. Other design levels follow, in which further details, such as the choice of the positive or negative logic convention, calculations of the (maximum) propagation delay times of signals through the circuit, the partitioning of functions over ICs and their placement on boards, are filled in. A number of these details, such as timing, can be anticipated in the logic model, while others, such as layout, can hardly be anticipated. Of course, one has more flexibility when designing ICs than when designing with catalogue components.

5.3 Designing in NANDs or NORs

A designer prefers to use components onto which the operations used in the formulas can be mapped easily. Other aspects also play a role in the choice of components. In TTL technology, for example, the NAND (AND-NOT) is easier to realise than the AND. The same holds for the NOR (OR-NOT) with respect to the OR. Besides that, the NAND has certain advantages over the NOR. In other technologies a similar preference for certain operations is found. Chapter 6 will deal with the technical background.

To a manufacturer only the total costs of a product are of importance. Of these the design costs, especially the part concerning the mapping of a logic design onto hardware, are usually relatively low. The somewhat higher costs (design time) of a technology dependent preference are compensated by better performance (IC surface, propagation time of the circuit, decrease in dissipation, etc.). The designer must therefore be able to use components other than AND, OR and NOT. For the design process this means that the formulas specified in, for example, the AND-OR form must be transformed into a form that suits the available components better. Then every design can, for example, be completely realised in NAND gates. This has been done fairly recently. The same holds for NORs. Figure 5.9 shows how the logic operations AND, OR and NOT can be realised with NANDs. Direct application of this conclusion leads to circuits of, in principle, four or five gate levels. That it can also be done in two levels can be seen from the following 'recipes' for the mapping of formulas on NANDs or on NORs.

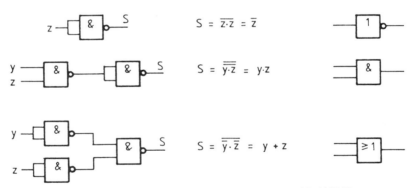

Figure 5.9 Realisation of AND, OR and NOT operations with NANDs.

Recipe for NANDs

— Start with a formula in the *sum-of-products form*.

— Negate the formula twice. This does not change the formula.

— Work out one negation. This results in a form that can be mapped directly onto NANDS.

Example
The formula

$$S = wx + wz + yz \qquad (5.3)$$

is given in the *sum-of-products form*. With a double negation this becomes

$$S = \overline{\overline{wx + wz + yz}}.$$

It follows from DeMorgan's laws that

$$S = \overline{\overline{wx} \cdot \overline{wx} \cdot \overline{yz}}.$$

In this form we see the operation NAND four times:

$$S = \text{NAND}[\text{NAND}(w,x), \text{NAND}(w,z), \text{NAND}(y,z)].$$

The corresponding logic diagram is shown in Figure 5.10. □

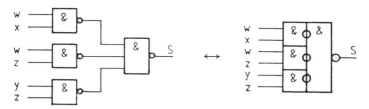

Figure 5.10 Realisation with NANDs.

The above recipe for a realisation with NANDs is applicable to any function, written in the sum-of-products form. This form is found by reading out the 1s in a Karnaugh map. There is also a recipe for NORs.

Recipe for NORs

– Start with a formula in the *product-of-sums form*.

– Negate the formula twice. This does not change the formula.

– Work out one negation. A realisation with NORs can then be seen immediately.

Example
The direct application of this recipe has been worked out for (5.3) in Figure 5.11. □

The recipes for NAND and NOR realisations are based on respectively the sum-of-products form or the product-of-sums form of the formula. No restrictions on the number of inputs of a gate are assumed, even though there usually are some. For NANDs special gates, called *expanders*, are available. These can be used to increase the number of inputs of the AND part (Figure 5.12). Decomposition of the formula can also be applied. Section 5.5 will discuss this possibility in more detail.

5.4 AND-NOR gates

In TTL technology the input transistors of a NOR gate are sometimes replaced by multi-emitter transistors. Every multi-emitter transistor then forms an AND gate (positive logic). The other part of the circuit remains a NOR gate. The modified circuit effectively realises two logic levels in one physical gate level. Such a circuit is called an AND-NOR or an

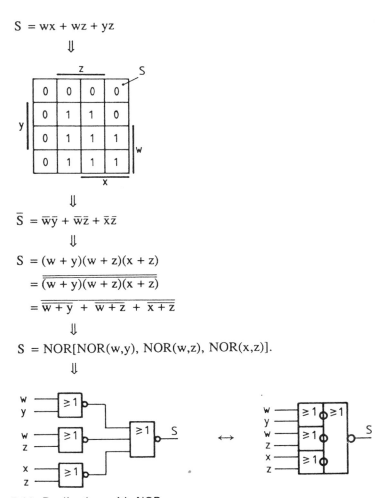

Figure 5.11 Realisation with NORs.

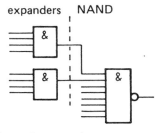

Figure 5.12 14-Input NAND through expanders.

AND-OR-INVERT gate. AND-NORs are available in several types. As an example Figure 5.13 illustrates circuit SN7454. In positive logic its logic function is

$$S = \overline{ab + cd + ef + gh}.$$

To be able to map a formula onto an AND-NOR gate it must first be transformed into a

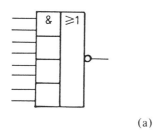
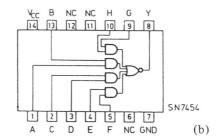

(a)   (b)

Figure 5.13 AND-NOR gate (SN7454).

suitable form. We can do this by reading out the function S, in the sum-of-products form, from a Karnaugh map. A negation gives S in a form that can be mapped directly.

The method with Karnaugh maps is not suitable for functions in many variables. One often has to resort to negating the function S, given in the sum-of-products form, twice, of which one negation is worked out. Another possibility may be to map S, given in the sum-of-products form, directly onto an AND-NOR gate. This yields $\overline{S}$, a function which can be converted into S with one negator. The circuit then consists of two gate levels, so that the advantage of higher speed is lost. Chapter 7 gives some examples of applications of AND-NOR gates, for instance in the realisation of fast adders.

5.5 Designing through decomposition

Circuits with contacts are often designed with the help of *decompostion* based on (5.4):

$$S = f(x_1, x_2, ..., x_k) \qquad (5.4)$$
$$= \overline{x}_1 \cdot f(0, x_2, ..., x_k) + x_1 \cdot f(1, x_2, ..., x_k).$$

Repeated decomposition finally leads to a complete design (see Section 4.3).

Decomposition can also be applied in the design of electronic combinational logic circuits. Figure 5.14 shows how decomposition can be realised with gate circuits, according to (5.4). The circuit *selects*, depending on x_1, either $f(0, x_2, ..., x_k)$ (when $x_1 = 0$) or $f(1, x_2, ..., x_k)$ (when $x_1 = 1$) and assigns the selected value to the output signal S. This *module* is therefore referred to as *selector*.

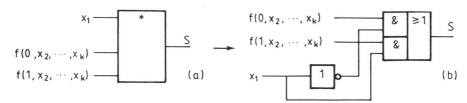

Figure 5.14 Decomposition to one variable, basic module.

Based on the circuit of Figure 5.14 repeated decomposition is possible. The result is a design, in identical modules, with variables x_1 through x_n as input signals on the different levels. On the last level the input signals are constants, equal to the residues $f(0, 0, ..., 0)$ through $f(1, 1, ..., 1)$, the function values f_0 through f_{2^k-1}.

Note
The procedure described here can be refined by applying reduction techniques, by combining residues and also by taking variables of the function as input signals. The interested reader is referred to the References at the end of this chapter. □

Decomposition to two variables at the same time is also possible:

$$S = f(x_1, x_2, ..., x_k) \quad (5.5)$$
$$= \bar{x}_1\bar{x}_2 \cdot f(0, 0, x_3, ..., x_k) + \bar{x}_1 x_2 \cdot f(0, 1, x_3, ..., x_k)$$
$$+ x_1\bar{x}_2 \cdot f(1, 0, x_3, ..., x_k) + x_1 x_2 \cdot f(1, 1, x_3, ..., x_k).$$

The residues $f(0, 0, x_3, ..., x_k)$ through $f(1, 1, x_3, ..., x_k)$ are then functions in at most $k-2$ variables. Figure 5.15 shows the circuit which can be used for decomposition to two variables.

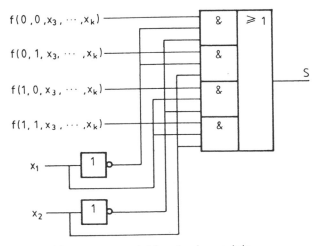

Figure 5.15 Decomposition to two variables, basic module.

Modules of a similar structure can decompose directly to any number of variables. (Note that the input signals cannot be interchanged freely, as was the case with inputs of ANDs and ORs and of NANDs and NORs.)

Application

Decomposition is used frequently. In fact we have already used it in realising n–input AND gates and OR gates from smaller ones (Section 5.2). In the following example two realisations of a given function are worked out, using NANDs with a maximum of four inputs. The realisation of any function with 4–input NANDs involves in the first place the selection of a maximum of four subfunctions S_0, S_1, S_2 and S_3, in such a way that (5.6) holds:

$$S = \text{NAND}(S_0, S_1, S_2, S_3) = \overline{S_0 S_1 S_2 S_3}. \quad (5.6)$$

After all, on the output level there is only one NAND that emits the signal S, which is only possible if (5.6) has been satisfied. This, too, is a form of *decomposition*, although in this case it is decomposition to a given component/logic function. The problem now is how to find subfunctions S_0 through S_3 such that

– condition (5.6) is true;

– the subfunctions can (economically) be realised with 4–input NANDs.

All logic operations can be performed with NANDs. This means that the decomposition problem can be solved, in principle. However, the degree of freedom in such a decomposition is very large. Moreover, the residues can usually not be determined unambiguously.

Example
Realise the function

$$S = w\bar{x}yz + w\bar{y}\bar{z} + \bar{w}\bar{x}y\bar{z} + \bar{w}xz + wx\bar{z} \tag{5.7}$$

with NANDs. These have at most four inputs.

Formula (5.7) contains five product terms, one too many when the *standard recipe* is used. However, it holds that

$$S = (w\bar{x}yz + w\bar{y}\bar{z}) + \bar{w}\bar{x}y\bar{z} + \bar{w}xz + wx\bar{z}$$

$$= \overline{\overline{(w\bar{x}yz + w\bar{y}\bar{z}) + \bar{w}\bar{x}y\bar{z} + \bar{w}xz + wx\bar{z}}}$$

$$= \overline{\overline{(w\bar{x}yz + w\bar{y}\bar{z})} \cdot \overline{\bar{w}\bar{x}y\bar{z}} \cdot \overline{\bar{w}xz} \cdot \overline{wx\bar{z}}}$$

$$= \overline{S_0 \cdot S_1 \cdot S_2 \cdot S_3}.$$

Of the four residues only S_0 cannot be realised directly with one NAND. The others can. For S_0 we find

$$S_0 = \overline{w\bar{x}yz + w\bar{y}\bar{z}}$$

$$= \overline{\overline{w\bar{x}yz} \cdot \overline{w\bar{y}\bar{z}}}$$

$$= \overline{\overline{\overline{w\bar{x}yz} \cdot \overline{w\bar{y}\bar{z}}}}$$

$$= \text{NOT}\{\text{NAND}[\text{NAND}(w,\bar{x},y,z), \text{NAND}(w,\bar{y},\bar{z})]\}.$$

Since the negator (NOT) can also be realised with a NAND the complete function can be mapped onto 4-input NANDs (see Figure 5.16).

In fact the standard recipe for NANDs has been applied twice in Figure 5.16. In general, refinements are possible. One of these is shown in the following example. □

Example
Formula (5.7) may be written as

$$S = w\bar{x}yz + w\bar{y}\bar{z} + \bar{w}\bar{x}y\bar{z} + \bar{w}xz + wx\bar{z}$$

$$= w\bar{x}yz + \bar{w}\bar{x}y\bar{z} + \bar{w}xz + w\bar{z}(\bar{y} + x).$$

When we now apply the standard recipe, we find

$$S = \overline{\overline{w\bar{x}yz} \cdot \overline{\bar{w}\bar{x}y\bar{z}} \cdot \overline{\bar{w}xz} \cdot \overline{w\bar{z}(\bar{y} + x)}}.$$

This form can be mapped onto Figure 5.17. □

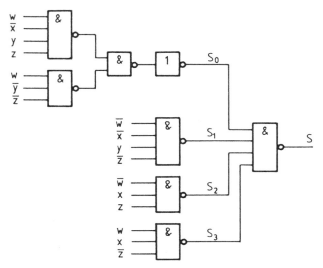

Figure 5.16 Circuit, realised through decomposition.

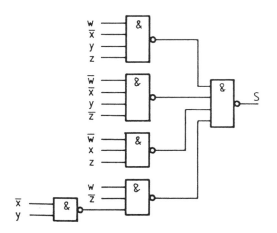

Figure 5.17 Circuit, realised through decomposition.

Several systematic methods for mapping functions onto hardware through decomposition are known from the literature. One of the best known is the Reed–Muller decomposition, which maps directly onto EX-OR gates. Methods for decomposition to any logic gate or function have been published, for example, by Ashenhurst, Kjelkerud and Shen. These methods are difficult and not suited to manual use. A computer is necessary and they are not used frequently. The References provide additional information.

5.6 Some practical aspects

When mapping logic functions onto gates the designer has to deal with restrictions such as

– the number of inputs per gate;

– fanout restrictions;

– the maximum time available for an operation;

and many others. Except for these, the designer has a lot of freedom.

Example

Figure 5.18 shows two realisations of an EX-OR gate with NANDs. The second realisation uses one gate less. This is based on the laws

$$y \cdot \overline{yz} = y \cdot (\overline{y} + \overline{z}) \qquad \text{and} \qquad z \cdot \overline{yz} = z \cdot (\overline{y} + \overline{z})$$
$$= y \cdot \overline{y} + y \cdot \overline{z} \qquad\qquad\qquad = \overline{y} \cdot z + z \cdot \overline{z}$$
$$= 0 + y \cdot \overline{z} \qquad\qquad\qquad = \overline{y} \cdot z + 0$$
$$= y \cdot \overline{z} \qquad\qquad\qquad\qquad = \overline{y} \cdot z. \ \square$$

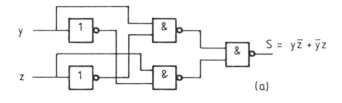

(a)

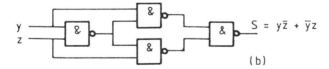

(b)

Figure 5.18 EX-OR realisations with NANDs.

Situations such as those depicted in Figure 5.18 occur frequently, and one has to be able to recognise them. They are based on the facts that

– in a product, variables can be added (in a sum) to each factor which occurs negated in the other factors with respect to the added variables.

– in a sum, variables can be added (in a product) to each term which occurs negated in the other terms with respect to the added variables.

Example

$$S = w\overline{x}yz \qquad\qquad S = w + \overline{x} + y + z$$
$$= w(\overline{x} + \overline{w})yz \qquad\qquad = w + \overline{w}\overline{x} + y + z$$

$$= w(\bar{x} + \bar{w})(y + \bar{z})(x + z) \qquad = w + \bar{w}\bar{x} + y\bar{z} + xz$$
$$= \ldots \qquad\qquad\qquad\qquad = \ldots . \square$$

In many larger circuits it is possible for parts of a function to have already been realised in other places, although they are not always in the minimal form specified for that function. In that case the function is first realised on available parts and then supplemented by other product or sum terms. This is only useful when the savings in the number of gates compensates for the extra connections that have to be made throughout the circuit. In an IC layout this latter aspect certainly plays a role. □

Example
Figure 5.19 specifies three functions. A realisation based on the minimal forms of the formulas requires seven product terms.

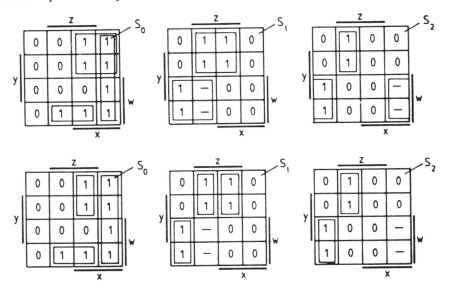

Figure 5.19 Multiple output specification.

On the basis of combined use, the realisation requires five different product terms. This aspect is especially important in designs with programmable logic (Chapter 8). □

The above merely serves to give an impression of the freedom that the designer still has after a logic specification has been formulated. It is by no means meant to be an exhaustive treatment of all of the techniques available to optimise circuits. The availability of programmable logic such as PALs and PLAs (Chapter 8) certainly simplifies the whole process.

References

1. R.L. Ashenhurst, *The Decomposition of Switching Functions*, Annals Computation Lab., Harvard Univ., Vol. 29, Cambridge, Mass., 1959, pp. 74–116.
2. T.C. Bartee, *Computer Design of Multiple-Output Logical Networks*, IRE Trans. Electron. Comp., Vol. EC-10, 1961, pp. 21–30.

3. P.W. Besslich and P. Pichlbauer, *Fast Transform Procedure for the Generation of Near Minimal Covers of Boolean Functions*, IEEE Proc., Vol. 128, Pt.E, no. 6, pp. 250–254.
4. R.K. Brayton and C. McMullen, *The Decomposition and Factorization of Boolean Expressions*, Proc. Int. Symp. on Circuits and Systems, Rome, 1982, pp. 49–54.
5. R.K. Brayton and C. McMullen, *Synthesis and Optimization of Multistage Logic*, Proc. Int. Symp. on Circuits and Systems, Rye N.Y., 1984, pp. 23–28.
6. R.K. Brayton, *Logic Minimization Algorithms for VLSI Synthesis*, Kluwer Academic Publishers, Hingham, 1984.
7. S.G. Chappell, *Simulation of Large Asynchronous Logic Circuits Using an Ambiguous Gate Model*, Proc. Fall Joint Computer Conference, 1971, pp. 651–660.
8. H.A. Curtis, *Design of Switching Circuits*, Van Nostrand, Princeton N.J., 1962.
9. R.B. Cutler and S. Muroga, *Derivation of Minimal Sums for Completely Specified Functions*, IEEE Trans. On Comp., Vol. C-36, 1987, pp. 277–292.
10. E.S. Davidson, *An Algorithm for NAND Decomposition under Network constraints*, IEEE Trans. Comp., Vol. C-18, 1969, pp. 1098–1109.
11. M. Davio a.o., *Digital Systems with Algorithm Implementation*, John Wiley & Sons, New York, 1983.
12. E.L. Dietmeyer and Y.H. Su, *Logic Design of Fanin Limited NAND Networks*, IEEE Trans. Comp., Vol. C-18, 1969, pp. 11–22.
13. B. Dormido and C. Canto, *Systematic Synthesis of Combinational Ciruits using Multiplexters*, Electron. Lett., Vol. 154, no. 18, 1978, pp. 588–590.
14. D.T. Ellis, *A synthesis of Combinational Logic with NAND and NOR Elements*, IEEE Trans. on Electr. Comp., Vol. EC-14, no. 5, Oct. 1965, pp. 701–705.

6
Logic families and their gate models

6.1 Introduction

Circuits perform their logic function only under certain conditions. When transforming a logic design into hardware the designer is confronted with the question of the extent to which the specified logic behaviour is actually realised in the circuit and under which conditions. This chapter deals with a *qualitative study* of the rules and conditions necessary for a circuit to function correctly. These conditions, as it will appear, are partly technology dependent.

A number of design rules should already be anticipated in the logic design level. These rules will be treated. Other rules, e.g. those dealing with the layout of printed circuit boards and ICs, will be left out of the discussion. These problems are primarily of a topological nature. For a number of design rules we will try to frame *simple models* so that the designer will not have to bother about details every time. This attempt at modelling is only successful if the model sufficiently covers the physical phenomena it describes. On the other hand, a model that is too detailed is unmanageable because, for example, computer simulations of the logic circuits will be very time consuming. The updating of all parameters of a detailed model takes up a lot of time. Where necessary we will bring the limits of our models up for discussion and indicate possible refinements.

We first introduce a number of concepts to be used when discussing the properties of the different logic families. After that we will return to the subject of modelling.

Logic levels and noise

In the static state all logic signal levels are within the specified ranges for H and L. When signal transitions occur this state temporarily ends. Not only do intended signal transitions take place, but also signal transitions are found on other lines. *Cross talk* between signal lines and through the power supply often occurs in digital circuits. This phenomenon results in *noise*. From now on we will indicate any interference which has a negative influence on the quality of a signal as 'noise'. Noise usually has a random nature (random noise).

There are several approaches to restricting the effects of noise, such as suppression using components that are less susceptible to noise, or waiting until the noise has faded away. Suppressing noise is difficult, makes circuits more expensive and is certainly not fully 100 % possible. To avoid noise problems other logic families, which are less sensitive to small variations in signals, may also be used. The noise then has less of an effect. In digital design the noise problem is usually approached differently. A signal is only tested or used when all transient phenomena have faded away.

There are two types of noise, *static* or *DC noise* and *dynamic* or *AC noise*. Noise is called static when the duration of the phenomenon is relatively long with regard to the response time of the circuit. Dynamic noise is noise with a duration comparable to the response time of a circuit after an input transition. In addition to the duration of the phenomenon, the degree to which the noise pulse exceeds the limits for H and L, $V_{IH(min)}$ and $V_{IL(max)}$, is important. In this context one also speaks in terms of the *power density* of a noise pulse.

Because of noise there has to be a certain distance between the range in which the threshold voltage lies and the defined limits $V_{IH(min)}$ and $V_{IL(max)}$ of H and L. As a measure of the *DC noise margin* for the H level one usually takes the distance between $V_{OH(min)}$ of an output and $V_{IH(min)}$ of an input driven by it, and for the L level $V_{OL(max)}$ of an output with regard to $V_{IL(max)}$ of a driven input. See Figure 6.1. The greater the margins, the better the noise immunity for, e.g. DC noise. The reference for the DC noise margin is not the threshold region, partly because there might also be AC noise in combination with DC noise.

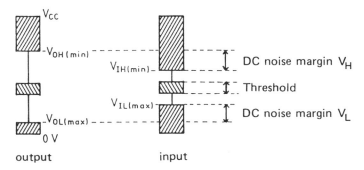

Figure 6.1 DC noise margins.

Within a *logic family* there are no great differences between the DC noise margins. Reliable data transfer is usually possible between components from one family. This is not the case where different logic families are concerned. Figure 6.2 shows the relevant margins in the case of data transfer from a TTL IC to a CMOS IC. Reliable data transfer is not possible for the H level. The DC noise margin is actually negative. Data transfer from the CMOS IC to the TTL IC is possible. The margins are in fact greater than for a design consisting of all TTL ICs or all CMOS ICs.

Most catalogues supply reasonable information about the DC noise margins. Far less information is given about AC noise margins and particularly about the allowed power density of a noise pulse. In 7400 TTL circuits, for example, AC noise pulses usually do not cause output transitions if they last less than about 60 % of the propagation delay time of the gate, even when they exceed the H ↔ L threshold.

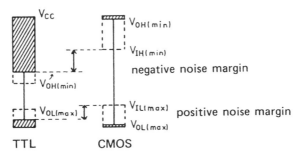

Figure 6.2 DC noise margins between TTL and CMOS.

Schmitt-trigger inputs

In order to reduce noise susceptibility substantially some ICs have Schmitt-trigger inputs. The L→ H threshold of these ICs is at least 0.4 V higher than the H → L threshold. This difference increases the noise margins proportionally. Schmitt-trigger inputs are used especially on inputs that are driven by *external input signals*, e.g. bus receivers.

Power supply

The *de facto standard* for the *power supply voltage* V_{CC}/V_{DD} of integrated circuits is +5 V to ground. This is the power supply voltage of the TTL family, the oldest and at the moment the most extensive family of integrated logic circuits. The other families, of a more recent date, are adapted to it and can usually operate with a 5 V supply, or have input and output adapters.

The characteristics required of a good power supply are not only concerned with the output voltage (+5 V ± ΔV) and the current to be supplied. The *dynamic high frequency behaviour* of a power supply, the ability to compensate fast changes in current, is just as important. The high frequency behaviour determines the amount of cross talk through the power supply.

In larger designs the power supply is often stabilised on each board separately and subsequently distributed on that board. Usually an electrolytic capacitor then absorbs the greater current changes. For good *high frequency behaviour* tantalous or ceramic *decoupling capacitors* are necessary, one per one or two ICs. They have to be connected as closely as possible to the power pins of the IC. Information on this can be found in the application notes of the various logic families.

A well-designed circuit usually operates within greater margins than a critical circuit. Testing with somewhat larger voltage variations, and if possible without decoupling capacitors, usually results in a quick and often correct impression of the quality of the circuit under design, at least as far as this point is concerned.

A number of IC parameters, e.g. the signal propagation time and the noise margins, are influenced by the power supply. This influence is reduced as the power supply is better stabilised. In any case these variations have to be incorporated in the model parameters.

Open inputs

Unused inputs are never allowed to remain open, but have to be provided with an H or L voltage level. *Open inputs are very susceptible to noise.* For the H level an unused input is usually connected to V_{CC} via a 1 kΩ resistor. One resistor can drive 10-20 inputs. The inputs of some IC families, e.g. 7400LS and CMOS, can be directly connected to V_{CC}. The resistor is used if V_{CC} can become higher than +5 V (consider switching power on). Some inputs have to be protected against this.

To force inputs to the L level one may ground them. Sometimes a resistor or *clipping diode* is necessary, to protect inputs against negative voltages. This is also technology dependent.

Timing parameters

Transition time t_T

The timing behaviour of logic circuits is described by several parameters. One of them is the *transition time* t_T, the time a signal needs to change from a high level voltage H to a low level voltage L (t_{THL}), or from L to H (t_{TLH}). This time is measured between 10 % and 90 % of the difference between the old and the new logic levels. See Figure 6.3.a.

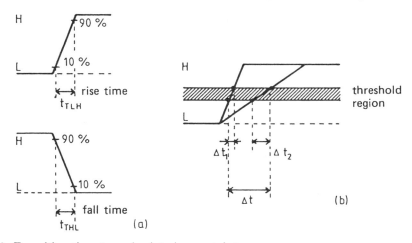

Figure 6.3 Transition time t_T and related uncertainty.

At first sight the parameter t_T is of little importance. After all, an input reacts to the passing of the H $\leftrightarrow$ L threshold, and hardly notices anything of the slope of the driving signal (TTL and to a certain extent also standard CMOS). Figure 6.3.b explains the importance of t_T. To begin with, the moment the threshold is passed is delayed by increasing t_T. Besides, the size of the threshold region also influences the uncertainty in the detection moment. The uncertainty between Δt_1 and Δt_2 increases as t_T increases. When the transition times in a circuit lie between the two extremes in Figure 6.3.b, the period denoted as Δt corresponds to the *maximum uncertainty* in the moment at which an input change is detected. This uncertainty, which may be some nanoseconds, has to be incorporated in the *timing model*.

Example

Parameters which have a large influence on t_T are the *supply voltage* of the IC, its temperature and the *capacitive load* of an output. Figure 6.4 shows, for example, the influence of V_{CC} and the capacitive load on t_T for HCMOS ICs. □

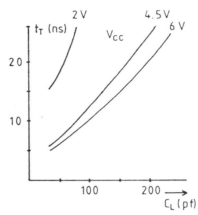

Figure 6.4 Transition time t_T for HCMOS ICs.

Propagation delay times

The time a gate or combinational circuit needs to react to an input transition is called the *propagation delay time* t_{PD} of a gate. Figure 6.5 illustrates how t_{PD} is measured (TTL). As reference level the 1.5 V level is taken (sometimes 1.4 V). After all, the real thresholds vary. The difference between V_{ref} and the threshold introduces a small uncertainty. This must also be incorporated in the timing model.

In data books t_{PD} is usually specified for a standard load at the output of the gate/circuit under test. The actual load may deviate from this. With a higher *capacitive load*, for example, the slope of the signal diminishes. The threshold of the driven gate is then reached later. The load of every output in a circuit can be measured or calculated, which is a tedious job, or its (maximum) effect can be estimated and incorporated in the timing model. The latter is simpler.

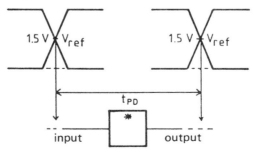

Figure 6.5 Definition of t_{PD}.

A well-known phenomenon of TTL gates, and others, is that the gate reacts faster as *more inputs are driven in the same direction*. This situation occurs especially when unused inputs are interconnected and it is also to be expected when a circuit is driven by different sources. This effect may result in a minimum *propagation delay time* of a gate, $t_{P(min)}$, which may be less than the minimum propagation delay time measured between one input and the output, with a single input transition. The reduction may be as high as 30 % to 40 %.

Furthermore, measurements indicate that there is always a difference between t_{PHL} and t_{PLH} (referring to the direction of the output transition). This depends on the *output driver circuit* of a gate. See Section 6.3. Usually the so-called *global timing model* is used, i.e. measurements are made under the most (un)favourable conditions and a $t_{P(min)}$ and a $t_{P(max)}$ are specified for a gate or circuit. When different values are expected or measured for different inputs, the $t_{P(min)}$ of the fastest input is taken as the $t_{P(min)}$ of the circuit and the $t_{P(max)}$ of the slowest input is taken as the $t_{P(max)}$ of the circuit. If the designed circuit turns out to function correctly according to this timing model, the verification of the timing behaviour is complete. If the circuit as designed turns out not to function correctly, but the differences between requirements and performance are not great, a *detailed timing model* may be a solution. One relevant aspect in that case is that the t_{PHL} and t_{PLH} of the input signals are used instead of

$$t_{P(min)} = \min\{t_{PHL}, t_{PLH}\}, \quad t_{P(max)} = \max\{t_{PHL}, t_{PLH}\}.$$

Sometimes this may be advantageous.

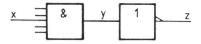

Global model: $t_{P(max)} = t_{P(max)}(AND) + t_{P(max)}(NOT)$
Detailed model: $t_{P(max)} = \max\{t_{PLH}(AND) + t_{PHL}(NOT) / t_{PHL}(AND) + t_{PLH}(NOT)\}$.

Figure 6.6 Analysis according to the global and the detailed timing model.

In Figure 6.6 the *direction of a signal transition* at point y is always opposite that of point z. If we include this factor the estimated $t_{P(max)}$ will be lower. In the detailed timing model one also distinguishes between the t_{PD}s of the various inputs to the output. Doing this allows the propagation delay time of each path to be estimated more accurately. In the detailed timing model some *topological data* of the circuit must also be incorporated, e.g. whether or not inputs are interconnected. Furthermore, the timing can also be *data dependent*. This makes a computer simulation very time consuming and thus expensive, if not impossible.

Which aspects and parameters are incorporated in the timing model depends to a large extent on the *application*. Medical and military equipment must satisfy very high demands with respect to reliability. The timing model must then provide absolute certainty that the specifications will be met. Consumer electronics usually have lower demands, even though in this area as well demands get stricter as the integration density increases. Otherwise the costs of redesigning ICs become astronomical.

From the above it must be concluded that many parameters influence the timing behaviour. When manufacturers specify the parameters of their ICs they usually gloss over certain effects. As a rule an adjustment of the specified values of $t_{P(max)}$ is necessary. An increase of about 30 % is usually sufficient. A small increase or none at all requires a very thorough verification.

The fanout factor

Each gate can only drive a restricted number of other gates (of the same logic family). In most standard ICs the *fanout* is 10 to 20. Correct operation is guaranteed up to that limit. To prevent the designer of logic circuits from having to make *load calculations*, a standard load has been defined. The fanout of a gate output then expresses how many standard inputs this gate output is able to drive. Exceeding the fanout of a gate output does not necessarily mean that the entire circuit will not function properly. Propagation delay times, for example, do increase, but in applications with few timing constraints this is not a problem. *In general, it is advisable to respect all operating conditions.*

Special versions of some ICs are available with a higher fanout. A triangle placed after the qualifying symbol indicates that the IC has buffered outputs. See Figure 6.7.

Figure 6.7 Gates with buffered outputs.

Using a standard fanout means that the actual load of an input has to be related to some standard load. If an input constitutes a higher load than normal, this is indicated by a fanin of, for example, 2 or 3. The introduction of the concepts of fanin and fanout has resulted in a load verification being reduced to a *weighted counting* of the number of inputs.

The above discussion of the fanout holds for the use of standard ICs from the same series of integrated circuits. Similar rules apply between different series of the same logic family. The 7400 TTL series can drive about 30 7400LS TTL inputs. The other way round the fanout is 3 or 4. Between different logic families adapted rules apply. Sometimes direct data transfer is not at all possible, or the noise susceptibility is higher. In the discussion of the logic families we will give some examples. The designer of (VLSI) ICs has more freedom with regard to fanout than the user of standard components. His design software automatically adjusts the layout of transistors.

Note

The concepts of fanin and fanout are sometimes used in a wider context, such as indicating the number of inputs of a gate (4-input NAND $\Rightarrow$ fanin = 4) or the number of outputs of a gate (see I^{2}L logic). The load factor is then left out of discussion. □

Speed-power product

In many applications a low dissipation is essential, measured in Watts (μWs) per gate equivalent. In ICs it is possible to obtain shorter propagation delay times, at the expense of a higher dissipation. In this context one often uses the concept of the *speed-power product*, the product of the power consumption of a gate and its propagation delay time. In addition, the clock frequency has a considerable influence on the dissipation. See Section 6.8.

6.2 The TTL logic family

Before ca. 1965 electronic logic circuits were built with discrete diodes and transistors, complemented with resistors and capacitors for the DC bias. Well-known families were the RTL family *(Resistor-transistor-logic)* and the DTL family *(Diode-transistor-logic)*. The complexity of the logic modules was small and hardly exceeded the gate level. Miniaturisation of the discrete components and their mounting on a small PCB (Printed Circuit Board) led to manageable building blocks. Since the introduction of the bipolar TTL family *(Transistor-transistor-logic)* in 1965 (Texas Instruments) RTL and DTL have faded into the background.

At this moment TTL is the most comprehensive logic family. Most manufacturers of semiconductors carry an extensive TTL series. A great variety of functions is available, varying from gates (SSI, Small Scale Integrated Circuits), the register level (MSI, Medium Scale) to complex logic functions (LSI, Large Scale, ≥ 100 gate equivalents). The market for very large scale (VLSI) designs is dominated by MOS technology because of its high integration density and low dissipation per gate equivalent.

Basic component in TTL technology

The basic component in the bipolar TTL technology is the NAND circuit. Table b in Figure 6.8 describes the external operation of the NAND circuit in Figure 6.8.a. In positive logic this circuit realises the logic function

$$S = \overline{x \cdot y \cdot z},$$

which explains its name.

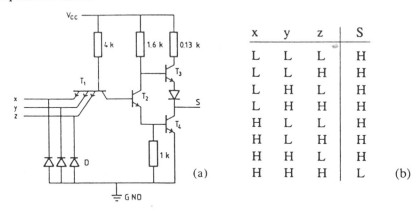

Figure 6.8 Basic component TTL integrated circuits: NAND (positive logic).

The circuit is designed as follows. The input transistor is a multi-emitter transistor. This transistor realises the AND function. The output stage follows T_1. Its function is to amplify the signal obtained from T_1 and to produce a signal S with an H or L voltage level which lies within the predefined ranges. Depending on the bias voltage (provided by T_1) transistor T_2 either drives T_4 or T_3 into saturation. If T_3 conducts, then the output voltage of S is at the H level. If T_4 is conducting the output level is L. Transistor T_2 is therefore called a *phase-splitter*. An output stage constructed in this way is called a *totem pole*. Its output S is either H or L. Other output constructions are discussed in Section 6.3.

The resistor values given correspond to those of the standard TTL series. From now on they will be referred to as 'N'. Higher resistor values result in a lower dissipation and lower speed, and lower values in a higher speed. This property has led to different IC series within the TTL family.

The diodes (*clipping diodes*) in Figure 6.8 protect the inputs against negative input voltages. These may arise as a result of cross talk or because of signal reflections. The diodes clip the input voltage at ca. -0.4 V.

The output section of the circuit in Figure 6.8 inverts the signal from the input transistor T_1, which drives T_2. The logical interpretation is a *negation*. It can be compensated by adapting the circuit as shown in Figure 6.9.

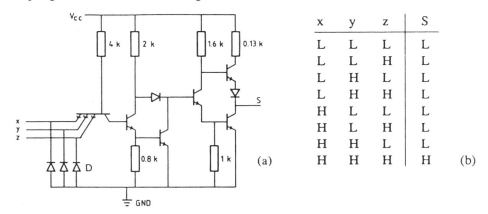

x	y	z	S
L	L	L	L
L	L	H	L
L	H	L	L
L	H	H	L
H	L	L	L
H	L	H	L
H	H	L	L
H	H	H	H

(a) (b)

Figure 6.9 TTL AND gate (positive logic).

In positive logic this circuit realises the AND function. The preference for the NAND in TTL technology follows directly from a comparison between the circuit diagrams. (Chapter 5 has already shown that, in principle, any circuit can be built with NANDs).
Figure 6.10 shows the TTL NOR circuit diagram. Three pairs of two transistors realise the 3-input OR function, and the resulting signal is buffered by the inverting output stage. A NOR circuit is more complex than a NAND circuit so, when faced with a choice between NAND and NOR, the NAND is preferable. If the input transistors of a NOR are implemented as a *multi-emitter transistor* (see the dashed additional inputs) a gate with the logic function AND-NOR, in the positive logic convention, is formed. This component effectively realises two logic levels in one physical gate level. This property of the AND-NOR is, for example, used in the realisation of fast adder circuits (Chapter 7).

TTL family

The TTL family consists of several series, viz.:

N   Standard TTL. Propagation delay time ca. 10 ns per gate. See Table 6.1 for further characteristics (7400 series).

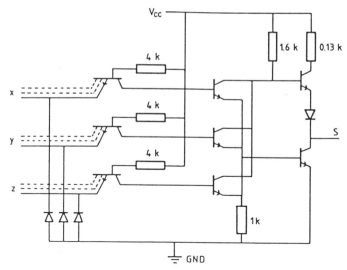

Figure 6.10 TTL NOR (AND–NOR) (positive logic).

L   Low-power TTL. Resistance values have been chosen higher than in the N series. The dissipation is lower, and the propagation delay time increases (74L00 series).

H   High-power TTL. Lower resistance values. Greater dissipation. Less susceptible to noise (74H00 series).

S   Schottky-clamped TTL. Schottky diodes prevent the transistors from going into saturation, making a higher switching frequency possible (74S00 series).

LS Low-power Schottky TTL. A combination of the L and S series. Has superseded the standard series (74LS00 series).

AS Advanced Schottky TTL. Recent development with greater switching speeds compared to the S series. Lower dissipation (74AS00 series).

ALS Advanced low-power Schottky TTL. An improved version of the LS series (74ALS00 series).

The series are available with different specifications: the 5400 and 7400 series. The temperature range in the 5400 series is roughly from −55° C to 125° C. The 7400 series goes from 0° C to 70° C.

Figure 6.11 gives an overview of the H and L ranges of the TTL series. The series are fully compatible, although there are small differences in, for example, the noise margins (DC noise) and fanin and fanout.

Table 6.1 shows the mean value of some parameters of the TTL series. The maximum flip-flop frequency is related to the clock frequency of flip-flop memory elements. See Chapter 10.

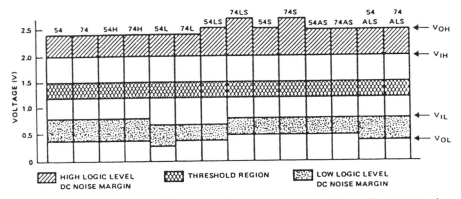

Figure 6.11 Series 5400/7400 TTL family, DC noise margins (Texas Instruments).

TTL series	Propagation delay (ns)	Dissipation /gate (mW)	Speed-power product (pJ)	Maximum FF freq. (MHz)	Fanout capability
N	10	10	100	35	10
L	33	1	33	3	20/10
H	6	22	132	50	10
S	3	19	57	125	10
LS	9	2	18	45	20/10
AS	1.7	8	13.6	200	40
ALS	4	1.2	4.8	70	20

Table 6.1 Parameters of TTL series.

With TTL gates having an input-output structure as shown in Figure 6.8, for example, an excellent *digital design kit* can be constructed. Input and output impedances are decoupled well, and there is little interaction between gates. This means that the designer can concentrate mainly on the *functional part* of the design. The restrictions imposed on the designer by the components do not dominate.

Development of the standard TTL series has practically stopped. New types are primarily designed in the ALS and the AS series.

6.3 Output drivers in the TTL family

TTL gates with *totem pole outputs* are capable of driving 10 to 20 inputs of other gates of the same type. The designer can therefore restrict himself to counting the number of driven inputs, with their fanin if it deviates from the standard. Other *output constructions* are also used. Their treatment requires a short introduction to the transistor model.

The semiconductor diode

Figure 6.12 shows the volt-ampere (V-I) characteristic of a semiconductor diode. A diode is an element that has a low resistance in one direction and a high resistance in the other. As a rough estimate the forward-biased resistance may be taken as 0 Ω and the reversed-

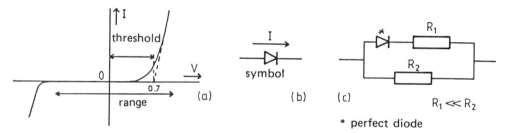

Figure 6.12 I-V characteristic of a semiconductor diode.

biased resistance as infinitely high. This is called a *perfect diode*.

In digital design a diode is usually used in only *two modes*: reverse-biased or conducting far above its threshold. A model for this application is shown in Figure 6.12.c. In the forward direction the resistance is low. A perfect diode and resistor R_1 describe this behaviour. In the non-conducting mode R_2 determines the effective resistance. For a proper diode it holds that $R_2 \gg R_1$. The current through R_2, in the *reverse* mode, is the *leakage current*.

The bipolar transistor

The serial interconnection of two reverse-biased diodes forms a bipolar transistor. There are two types, NPN and PNP transistors, which are shown in Figure 6.13.

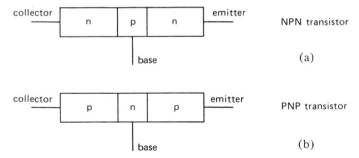

Figure 6.13 Bipolar transistors.

The NPN transistor consists of a thin layer of p-type semiconductor between two layers of n-type semiconductor. Depending on the collector-emitter voltage one diode is forward-biased, while the other is reverse-biased. PNP transistors are complementary to NPN transistors. For that reason we restrict ourselves to a short introduction to NPN transistors.

A positive collector-emitter voltage allows a current I_{ce} to flow. This current is controlled by the base-emitter voltage and the base current I_b. With a negative base-emitter voltage the transistor is in the non-conducting mode. A base-emitter voltage above a certain threshold voltage induces a current I_{ce}, which, to a great extent, is linearly proportional to I_b,

$$I_{ce} = \alpha \cdot I_b,$$

with
$$10 < \alpha < 400.$$

Above a certain value of I_b the transistor *saturates* and I_{ce} does not increase further. Bipolar transistors are current-controlled. With a base-emitter voltage below the threshold the transistor is non-conducting, in which case only a small leakage current flows. The leakage current is much smaller than the forward-biased current. It is omitted in most of our models.

Symbols and simple models of bipolar transistors are given in Figure 6.14. In digital design transistors are mainly used in one of two modes: (near-)saturation or non-conducting (with a certain leakage current). For its behaviour as a *switch* the models as shown are sufficient. A bipolar transistor then has a conducting path (low resistance) between the emitter and the collector when $V_b < V_e$ (PNP) or when $V_b > V_e$ (NPN). (For convenience we will neglect the base-emitter threshold voltage of ca. 0.5 V.) The minimum voltage drop between the emitter and the collector is about 0.2 V. A small base current also flows in a conducting transistor. We will omit this current from our models (see Section 6.5). When $V_b > V_e$ (PNP) or $V_b < V_e$ (NPN) the transistor is cut off. The collector-emitter resistance is then high and the voltage drop over the transistor may be much higher, even of the order of V_{cc}.

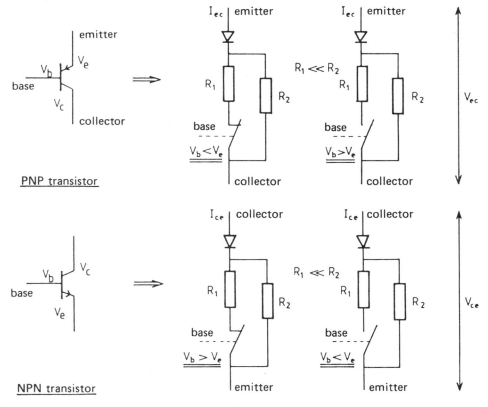

Figure 6.14 Substitution schemes of transistors.

With the transistor models in Figure 6.14 one can understand the external operation of the following output circuits. For the logic designer this is sufficient. (The designer of ICs must, of course, use a much more detailed model.)

Output drivers
Totem pole output

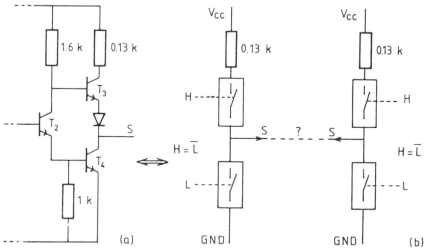

Figure 6.15 Totem pole output and substitution scheme.

The standard output driver stage of TTL integrated circuits is the totem pole. Depending on the input voltages of the gate one of the two output transistors is conducting. The output level is pulled to V_{CC} (T_3 conducting) or to ground (T_4 conducting). Figure 6.15.b shows a model with two 'switches'. The switches are coupled in the totem pole. One of them is closed.

Clearly two totem pole outputs must never be allowed to be interconnected directly. If T_3 of one circuit and T_4 of the other is conducting, then a *short circuit* is created from the power supply V_{CC} to ground. Circuits cannot normally withstand this for very long. Besides, the voltage level at such a point is not well-defined, compared to the specifications for H and L.

Open-collector output

It is possible to omit one of the two output transistors. See Figure 6.16. In TTL one generally omits T_3. As the collector of transistor T_4 is then open, this output construction is called the *open-collector output*. When T_4 is cut off the output is floating. In order to get a well-defined H output level an external resistor R to V_{CC} is necessary (*pull-up resistor*). In many ICs the resistor R is added internally. The output is then called 'active low/passive high' (passive pull-up).

With open-collector outputs it is permissible to interconnect a number of IC outputs in parallel. See Figure 6.17. The value of the resistor R has to be adjusted to the number of outputs connected in parallel. A number of inputs can be driven with the resulting signal.

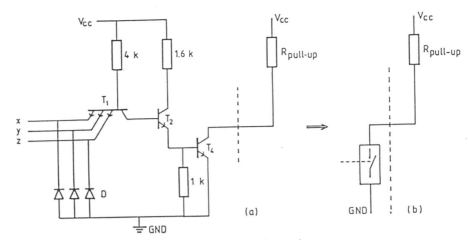

Figure 6.16 Open-collector output and substitution scheme.

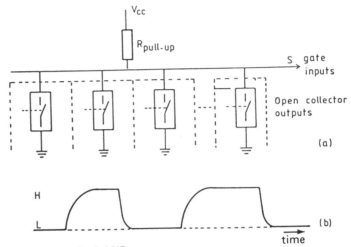

Figure 6.17 Open-collector wired-AND.

Figure 6.17.b shows the typical oscillogram of an open-collector output line. The rise time t_{TLH} is high (passive high), while the fall time t_{THL} is low (active low).

Note
A construction with several open-collector outputs wired in parallel is often called a *wired-AND*. The construction in Figure 6.17.a does indeed realise the AND function (in positive logic). If one of the output transistors T_4 is conducting, the voltage level at point S is low. In principle, the name *active-low wiring* would be more correct for a number of electrically interconnected open-collector outputs. □

3-State output

The principle of 3-state outputs is described in Figure 6.18. An *output enable* signal EN determines whether the output has its *normal effect* externally (EN = 1) or whether it is in

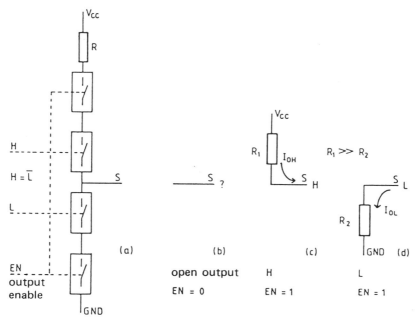

Figure 6.18 3-State output.

the *high-impedance state* (EN = 0). A circuit with 3-state outputs can work normally internally, although this cannot be observed at the output pins. This is schematically shown in Figures 6.18.b through 6.18.d.

3-State outputs can be interconnected directly. Each output would then be capable of determining the output level of the line (EN = 1), provided that the outputs of the other circuits were disabled (EN = 0). With 3-state outputs two circuits are *not allowed* to be enabled at the same time. This might cause a short circuit between V_{CC} and ground. Most 3-state outputs cannot withstand this for long. Moreover, the signal level is not defined correctly in this case.

A comparison between 3-state and open-collector outputs reveals that both have pros and cons. Because the 3-state output switches between 'active High' and 'active Low' the signal wave form at the interconnected outputs is better than with interconnected open-collector outputs, and signal transitions are quicker (at an L $\rightarrow$ H transition). The open-collector output, however, is protected against short circuits.

Symbols

Figure 6.19 lists the symbols used to indicate open-collector and 3-state outputs. In other technologies as well output constructions are found that correspond to the above TTL-oriented constructions.

The application of the output constructions that have been introduced here will be discussed in Chapter 7.

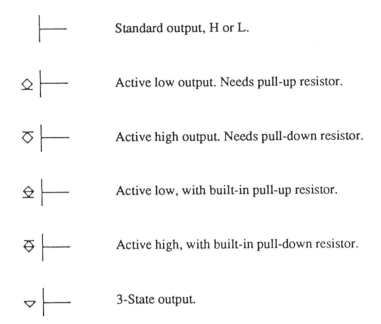

Figure 6.19 Output symbols.

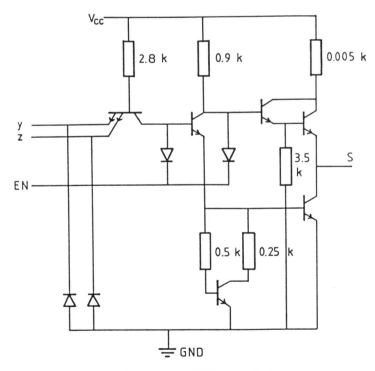

Figure 6.20 Circuit diagram of a 2-input NAND with 3-state output.

Figure 6.20 describes the circuit diagram of a 2-input NAND with an output enable input EN. For a further discussion of the properties of wired logic see Kambayashi and Muroga.

6.4 The I²L logic family and other bipolar families

Besides TTL there are several other bipolar logic families. One of them is the I²L family (Integrated-injection-logic), which has been under development since 1972. The supply voltage is ca. 0.8 V and the dissipation is very low. I²L permits a high gate density, owing to the structure of the cell.

Basic I²L cell

The *basic I²L cell* consists of a PNP and an NPN transistor. This combination (Figure 6.21.a) can be mapped very efficiently onto a cell layout. The PNP transistor serves as a *current source (load)*. Within an I²L environment the cell functions as an *inverter*.

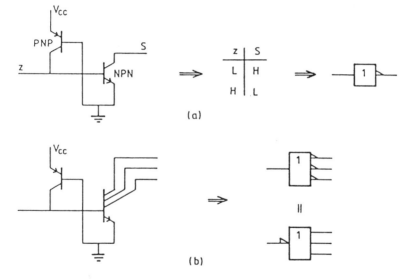

Figure 6.21 Basic cell I²L technology.

At the 'silicon level' the collector of the PNP transistor is a small island of n-material in a P-type region. The collector can therefore easily have multiple implementations (Figure 6.21.b). The cell is then a multiple-output inverter. In logic diagrams the polarity indicator is usually drawn at the input of the symbol. When outputs of different cells are interconnected in parallel, an active low wiring is formed. In positive logic an active low wiring realises the AND function. Figure 6.22 illustrates how the various logic operations in I²L can be realised at the gate level.

Logic functions with I²L

The basic components in an I²L environment are, as we have already said, the multiple-output inverter and the wired-AND. When mapping a logic function onto an I²L structure the designer preferably starts from a somewhat higher level, for example from NANDs or NORs. (How a random logic function can be transformed into these operations has been discussed in Chapter 5.) NANDs and NORs in I²L are shown in Figure 6.23. The schemes follow directly from Figures 6.21/22.

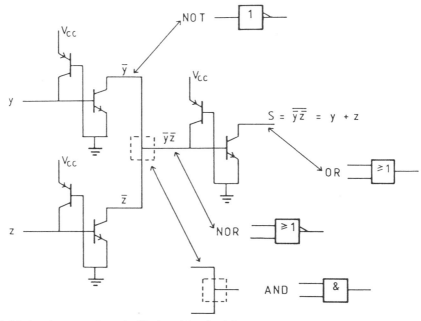

Figure 6.22 Logic operations in I²L (positive logic).

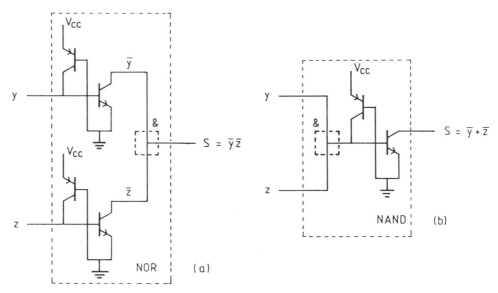

Figure 6.23 2-Input NOR and 2-input NAND in I²L.

An advantage of a NOR is that external signals immediately enter the I²L cells, before the wired-AND is performed. A NAND realisation assumes that the inputs come from other I²L cells, in other words, from collectors on the bottommost NPN transistor. External signals have to be adapted. The 'NAND' cell generally results in simpler circuits after mapping than the 'NOR' cell, even when input buffers are necessary.

Example *EXOR gate in I²L (NANDs)*
The formula for an EXOR gate is

$$S = \bar{y}z + y\bar{z}$$
$$= \overline{\overline{\bar{y}z + y\bar{z}}}$$
$$= \overline{\overline{\bar{y}z} \cdot \overline{y\bar{z}}}$$
$$= \text{NAND}(\text{NAND}(\bar{y},z), \text{NAND}(y,\bar{z})).$$

Figure 6.24 describes the result after mapping onto an I²L structure. □

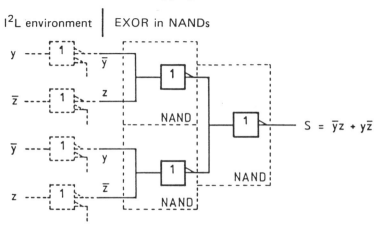

Figure 6.24 NAND implementation of an EXOR gate in I²L (positive logic).

Example *EXOR gate in I²L (NORs)*
The formula for an EXOR gate is

$$S = \bar{y}z + y\bar{z}.$$

The NOR-NOR form of this formula (procedure Section 3.4) is

$$S = \overline{\overline{\bar{y} + \bar{z}} + \overline{y + z}}$$
$$= \text{NOR}(\text{NOR}(\bar{y},\bar{z}), \text{NOR}(y,z)).$$

Figure 6.25 describes the result after mapping onto I²L structures. The input signals $\bar{y}$ and $\bar{z}$ have been made with extra collectors on the input cells. If $\bar{y}$ and $\bar{z}$ are available externally these may be used. □

From this example it follows that the *concept 'gate'* as it has been used in Chapter 5 in the discussion about mapping logic functions onto hardware has to be replaced by the concept *'layout structure'* at the IC level. For every function a structure has to be determined

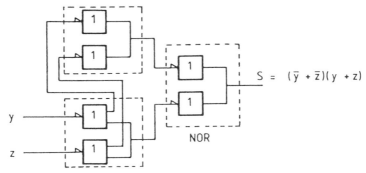

Figure 6.25 NOR implementation of an EXOR gate in I²L (positive logic).

which, within the agreed logic convention, realises the desired function externally. This mapping process is certainly not always one-to-one during the design of ICs, as is demonstrated by the I²L structures shown.

Properties

I²L is, in view of the very low dissipation and the high integration density, a technology that is very suitable for LSI/VLSI applications. In SSI/MSI practically no additional standard components are available. Most I²L circuits therefore have TTL-I²L adapter circuits at the inputs and I²L-TTL adapter circuits at the outputs. Depending on the process, the propagation delay times of I²L vary between 50 ns to ca. 20 ns. The technology cannot be considered very fast (yet), but it has a very low speed-power product (up to ca. 0.5 pJ).

ECL logic

We will deal very briefly with the remaining bipolar families. We only mention in passing the ECL family (*Emitter-coupled logic, Current-mode logic*). This family is very fast, with propagation delay times of up to 0.5-2 ns/gate. These switching times are obtained at the expense of a high dissipation and, as a result, a low integration density. The dissipation per gate equivalent is 3–10 times higher than with TTL.

ECL logic is used in *fast computers*. Such computers consist of many ICs (heat sink is a problem). ECL logic is also applied in the input stages of logic devices, for example, to sample high-frequency analog signals (prescalers).

The basic element for mapping logic functions onto ECL is the NOR. Output wiring is permitted and results in a wired-OR (positive logic). Because of high switching frequencies ECL requires a lot of attention with regard to HF shielding and transmission line effects. Consequently, the 'electrical design' of an ECL circuit is difficult, as is the layout.

6.5 MOS field-effect transistors

Bipolar transistors are based on *current amplification*. The collector-emitter current I_{ce} is proportional to the base current I_b,

$$I_{ce} = \alpha I_b,$$

with α between approx. $10 < \alpha < 400$. The impedances in ICs based on bipolar transistors are low. The currents and consequently the power consumption are relatively high, in both the *static state* and the *dynamic state*. Additionally, the IC area used per transistor/gate is considerably larger than in MOS. Bipolar technologies are therefore less suitable for VLSI designs.

Nevertheless, bipolar ICs are frequently used. This is because of their high switching frequency, up to 10^8 MHz or more. This frequency is considerably higher than the obtainable standard in MOS. For the rest, bipolar technology is often used for historical reasons, as it was the first technology to yield workable ICs.

MOS transistors (Metal-Oxide Semiconductors) belong to the family of *field-effect transistors* (FETs), in which the output current is *voltage controlled* (typically 0.25 mA/V). They have an extremely high input impedance, which is capacitive by nature. This input capacitor restricts the frequency of MOS ICs to about 25 MHz, although the frequency range can be extended to approx. 100 MHz by special 'low-capacitance' technologies, including CMOS-SOS (Silicon-On-Sapphire).

MOS has many advantages over bipolar technologies. Besides its very low power consumption (static), a MOS transistor/gate requires much less area on an IC. Because the yield of a process depends to a great extent on the IC's area, VLSI designs require a high gate density. Another advantage of MOS over bipolar is the fact that the input impedance of a MOS transistor is capacitive. The resistance of this input capacitor is very high. An electrical charge can therefore be stored for a longer period of time. *Dynamic logic*, which has a simpler layout structure than *static logic* (memories), is based on this principle.

Because the processing of MOS ICs is simpler than that of bipolar ICs, higher yields can generally be obtained. This advantage disappears as the sub-micron area is approached. Then the MOS process becomes very complex as well.

In field-effect transistors the current through the transistor is based on the *majority carriers* in the conducting channel. In contrast, in bipolar transistors both majority and minority carriers contribute to the transport. Because the mobility of electrons is much higher (2.5 times with the present internal field strengths) than that of holes NMOS is usually preferred. The frequency of NMOS ICs is higher than that of PMOS ICs. For entirely different reasons, to be explained later, a combination of both, CMOS, is preferred at the moment.

The structure of a MOS transistor

Figure 6.26 shows the structure of an NMOS transistor. In a p-type silicon substrate two small n-type areas, '*source*' and '*drain*', are diffused. (Because of the symmetrical structure the source and drain are interchangeable, in principle. The 'output terminal' is usually called the 'drain', and the common terminal the 'source'.) The area between the source and the drain is called the '*channel*'. An insulated conductor is deposited over the channel, the '*gate*', which is made of aluminium or polycrystalline silicon ('poly'). The length L of the channel and the width W largely determine the characteristics of the transistor.

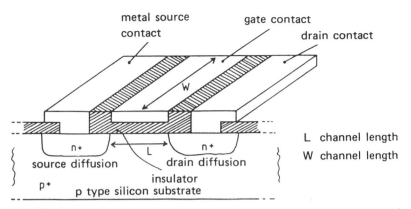

Figure 6.26 NMOS transistor.

The input resistance is very high, approx. 10^{10} Ω. The input current through the gate can be neglected in the static state. The transistor is then *voltage controlled*. The input capacitance is inversely proportional to the thickness of the insulator under the gate (approx. 0.5 μm). For that reason the input capacitance is relatively high. In the dynamic state some current will flow, charging/discharging the input capacitor. The *dynamic behaviour* of MOS ICs differs greatly from the *static behaviour*.

Because of the high input impedances one must take care to prevent *static electricity* from blowing up the MOS ICs during handling or assembling. For that reason modern ICs have a built-in protection circuit (diodes over the inputs).

The operation of a MOS transistor

NMOS transistors have a p-type substrate, in which holes are the majority carriers. A *depletion layer*, in which the majority carriers are repelled by minority carriers (electrons), is formed around the source and drain junctions. The substrate in between them is of the p-type. No current flows between the source and drain, and the impedance between them is very high.

A *voltage* on the gate induces a difference in potential over the insulator between the gate and the substrate. A positive voltage on the gate induces a negative charge in the semiconductor substrate on the other side of the gate insulator. This negative charge repels the holes in the p-type substrate. The induced *surface charge* reduces the charge in the surrounding substrate. A *depletion layer* (a layer with a reduced charge density) is formed on the surface. With an increasing positive difference in potential across the gate-substrate insulator an equilibrium is created on the substrate surface between the concentration of majority and minority carriers (holes and electrons respectively). Holes from the substrate are then repelled from the channel.

With a higher gate potential the electron density in the channel exceeds that of the holes. The channel *inverts* and becomes of the n-type. In this way it bridges the gap between the n-type source and drain junctions. The transistor starts to conduct and the source-drain path behaves as a resistor with an impedance between 20 Ω and 10 kΩ. For a non-conducting channel the resistance is considerably higher.

The gate-substrate voltage above which the channel between the source and the drain conducts is called the *threshold voltage*, denoted as V_t or V_{th}. Normally V_t is denoted with respect to the source voltage (common part input/output circuit). With a gate-source voltage $V_{gs} > V_t$ the transistor is in the ON state, and with $V_{gs} < V_t$ it is in the OFF state. The threshold voltage V_t is an important parameter in a MOS transistor. The type of channel between the source-drain influences the value of V_t.

The n-channel enhancement-mode MOS transistor

Figure 6.27 shows the structure of an n-channel enhancement-mode MOS transistor. For $V_{gs} = 0$ V no conducting channel has been formed. The transistor is in the OFF state.

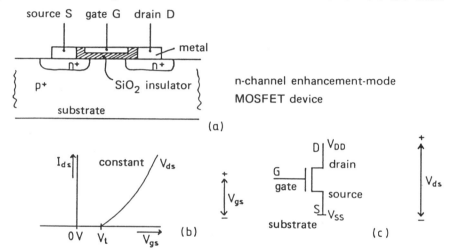

Figure 6.27 The n-channel enhancement-mode MOS transistor.

With a gate-source voltage $V_{gs} > V_t$ an n-type channel is formed between the source and drain. The transistor is then in the ON state. In many MOS processes V_t has a value of about

$$V_t \approx 0.2\, V_{DD}, \tag{6.1}$$

so that $V_t = 1$ V with $V_{DD} = 5$ V. This is not the place to discuss which and in what way process parameters influence V_t.

The n-channel depletion-mode MOS transistor

For $V_{gs} = 0$ V a depletion-mode MOS transistor already has a small n-type channel between the source and drain. As a consequence, a depletion-mode MOS transistor conducts for $V_{gs} = 0$ V. A negative voltage $V_{gs} < V_t$, which, depending on the process V_t may vary between -1.5 V and -5 V, forces the transistor into the non-conducting mode. One usually chooses $V_t = -0.8\, V_{DD}$. We note that the type of channel between the source and drain has to be realised during processing.

Figures 6.27 and 6.28 show the symbols which we shall use from now on to denote enhancement-mode and depletion-mode NMOS transisitors in the logic diagrams.

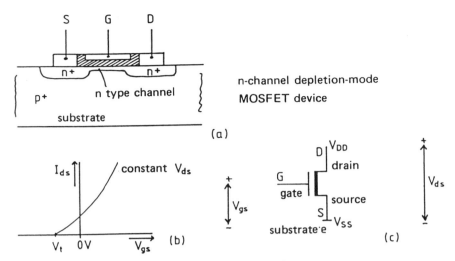

Figure 6.28 The n-channel depletion-mode MOS transistor.

Note
We have restricted the discussion to NMOS transistors. PMOS transistors are similar in behaviour. The I-V characteristics, however, are complementary with respect to the positive/negative values of currents and voltages. □

A MOS transistor is a near-perfect *voltage controlled switch* with a very high ON/OFF ratio (resistance in the conducting mode versus resistance in the non-conducting mode). In the *ON state* there is a small voltage drop across the transistor. Then, the resistance is 20 Ω – 10 kΩ. In the OFF state the source-drain path operates effectively as an open path, with a resistance higher than 10^{10} Ω. Except for its switching speed a MOS transistor is therefore a *perfect component* for logic circuits.

Dynamic behaviour

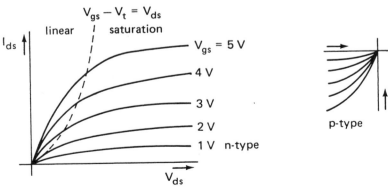

Figure 6.29 I_{ds}-V_{ds} characteristics with constant V_{gs} (n-channel).

Figure 6.29 sketches the characteristics of a MOS transistor obtained experimentally (with constant V_{gs}). For a discussion of the physical phenomena leading to this form of the characteristics, we refer to the References (see e.g. Mavor et al.). The following simplified model, consisting of three regions, is used as a first-order description for MOS transistors:

1. $V_{gs} < V_t : I_{ds} = 0$. (6.2)

The transistor is non-conducting (cutoff region) because the gate-source voltage V_{gs} is less than the threshold value V_t. There is no conducting channel and the transistor is in effect an open path.

2. $0 < V_{ds} < V_{gs} - V_t : I_{ds} = \beta[(V_{gs} - V_t)V_{ds} - (V_{ds}^2/2)]$. (6.3)

The gate-source voltage exceeds the threshold voltage. The transistor is conducting. The source-drain voltage is low and smaller than the difference $V_{gs} - V_t$. This is the so-called *linear region* of the transistor, also called the *triode region*. Roughly speaking, the current I_{ds} is proportional to V_{ds}. The transistor behaves as a (non-linear) resistance.

3. $0 < V_{gs} - V_t < V_{ds} : I_{ds} = \frac{\beta}{2}(V_{gs} - V_t)^2$. (6.4)

The source-drain voltage is so high that the current is maximal. The transistor is in the saturation mode.

So, depending on its mode a MOS transistor has the character of an
- open interconnection, (6.2)
- (non-linear) resistance, (6.3)
- current source. (6.4)

These approximations apply to the way in which MOS transistors are used in digital design, where signal changes are generally large. We shall return to this subject when discussing the behaviour of NMOS and CMOS inverters.

Note
The proportional factor in (6.3) and (6.4) is

$$\beta = \varepsilon \mu \frac{W}{LD},$$ (6.5)

in which ε and μ,
where ε is the permittivity of the gate insulator (dielectric constant) and
μ the mobility of the carriers (holes/electrons),
are both *material constants* which depend on the process applied and in which
W the width of the channel,
L the length of the channel and
D the thickness of the gate insulator
depend on the *geometry* of the transistor. β is called the *gain factor* of the transistor. □

Note
For ε it roughly holds that

$$\varepsilon = 3 \cdot \varepsilon_o = 3 \times 8.85 \cdot 10^{-12} \text{ Farad/m},$$

in which ε_o is the permittivity in free space. The typical value of μ for electrons in silicon is

$$8 \cdot 10^{-2} \, m^2/Vs. \quad \square$$

6.6 MOS inverters

The ratioed inverter

There are two basic principles in realising inverters in MOS, *'ratioed'* and *'ratio-less'*. See Figures 6.30 and 6.31.

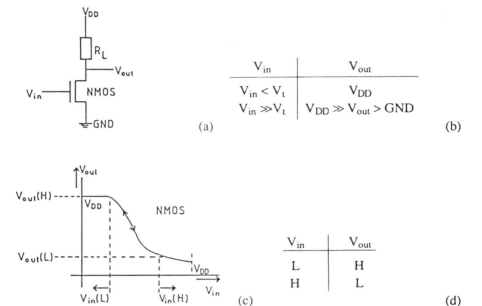

Figure 6.30 Principle of a ratioed inverter.

Figure 6.30 illustrates the principle of a *ratioed* inverter, based on an NMOS transistor and a pull-up resistor R_L (we will see that for R_L a transistor is used as well). For a cutoff transistor ($V_{in} < V_t$) the output voltage is approximately equal to V_{DD}, and, depending on the output load, possibly somewhat lower. With an input voltage $V_{in} \gg V_t$ the transistor conducts. An equilibrium then starts to be set between the voltage across the conducting transistor (resistance some kΩs) and the pull-up resistor R_L. Consequently, in this state V_{out} remains well above the zero potential (ground).

Figure 6.30.c describes the *transfer characteristic* of the inverter in Figure 6.30.a. In this characteristic the *logic interpretations* of the voltage levels have also been indicated. The corresponding logic behaviour is described by Table d in Figure 6.30.

A characteristic of a *ratioed inverter* constructed in this way is that the 'quality' of the H signal is obviously better than that of the L signal. One sometimes talks about a *soft* logic 0 (positive logic). The ratio-less inverter does not have this difference.

The ratio-less inverter

A ratio-less inverter is composed of a PMOS and an NMOS transistor. See Figure 6.31.a. The transistors have been designed such that the threshold voltage of the upper PMOS transistor is approx. 0.8 V_{DD} ($V_{in} < V_t$ → conducting) and that of the lower NMOS transistor approx. 0.2 V_{DD} ($V_{in} < V_t$ → non-conducting). A far more favourable transfer characteristic results, as can be seen in Figure 6.31.c.

With *complementary* transistors (PMOS and NMOS) both the L level and the H level are as close as possible to respectively the level of V_{SS} (GND) and the level of V_{DD}. The *logic swing* is then as large as possible. The CMOS technology to be introduced below is based on this fact. We shall first discuss *NMOS logic* (ratioed logic).

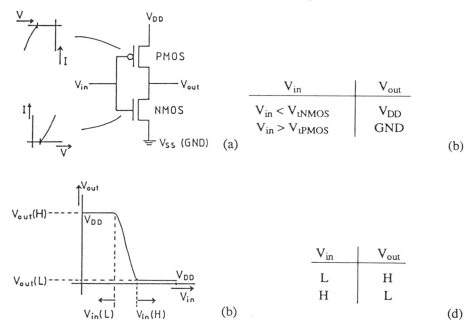

Figure 6.31 Principle of a ratio-less inverter.

NMOS inverters

Resistors require a relatively large area on ICs. Therefore, a so-called *pull-up transistor* is generally used as load R_L (Figure 6.30). By adapting the W/L ratio the resistance of a conducting transistor can be set within a certain range. For an enhancement-mode pull-up transistor, the gate and drain are interconnected (Figure 6.32.a). With a depletion-mode pull-up transistor, the gate and source are interconnected (Figure 6.32.b).

For an *enhancement-mode pull-up transistor* it holds that

$$0 < V_{gs} - V_t < V_{ds} \quad (V_t > 0). \tag{6.6}$$

The transistor then operates as a *current source*. A disadvantage of this circuit is that V_{out} is not able to exceed the voltage $V_{DD} - V_t$. Above this voltage level the transistor is cut off and the voltage drop across it automatically increases to V_t. This balances to

$$V_{out} < V_{DD} - V_t. \tag{6.7}$$

This phenomenon restricts the 'logic swing' of the output H level. See Figure 6.33.a for the transfer characteristic ($V_{in} - V_{out}$) of the inverter with enhancement-mode pull-up. For an input voltage $V_{in} < V_{inv}$ the output voltage is $V_{out} > V_{inv}$ and conversely, for $V_{in} > V_{inv}$ it is $V_{out} < V_{inv}$. The voltage V_{inv} may be considered as the *logic threshold voltage* (for which $V_{in} = V_{out}$).

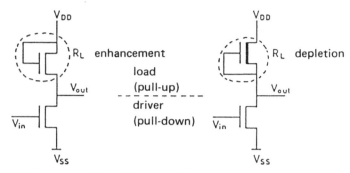

Figure 6.32 MOS inverters with pull-up transistors.

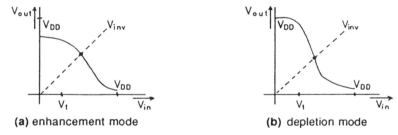

(a) enhancement mode (b) depletion mode

Figure 6.33 Transfer characteristics.

A *depletion-mode pull-up transistor* does not have this disadvantage. Here gate and source are interconnected. Since V_t is negative (for $V_t = 0$, the transistor conducts) it holds that

$$0 < V_{ds} < V_{gs} - V_t, \tag{6.8}$$

so that the transistor is used in its linear mode, up to (near) saturation with a high input voltage V_{in}. A depletion-mode pull-up transistor also has a somewhat steeper transfer characteristic than an enhancement-mode pull-up transistor. Depletion-mode pull-up transistors are therefore preferred.

In addition to the properties mentioned here the W/L ratios of the transistors of an inverter must have a certain relation to each other. These layout details will be discussed in Section 6.8.

6.7 NMOS logic circuits

The NMOS NAND

With two or more transistors in series, a NAND circuit (positive logic) is formed. See Figure 6.34.

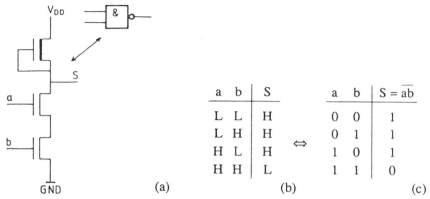

Figure 6.34 MOS NAND (positive logic).

In Table b, an input voltage L should be interpreted as a voltage below V_t of the input transistors and H as an input voltage above V_t. For layout and other reasons the number of inputs on an NMOS NAND is limited to 2 or 3 (see also Section 6.8).

The NMOS NOR

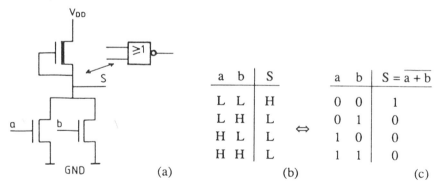

Figure 6.35 MOS NOR (positive logic).

A configuration of two or more transistors in parallel results in a NOR circuit (positive logic). See Figure 6.35. This can be extended to form NANDs/NORs with several inputs.

The NMOS AND and the NMOS OR

In order to realise an AND (positive logic) it is necessary to compensate for the inversion, which by nature is present in MOS NANDs. This can be done by extending the NAND with an inverter (NOT). Figure 6.36 shows the resulting diagram.

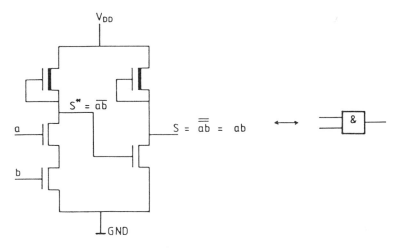

Figure 6.36 MOS AND (positive logic).

For an NMOS OR a similar diagram holds. It will be clear that logic designs are preferably made with the operators NAND/NOR instead of AND/OR.

NMOS AND-NOR circuits

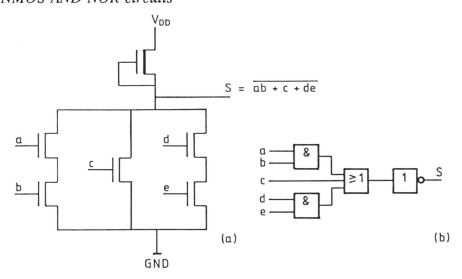

Figure 6.37 MOS AND-NOR circuit (positive logic).

A series of MOS transistors can easily be interconnected in parallel. Figure 6.37 shows how the logic function

$$S = \overline{ab + c + de} \tag{6.9}$$

can be realised. A series interconnection of transistors realises the logic AND, followed by a wired OR which realises the OR operation in the formula. For the logic designer this means that for the realisation of a given function S one has to start with $\overline{S}$, in the sum-of-products form (Section 3.4).

Figure 6.37.b shows the gate diagram of (6.9) in the AND-OR-NOT form. We see that at the transistor layout level the concept 'gate', as it was introduced in Chapter 5, becomes vague. All branches have a common pull-up transistor. A pure realisation based on the concept 'gate' would have different pull-up transistors. So when mapping onto, for example, NMOS one has to start with the logic formula instead of with the gate level.

6.8 Switching behaviour of ratio-type NMOS logic

Static behaviour

In MOS transistor logic there is a significant difference between the circuit's behaviour in the static state and in the dynamic state. In the *static state* all levels have been set and all capacitances in the circuit are charged/discharged. The input resistance of the gate of a subsequent transistor is very high ($> 10^{10}\ \Omega$), so that hardly any current flows between the successive transistors. We shall investigate what this means for NMOS inverter logic.

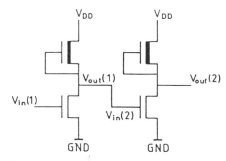

Figure 6.38 Series of two inverters.

Figure 6.38 shows a series of two inverters, each of which is realised with a 'depletion' pull-up transistor. Inverter no.1 conducts if $V_{in}(1) > V_t$ of transistor T_1. The corresponding logic input voltage is H; usually $H \approx V_{DD}$. The output voltage $V_{out}(1)$ then has the logic level L. This level is *only recognised as L* by the second inverter if $V_{in}(2) < V_t$ of transistor T_2. For standard NMOS V_t is approx. $0.2\ V_{DD}$, so for $V_{out(1)}$ to be a Low (L) output level it must hold that

$$V_{out}(1) \leq 0.2\ V_{DD}. \tag{6.10}$$

In the static state the current to the second inverter is practically zero. The currents through transistor T_1 and its pull-up transistor are then equal. The resistive part of the impedance of a MOS transistor is proportional to its length/width ratio Z. With Z_{pu} and Z_{pd} being the length/width ratio of respectively the pull-up and pull-down transistor, it must hold in the static state that:

$$\frac{Z_{pd}}{Z_{pu} + Z_{pd}} V_{DD} \leq V_t = 0.2\ V_{DD}. \tag{6.11}$$

After conversion it follows that

$$Z_{pu} \geq 4\, Z_{pd}. \tag{6.12}$$

The $Z_{pu}:Z_{pd}$ ratio is called the *inverter ratio*. In practice a 4:1 ratio is applied. The *logic threshold* V_{inv} is then approx. 0.5 V_{DD}.

From (6.12) it follows that:

$$(L/W)_{pu} > 4\,(L/W)_{pd}. \tag{6.13}$$

The interrelation between the ratios of the transistors has now been found.

Similar layout rules hold for NANDs or NORs to ensure that a driven gate reaches the appropriate voltage levels. A series of transistors (NAND) requires the L/W ratio to decrease proportionally, so that the transistors become increasingly wider or decreasingly shorter. With NMOS one therefore limits the number of transistors connected in series to two or three in practice.

Dynamic behaviour

In the *dynamic state* the capacitive part of the impedances strongly dominates the behaviour, especially the capacitance C_{gs}. The capacitance of the interconnections, $C_{interconnect}$, and of the substrate, $C_{substrate}$, may not be neglected either, as is commonly done with the calculations of propagation delays in TTL logic, for example. Charging/discharging (parasitic) capacitances is relatively slow for MOS, because of the relatively high impedance levels. This expresses itself in relatively long propagation delays for signal transitions. Because of this influence on the propagation delay the fanout of MOS gates is restricted in practice, although they have a very high fanout in the static state. (In principle the designer has some margin here; with fast logic the fanout is more restricted than with low-speed logic!)

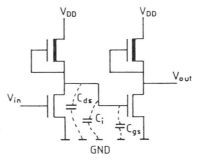

Figure 6.39 Series of inverters with some parasitic capacitances.

Figure 6.39 shows some of the parasitic capacitances that play a role in signal transfer. Loading is performed by the depletion-mode pull-up transistor, and discharging by the enhancement-mode pull-down transistor. We have seen that the L/W ratio of these transistors is related as pull-up:pull-down = 4:1. This has consequences for the speed of charging/discharging. The charge/discharge time t_{on}/t_{off} is also roughly related as 4:1. This can clearly be seen in oscillograms of MOS transistor logic. (For MOS the delay time is usually defined in relation to 0.5 V_{DD}, being V_{inv}.) The ratio of 4:1 is characteristic of *ratio-typed logic*.

Behaviour of NANDs/NORs

The above observations also apply to NANDs and NORs, with some restrictions. A series of two transistors of width W and length L roughly corresponds to one single transistor with length 2L. To maintain V_{inv} at approx. 0.5 V_{DD} the pull-up transistor must also be doubled in length. A 2-input NAND will get a rise/fall time that is twice as long as that of an inverter. For that reason signal transitions last longer. For three and more transistors connected in series the propagation delay increases proportionally (with otherwise unchanged size).

The effects of timing for a NOR depend on the number of inputs which simultaneously change to a high voltage. As this number gets larger the effective Z_{pd} will be smaller. For similar reasons the falling output transition will become faster and faster as more inputs simultaneously become high. The L/W ratio and the number of switching inputs also has consequences for V_{inv}, the voltage for which $V_{in} = V_{out}$.

6.9 Pass transistor logic

In the circuits discussed so far the transistors switch between V_{DD} and V_{SS} (GND). Figure 6.40 shows a different structure, in which a number of transistors are connected as a *pass transistor*. The type of switching between V_{DD} and V_{SS} discussed previously is called *restoring logic*, because after each gate *the logic voltage levels are restored*. The type of logic shown in Figure 6.40 is sometimes called *steering logic*.

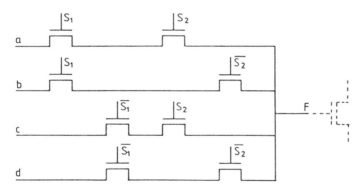

Figure 6.40 Selector with pass transistors.

The signals S_1, $\overline{S_1}$, S_2 and $\overline{S_2}$ drive the gates of the pass transistors in Figure 6.40 and determine whether the relevant transistor conducts or not. The uppermost chain conducts if $S_1 = S_2 = H$. The control of the selector is such that only one path is conducting, and the others are not.

Pass transistor logic requires more precautions than *restoring logic*. In applying the selector of Figure 6.40 one has to take care, for example, that points a through d control point F instead of the other way round. Moreover, attention must be paid to the removal of *hidden loops* (cf. Section 4.4). An input signal could, theoretically, influence another input signal via a path with F as intermediate node (out of the question here). The strength of *steering logic* lies in the large complexity of functions that can be realised with a relatively small number of transistors.

Besides static logic, pass transistors offer the possibility to realise *dynamic logic*, exemplified in Figure 6.41. When the value of S is High, the gate of transistor T_2 is set to the voltage V_{in} of the circuit and the gate-source capacitance is charged/discharged. For a non-conducting T_1 transistor the charge of this capacitor is stored and is able to keep T_2 open/closed for a longer period of time, independent of V_{in} to be precise. In this way *dynamic memories* can be realised.

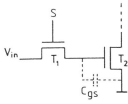

Figure 6.41 Dynamic memory function with pass transistors.

The dynamic behaviour of chains of pass transistors

A disadvantage of the use of pass transistors emerges when a timing model of Figure 6.42.a is made. In the conducting mode this chain roughly reacts as an RC chain.

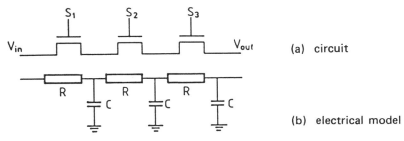

Figure 6.42 Model of pass transistor logic.

Let us assume that $S_1 = S_2 = S_3 = H$, where V_{in} switches from L → H. The input voltage is distributed over the pass transistors. As a result they are in the *resistive region*. The RC chain is set. It is known that its propagation delay time (RC time) is proportional to RCn^2, with n being the number of pass transistors. This is grounds for restricting the number of pass transistors which can be connected in a series.

As the signal V_{out} in Figure 6.42 becomes higher the voltage over the pass transistors decreases. The voltage V_{ds} of a pass transistor decreases and consequently V_{gs} as well. The transistors reach the cutoff region. The minimum voltage drop to which a transistor is set depends on its predecessor in the chain, but also on its threshold voltage V_t. For $V_{in} = V_{DD}$, the voltage drop over the first transistor is

$$V_{DD} - V_t,$$

so that V_{out} clearly remains below V_{DD} or the H level. A low voltage level (L) on V_{in}, however, is transmitted better.

One of the possible ways to limit these effects for pass transistors is to adjust the L/W ratio of the transistors driven by this chain of pass transistors, which decreases the voltage drop. This will not be discussed.

In practice, one usually applies a *complementary set* of pass transistors. An n-channel enhancement transistor poorly conducts an H level, and a p-channel enhancement poorly conducts an L level. The transistors do not have problems with the opposite voltage levels. The combination of both types forms a 'perfect' switch (Figure 6.43).

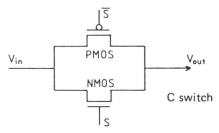

Figure 6.43 CMOS pass transistor switch.

To control a CMOS switch two complementary signals are required, often denoted as Φ and $\overline{\Phi}$ (anti-phase clocks).

Note
Pass transistors are always of the enhancement type. After all, depletion-mode transistors still conduct with a gate voltage of 0 Volt and can therefore not be cut off in this way. □

Conclusion
Pass transistors are a type of MOS logic that is somewhat difficult to handle, although it is able to realise some functions (selection) very efficiently. They are not often encountered with a low supply voltage V_{DD} (up to 2 V); in that case one uses 'restoring logic' only. □

6.10 CMOS logic circuits

In Section 6.7 the elementary NMOS circuits have been discussed. In that and the following sections a number of conclusions have been drawn with regard to these circuits:

- The depletion-mode transistor is clearly preferred as pull-up transistor within NMOS. This transistor has the lowest voltage drop and does not enter the cutoff region.

- If the pull-down transistors are conducting, there is a conducting path from V_{DD} to V_{SS} (GND): there is some dissipation. If the pull-down transistors are cut off, the dissipation is very low.

- The number of transistors connected in series/parallel influences the logic threshold voltage V_{inv}, so that the layout (L/W-ratio) must then be adjusted.

- There is a great difference between the charge and the discharge times of the gate capacitance of the next gate, which causes different delay times t_{PHL} and t_{PLH}.

The disadvantages of NMOS logic are compensated for within CMOS (Complementary MOS) logic.

The CMOS inverter

The CMOS inverter is of the 'ratio-less' type. We have already introduced its principles in Section 6.6. An enhancement-mode PMOS transistor is used as pull-up transistor. The pull-down transistor is of the enhancement-mode NMOS type. See Figure 6.31. The CMOS inverter has a number of obvious advantages:

- Because in the static state either the PMOS or the NMOS transistor is conducting there is no conducting path between V_{DD} and V_{SS} (GND) in either case (V_{in} = L or V_{in} = H). As a consequence, the dissipation of CMOS circuits is extremely low.
- In the static state no current flows between the gates (the same as in NMOS), nor in the gate itself. In the dynamic state such capacitances as C_{gs} and the like are charged/discharged.
- CMOS inverters have a very steep transfer characteristic, while V_{inv} remains approximately equal to 0.5 V_{DD}. The logic swing (difference between H and L) almost equals V_{DD}. The *DC noise margin* is approx. 2 V, compared to approx. 0.8 V with TTL. The DC noise margin is also better compared to NMOS.
- With CMOS the DC noise margin hardly depends on the L/W ratio of the transistors. The layout has far fewer restrictions.

In view of these characteristics NMOS is being replaced more and more by CMOS. The *quiescent power consumption* is approx. 10 nW/gate. The leakage current is some nAs. The switching speed of CMOS is somewhat lower than NMOS. In addition, CMOS requires more transistors, although they are smaller (see below). Nevertheless, at the moment CMOS is increasingly the technology of choice for VLSI ICs.

CMOS NAND and CMOS NOR

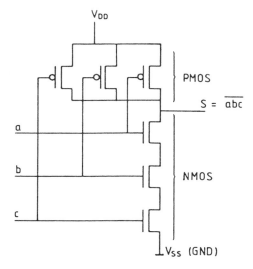

a	b	c	V_{out}
L	L	L	H
L	L	H	H
L	H	L	H
L	H	H	H
H	L	L	H
H	L	H	H
H	H	L	H
H	H	H	L

Figure 6.44 CMOS NAND (positive logic).

Figure 6.44 shows the transistor diagram of a CMOS NAND (positive logic). The design can be extended to a NAND with more than three inputs without difficulty. Figure 6.45 gives the transistor diagram of a CMOS NOR.

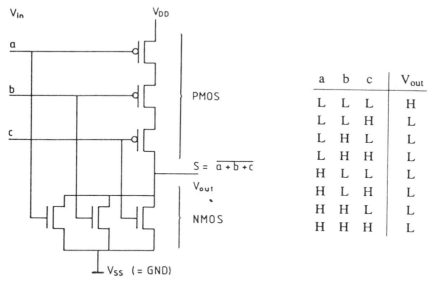

a	b	c	V_{out}
L	L	L	H
L	L	H	L
L	H	L	L
L	H	H	L
H	L	L	L
H	L	H	L
H	H	L	L
H	H	H	L

Figure 6.45 CMOS NOR (positive logic).

Dynamic CMOS (pseudo-NMOS) logic

A comparison between CMOS and NMOS shows that for the construction of a NAND or a NOR, for example, CMOS requires more transistors (with n inputs: 2n instead of n+1). This difference is partly compensated by the fact that in CMOS the W/L ratio is far less critical and that the transistors are on average smaller. All of the circuits introduced previously fit within a MOS environment called *static*.

To save IC area various MOS processes have been developed; these technologies are called *dynamic logic*. Its principle is shown in Figure 6.46.

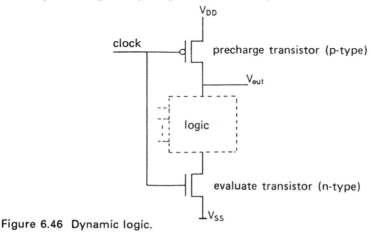

Figure 6.46 Dynamic logic.

Figure 6.46 shows a combinational logic circuit, comparable to the N part of CMOS (i.e. an NMOS construction without a pull-up transistor), interlocked between a p-type and an n-type enhancement-mode transistor. Depending on the phase of the signal 'clock' the following happens.

When the *'precharge transistor'* conducts V_{out} becomes approx. V_{DD}. All parasitic capacitances are charged accordingly. During the precharge phase the *'evaluate transistor'* does not conduct.

Subsequently, the clock signal switches to H. The evaluate transistor starts to conduct. When the logic function denoted as 'logic' is 'true', which corresponds to a conducting circuit, the voltage on V_{out} will drop. Otherwise V_{out} will remain approximately V_{DD}. V_{out} influences the next logic function.

Many problems are related to dynamic logic. The input signals may, for example, not only change during the precharge phase but also in the evaluation phase (cf. V_{out} in Figure 6.46). This causes problems with data transfer between gates, etc. Besides, the charge of the output capacitor will slowly leak away, so that V_{out} is only available for a limited period of time. Before the end of that period a new precharge phase must occur, if necessary a refresh cycle with the same data.

Dynamic logic more or less combines the advantages of both NMOS (few transistors) and CMOS (very low dissipation because the V_{DD}–V_{SS} path is always interrupted). For a detailed discussion see Weste, Chapter 5.

6.11 A comparison of MOS and bipolar

In the static state MOS ICs dissipate very little power (approx. 10 nW/gate in CMOS); the dissipation of NMOS is somewhat higher than with CMOS. A bipolar TTL gate then consumes approx. 1–10 mW, 10^6 times as much. In the dynamic state the power consumption of MOS ICs *increases proportional to the frequency*. One of the reasons for this can be attributed to the charging/discharging of the gate-substrate capacitance and other capacitances. Moreover, with CMOS one encounters the phenomenon that if the circuit switches, both the p-part and the n-part conduct for a short period. Then a much higher current than normal flows from V_{DD} to V_{SS} (GND).

Somewhere within approx. 1–20 MHz MOS ICs have the same power consumption as TTL ICs. This point is called the *cross-over frequency*. See Figure 6.47.

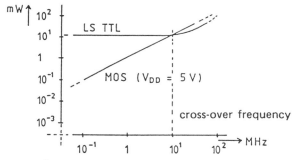

Figure 6.47 Cross-over frequency.

Clearly, both the timing mechanism and the switching activity in a design have a great influence on the definition and the position of the cross-over frequency. The clock pulse frequency (Chapter 10) can be ever so high, but if the number of signal transitions does not increase accordingly the power consumption will barely increase with the clock frequency. An often-used rule of thumb is that during one clock period about 50% of the signals change. For this assumption the cross-over frequency can be related to the clock frequency. Each timing mechanism has its own definition.

Power supply

MOS ICs usually operate for voltages between 3–15 V. The threshold voltage and the DC noise margins change proportionally. (So we must demand that with variations in the power supply the *entire* circuit must follow.) For TTL ICs the margins are much smaller, of the order of approx. 5 V ± 5%.
Although MOS ICs still operate correctly with a higher voltage, this is not without consequences for the *propagation delay time* of a gate. The propagation delay time is generally proportional to the supply voltage.

Fanout

With respect to TTL MOS is a high-impedance technology. It is therefore far more susceptible to *capacitive loads*. The time required to charge a parasitic capacitor is clearly longer for MOS than for TTL. This poses restrictions on the layout of ICs. The differences arise because bipolar logic generally uses *current control* and field-effect logic *voltage control*.
Because of its base current a bipolar transistor is a *continuous load*, allowing a single IC to drive a limited number of other ICs or transistors. A load considerably heavier than allowed according to the fanout rules not only interferes with the functional behaviour, but may even cause *permanent damage* through the great heat dissipation.
In a MOS IC capacitances are charged/discharged. An increase in the number of transistors (gates) to be driven increases the capacity to be charged/discharged, and consequently the RC time needed by the driver transistor. Its functionality is not affected.
As a result a very high load should preferably not be driven directly by a small transistor that is only able to conduct a small current. The delay will then increase more or less proportionally to the load.

The above discussion of bipolar and MOS families generally contains sufficient information for logic designers. Our intention was only to explain the backgrounds of possible technological restrictions and to convey to the logic designer *some understanding* of 'lower' design levels. An IC designer requires more technology-related know-how.

6.12 The gate model

Components have physical properties, which are interpreted logically. A component then operates logically 'correct' if certain conditions have been met, including:
– the supply voltage lies between $V_{cc} \pm \Delta V$;
– the temperature does not exceed the specified range;

- the fanin and fanout rules are respected;
- the input signals are within the H/L limits;
- the noise introduced by the environment is smaller than the noise margins allow.

All this was introduced in Section 6.1 and treated in more detail in Sections 6.2 through 6.11 for various technologies.

If the user requirements have not been met a circuit does not function correctly, or does not function reliably or the propagation delay times as measured are greater than the specified parameters indicate.

After an input transition the circuit enters the so-called *dynamic state*, the *collective name for all possible transient states* that may occur after an input transition. This state is left as soon as all internal and external signals have settled down again. According to the logic model the logic function of a circuit in the dynamic state is not specified. (In practice, the dynamic state does not necessarily lead to an output uncertainty.)

The fact that a circuit enters the dynamic state is not noticed at the output until a minimum time $t_{P(min)}$ has passed after the input transition. The dynamic state is left before a maximum time $t_{P(max)}$ after the last input transition.

$t_{P(min)}$:   – minimum propagation delay time;
- the time between an input transition and the earliest possible reaction of the outputs;
- until $t_{P(min)}$ the old output value is present.

$t_{P(max)}$:   – maximum propagation delay time;
- the time between the end of an input transition and the latest possible reaction of the outputs;
- after $t_{P(max)}$ the new output value is present.

The study of the timing properties of components and systems is based on the attributes $t_{P(min)}$ and $t_{P(max)}$ thus introduced. In data books $t_{P(typ)}$, the *typical value* of the propagation delay time, is often specified instead of $t_{P(min)}$. When the relation with $t_{P(min)}$ is not indicated, essential information is missing. See also Chapter 11.

Measurement conditions/operational conditions

Parameters of ICs are specified under '*parameter measurement conditions*'. These roughly correspond to the average application. The variations in the environment conditions are less than those allowed maximally for normal use (*operational conditions*). If necessary the user has to compensate for the parameter values as specified by making a correction. Application notes supply the required information. The reverse also occurs, i.e. ICs are used in an environment with smaller tolerances than the parameter measurement conditions. If necessary this can be used to one's advantage.

An example of a graph for the compensation of environmental influences is Figure 6.4. This kind of information is sparingly supplied by IC manufacturers, or rather it is scattered through the data books and is therefore inaccessible. Some manufacturers use a more pessimistic model than others. The authors' experience has taught them that an increase of up to 30 % of the propagation delay times specified is already necessary under normal operational conditions. Each path in a circuit with a margin of less than approx. 30 % must be thoroughly verified. By the way, problems with regard to timing verification do not only occur in *fast* logic circuits. See, for example, Chapter 11, clock skew.

The timing model of a gate

In the reflections on 'timing' we will restrict ourselves from now on to circuits with, as lower limit, the 'gate' as their smallest logic unit. Based on this we will derive the characteristics of larger logic circuits. For circuits composed of discrete components this is sufficient. For integrated circuits the model has to be extended. The model must then be made technology-dependent. A discussion of this extension lies outside the scope of this book.

In the following discussion we use the parameters $t_{P(min)}$ and $t_{P(max)}$ instead of the parameters t_{PLH}, t_{PHH}, t_{PHL} and t_{PLL}. Using the latter introduces a *data-dependent element* in the timing verification, which we would prefer to avoid for reasons of efficiency. Where necessary the restrictions of the timing model are indicated and suggestions for possible extensions given. In the following we assume that, depending on the application, a sufficiently accurate estimate of $t_{P(min)}$ and $t_{P(max)}$ is available.

The *reference level* to which the delay times are related differs greatly per technology. The reference level should indicate the *actual threshold level* of the gate as accurately as possible. With TTL the threshold is quite small. Limits are given in Figure 6.11. Usually the '*threshold area*' is defined by those input voltage levels for which the output voltage may leave the specified regions $V_{OH(min)}$ and $V_{OL(max)}$. With TTL, the *measured data* are generally related to 1.5 Volt.

With MOS one often applies 10/90 % of V_{DD} as reference levels (data dependent specification) or an input voltage V_{inv} for which $V_{in} = V_{out}$ (for an inverter) is taken as reference level. Normally speaking, MOS circuits are designed so that this level is 50 % of V_{DD}.

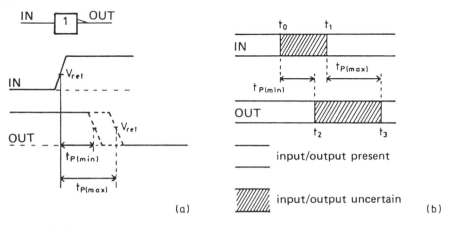

Figure 6.48 Timing model of an inverter.

Figure 6.48 describes the delay model of an inverter. The moment at which an input changes will usually be somewhat uncertain. See, for example, Figure 6.3. Figure 6.48.b illustrates how an uncertainty interval in the input signal affects the output behaviour. The following formulas hold:

$$t_2 = t_0 + t_{P(min)}, \qquad (6.14)$$

$$t_3 = t_1 + t_{P(max)}. \qquad (6.15)$$

The model described here does not say anything about the *distribution* of the moments at which the output changes. The model is a *minimum/maximum model* and only gives the extreme values. The distribution is certainly not uniform over the interval t_2–t_3. Some publications describe a statistical model to calculate gate delays. When the circuit is a series of gates these models result in smaller uncertainty intervals than the minimum/maximum model. ICs do not often conform to this distribution. The uncertainties in the production process are still too great for that. In our opinion the minimum/maximum model is more adequate for standard components.

For gates with more than one input the model of Figure 6.48.b must be somewhat extended. A well-known effect is that as more inputs simultaneously switch the propagation delay time can be considerably smaller than with one switching input (Section 6.1). This phenomenon affects the parameter specification, especially of $t_{P(min)}$. It should be chosen smaller than the delay for a single input transition. Know-how about this kind of effect is important for the modelling process.

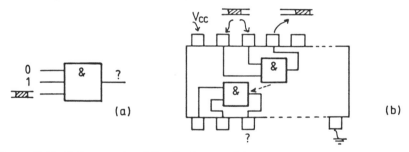

Figure 6.49 Refinements of the timing model.

Figure 6.49.a illustrates the situation of an AND gate having one input signal with an uncertainty interval, while at least one of the other inputs has a '0' value. Should the model prescribe an uncertainty for the output or not? Logically speaking this is not necessary, since one '0' input signal already completely determines the output value. We will see later that, depending on the internal structure of a design, there may be a difference between a logic (static) and a dynamic relation between inputs and dependent outputs.

The previous question brings up the following problem. What should be assumed for the outputs of the gates in one IC if an input of one of the gates switches (the situation in Figure 6.49.b)?

One might be of the opinion that the design contains a '*design error*' if, in this situation, the model prescribes an uncertainty interval for the outputs of the other gates. The gates of the IC have been insufficiently isolated from each other. Nevertheless, these effects have been observed in practice. If equipment has to meet very high reliability constraints, it is recommended that the uncertainty model be applied integrally, i.e. simultaneously over all gates in an IC and over all inputs of a gate. On printed circuit boards different rules hold. Here the ICs are, as a rule, isolated well enough from each other although 'crosstalk' does occur via the common supply lines. Decoupling capacitors may suppress much in such a case.

The model for a circuit composed of gates

We take Figure 6.50 as an example. For each gate in this circuit $t_{P(min)}$ and $t_{P(max)}$ are specified. This circuit can be considered in two ways:

– for each (group of) input(s) to the output separately;

– according to a more global model, which does not distinguish between the inputs.

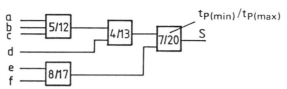

Figure 6.50 Gate circuit with propagation delay times.

Table 6.2 gives an overview of the propagation delay times, for each input-output combination. For a global model of the entire circuit we take the worst case situations. The propagation delay time of the entire circuit is then

11 ns $\leq t_{PD} \leq 45$ ns.

Inputs	$t_{P(min)}$	$t_{P(max)}$
a, b, c	16	45
d	11	33
e, f	15	37

Table 6.2 Overview of propagation delay times.

If we follow a more detailed model we may first of all differentiate according to the various groups of inputs. In addition, the direction of a signal transition also influences the delay (t_{PHL} and t_{PLH}). *Different levels of detail* may be observed. We usually use the global model, as introduced before.

In practice additions/corrections will always have to be made to this model. For example, the propagation delays increase because wiring delays have been neglected as compared to the delay of the gates. For printed circuit board designs that do not have to meet extremely high speed requirements this is acceptable. At the IC level it is not. Nor is this the case with a long interconnect line. The wiring then also requires a compensating propagation delay time.

A second correction is necessary when ICs from different logic families are used within one design. TTL inputs can be controlled directly by NMOS/CMOS ICs, but due to the difference in impedances (MOS high, TTL low) the transition times will strongly increase in comparison with a MOS or TTL design only. If we want to incorporate these effects as well, it is necessary to extend the model with an 'interconnect' and an 'interface delay', by which means the effect of the impedances can be compensated. For the case mentioned above 30 to 50 ns often have to be reserved for the interconnect, unless special interface circuits are used.

References

1. J.R. Armstrong, *Chip level Modeling of LSI Devices*, IEEE Tr. on CAD, Vol. CAD-3, 1984, pp. 288–297
2. H.H. Berger, *Modeling I²L/MTL Cells*, Microelectronics Journal, Vol. 9, 1979, pp. 16–23.
3. H.H. Berger and S.K. Wiedmann, *Merged-transistor logic – a low cost concept*, IEEE J. of Solid-State Circuits, Vol. SC-7, 1972, pp. 340–346.
4. W.N. Carr and J.P. Mize, *MOS/LSI Design and Application*, McGraw-Hill, New York, 1972.
5. R.H. Crawford, *MOSFET in Circuit Design*, McGraw-Hill, New York, 1967.
6. *Designers Guide High-speed CMOS*, Philips, Eindhoven, 1986.
7. M.I. Elmasry, ed., *Digital MOS Integrated Circuits*, IEEE Press, New York, 1981.
8. M.I. Elmasry, *Digital Bipolar Integrated Circuits*, John Wiley, New York, 1983.
9. R. Funk, *Interfacing HC/HCT QMOS Logic with Other Families and Various Types of Loads*, RCA Application Note ICAN-7325.
10. R. Funk, *Power-Supply Distribution and Decoupling for QMOS High-Speed-Logic ICs*, RCA Application Note ICAN-7329.
11. I. Gertreu, *Modeling the Bipolar Transistor*, Elsevier, Amsterdam, 1978.
12. L.A. Glasser and D.W. Dobberpubb, *The Design and Analysis of VLSI Circuits*, Addison-Wesley, Reading, Mass., 1985.
13. K. Hart and A. Slob, *Integrated Injection Logic, a New Approach to LSI*, IEEE J. of Solid State Circuits, Vol. SC-7, 1972, pp. 346–351.
14. A. Havasy and M. Kutzin, *Interfacing COS/MOS with other logic Families*, RCA Application Note ICAN-6602.
15. C.F. Hill, *Noise Margin and Noise Immunity in logic Circuits*, Microelectronics, 1986, pp 16–21.
16. S.K. Jain and V.D. Agrawal, *Modeling and Test Generation Algorithms for MOS Circuits*, IEEE Tr. on Computers, Vol. C-34, 1985, pp. 426–433.
17. Y. Kambayashi and S. Muroga, *Properties of Wired Logic*, IEEE Tr. on Computers, Vol. C-35, 1983, pp. 550–563.
18. Y.L. Levendel, P.R. Menon and C.E. Miller, *Accurate Logic Simulation Models for TTL Totempole and MOS Gates and Tristate Devices*, B.S.T.J., Vol. 60, 1981, pp. 1271–1287.
19. J. Logan, *Modeling for Circuit and System Design*, Proc. IEEE, Vol. 60, 1972, pp. 78–85.
20. J. Lohstroh, *Devices and Circuits for Bipolar (V)LSI*, Proc. IEEE, Vol. 69, no. 7, July 1981, pp. 812–826.
21. J. Mavor, M.A. Jack and P.B. Denyer, *Introduction to MOS LSI Design*, Addison-Wesley, London, 1983.
22. R.L. Morris and J.R. Miller, ed., *Designing with TTL Integrated Circuits*, McGraw-Hill, New York, 1971.
23. H.W. Ott, *Noise Reduction Techniques in Electronic Systems*, John Wiley, New York, 1976.
24. T. Poorter, *Electrical Parameters, static and dynamic response of I²L*, IEEE J. of Solid State Circuits, SC-12, 1977, pp. 443–449.

25. J.A. Scarlett, *Transistor-Transistor Logic and its Interconnections*, Van Nostrand Reinhold, New York, 1972.
26. Texas Instruments, *Advanced Schottky Family (ALS/AS) Application*, in: The TTL Data Book, Vol. 2, Texas Instruments, 1985.
27. J.T. Wallmark and H. Johnson, ed., *Field-Effect Transistors*, Prentice-Hall Inc., Englewood Cliffs, N.J., 1966.
28. R.L. Wadsack, *Fault Modeling and Logic Simulation of CMOS and MOS Integrated Circuits*, B.S.T.J., Vol. 57, 1978, pp. 1449–1474.
29. N. Weste and K. Eshragian, *Principles of CMOS VLSI Design, a Systems Perspective*, Addison-Wesley, Reading, Mass., 1985.

7
MSI combinational logic circuits

7.1 Introduction

Logic circuits are not always designed down to the lowest design level, the gate level (or in integrated circuits the layout level), nor is this always done down to every last detail. The more frequently an application or a problem occurs, the more one is tempted to give it extra attention and to design a special circuit for it. A great variety of such circuits for special applications have been designed and manufactured. A design is (partially) finished if a subproblem for which one or more *standard components* have been designed is recognised. It is clear that this method may save a lot of design time.

Roughly speaking the *design procedure* consists of the following steps:

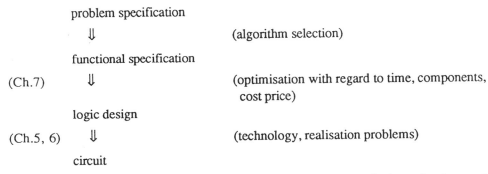

Upon every transition to the next level the most important aspects playing a dominant role in that stage of the design are given in parentheses.

The number of *degrees of freedom* needed to achieve a final design is great. From the many possible solutions the designer has to choose that solution which best meets the requirements. Examining these possibilities requires a lot of time, which may greatly influence the price of the final product. One is therefore often faced with the dilemma of having to make decisions early on the basis of a quick evaluation. Many solutions later turn out to be suboptimal. Clearly, the use of standard components may have a positive effect, certainly with prototypes or small series.

The function of a *standard component* has to be defined well. Its application must impose as few restrictions as possible on the designer. For that reason one often finds additional facilities in standard components, for example to restrict the fanin or extend the range of applications. The technology-dependent preference for certain logic operators also plays a role. In the following sections we shall treat some examples of component designs for standard functions. In so doing, we shall encounter a number of the aspects mentioned above.

7.2 The design of a 4-bit binary adder

Non-negative integers are notated in the binary system as

$$N = a_{n-1}a_{n-2} \ldots a_1a_0 \text{ BIN}$$
$$= a_{n-1}2^{n-1} + a_{n-2}2^{n-2} + \ldots + a_1 2^1 + a_0 2^0. \tag{7.1}$$

One of the operations often performed on numbers is *addition*. The rules of adding binary encoded numbers roughly correspond to those of decimal addition, on the understanding that in the binary system there is already a carry if the sum of the digits (bits) in a section is ≥ 2.

Example
Add A and B:

$$A = 1011_{BIN} \rightarrow \quad 1011$$
$$B = 0110_{BIN} \rightarrow \quad \underline{0110\ +}$$
$$\qquad\qquad\qquad\quad 10001 \quad \Rightarrow \quad S = 10001_{BIN}.$$

After examination it turns out that an adder for two numbers of four (n) bits may be composed of four (n) *identical sections*. Each section adds three bits, with a value of 0 or 1 and weight 2^i. These bits are:

– the bits a_i and b_i of the numbers to be added;
– the incoming carry C_i.

The least significant section does not have an incoming carry. We shall set it at $C_0 = 0$. Each section generates two outgoing bits:

– a sum bit S_i with a value of 0 or 1, with weight 2^i;
– an outgoing carry bit C_{i+1} with a value of 0 or 1, with weight 2^{i+1}.

Mathematically the operation of each section is specified by Equation (7.2.a),

$$\boxed{a_i 2^i + b_i 2^i + C_i 2^i = C_{i+1} 2^{i+1} + S_i 2^i,} \tag{7.2.a}$$

or by $\boxed{a_i + b_i + C_i = 2C_{i+1} + S_i.} \tag{7.2.b}$

In these formulas '+' is the plus sign, used in an *arithmetical sense*. Each of the variables has the value of 0 or the value of 1. □

Figure 7.1 shows the *structure of the full adder* for 4-bit numbers. Table 7.1 specifies the *arithmetical operation* of one of the sections.

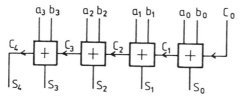

Figure 7.1 4-Bit full adder structure.

a_i	b_i	C_i	C_{i+1}	S_i	
0	0	0	0	0	
0	0	1	0	1	
0	1	0	0	1	
0	1	1	1	0	value of the bits, without weight.
1	0	0	0	1	
1	0	1	1	0	
1	1	0	1	0	
1	1	1	1	1	

Table 7.1 Specification of a full adder section.

In Figure 7.1 and Table 7.1 the indices i and i + 1 correspond to the power of the weight of the signal/bit, a power of 2.

Note

An adder section which can add three bits of equal weight is called a *full adder*. The least significant section of a binary adder only adds two bits, since $C_0 = 0$. An adder for two bits of equal weight is called a *half adder*. Use is not made of this possibility to reduce the circuit because

– C_0 is often necessary for other applications;

– the advantage of identical sections is lost. □

From arithmetic to logic

A *logic circuit* can now be designed for the full adder described here. To that end it is necessary to define a relation between the *arithmetical world* and the *physical world* in which the components and circuits are realised. This is done by logic, by means of *propositions*. These are:

A_i: 'Bit a_i has the value of 1'
B_i: 'Bit b_i has the value of 1'   } (and weight 2^i)
C_i: 'Bit C_i has the value of 1'

for the input signals and

C_{i+1}: 'Bit C_i has the value of 1'   } (and weight $2^{i+1}/2^i$)
S_i: 'Bit S_i has the value of 1'

for the output signals. The choice of the propositions is obvious in this case.

With Table 7.1 and these propositions truth Table 7.2.a can easily be formed. If we subsequently indicate the truth values of the propositions A_i through S_i by the logic variables a_i through S_i, in the usual way as 0 or 1, Table 7.2.b results. As the caption of Table 7.2 indicates, the information presented belongs to the *logic world*, the world of truth values.

The similarity between Table 7.2.b and Table 7.1 is a coincidence. This is generally not the case and the arithmetical specification clearly differs from the specification at the logic

A_i	B_i	C_i	C_{i+1}	S_i		a_i	b_i	C_i	C_{i+1}	S_i
F	F	F	F	F		0	0	0	0	0
F	F	T	F	T		0	0	1	0	1
F	T	F	F	T		0	1	0	0	1
F	T	T	T	F		0	1	1	1	0
T	F	F	F	T		1	0	0	0	1
T	F	T	T	F		1	0	1	1	0
T	T	F	T	F		1	1	0	1	0
T	T	T	T	T		1	1	1	1	1

(a) in truth values   (b) in logic 0 and 1

Table 7.2 Truth table of a full adder.

level. (In the above case the binary arithmetical specification is, of course, already close to the binary logic specification.)

Figure 7.2 shows the Karnaugh maps for C_{i+1} and S_i. These follow from Table 7.2.b. Equations 7.3 and 7.4 can now be easily formed. In these formulas the '+' sign indicates the *logic OR*. This concludes the *logic specification* of one section of a binary adder. Comparing the formulas in (7.2) shows that the *arithmetical formulation* and the *logic formulation* are clearly different.

$$\boxed{C_{i+1} = a_ib_i + a_iC_i + b_iC_i} \tag{7.3}$$

$$\boxed{S_i = \overline{a_i}\,\overline{b_i}C_i + \overline{a_i}b_i\overline{C_i} + a_i\overline{b_i}\,\overline{C_i} + a_ib_iC_i.} \tag{7.4}$$

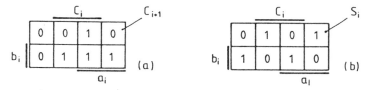

Figure 7.2 Derivation of logic formulas of a binary full adder.

Note

The previously introduced propositions belong to (7.3) and (7.4). Another choice of propositions usually results in other formulas. Therefore propositions have to be laid down clearly and unambiguously. □

From logic to circuit

Equations (7.3) and (7.4) can be converted into gate circuits in the usual way (Chapter 4). In this conversion several criteria for further optimisation may be used. A simple realisation with regard to the *number* of gates is found as follows:

$$S_i = \overline{a_i}\,\overline{b_i}C_i + \overline{a_i}b_i\overline{C_i} + a_i\overline{b_i}\,\overline{C_i} + a_ib_iC_i \tag{7.4}$$
$$= (\overline{a_i}\,\overline{b_i} + a_ib_i)C_i + (\overline{a_i}b_i + a_i\overline{b_i})\overline{C_i}$$

$$= \overline{(\overline{a_i}b_i + a_i\overline{b_i})}C_i + (\overline{a_i}b_i + a_i\overline{b_i})\overline{C_i}$$
$$= \overline{(a_i \oplus b_i)}C_i + (a_i \oplus b_i)\overline{C_i}$$
$$= (a_i \oplus b_i) \oplus C_i. \tag{7.5}$$

The formula for C_{i+1} can be written as (see Karnaugh map 7.2.a):

$$C_{i+1} = a_ib_i + a_iC_i + b_iC_i \tag{7.3}$$
$$= a_ib_i + a_i\overline{b_i}C_i + \overline{a_i}b_iC_i$$
$$= a_ib_i + (a_i\overline{b_i} + \overline{a_i}b_i)C_i$$
$$= a_ib_i + (a_i \oplus b_i)C_i. \tag{7.6}$$

Equations (7.5) and (7.6) result in the circuit of Figure 7.3.

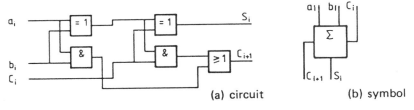

(a) circuit (b) symbol

Figure 7.3 Realisation of one section of a binary full adder.

Integrated adders are usually 4-bit oriented. Figure 7.4.a describes their internal structure, when composed of separate full adders. Figure 7.4.b shows the IEC symbol for 4-bit adders.

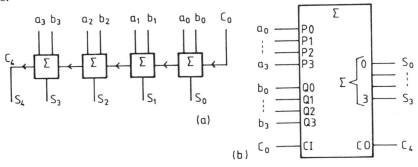

Figure 7.4 4-Bit full adder.

If 4-bit or 8-bit full adders in an integrated circuit form are available to the designer, the design process can stop as soon as the *applicability* of this component has been recognised. It is then not necessary to go into the internal structure. That this saves a lot of time and design effort will be made clear in the next two sections.

7.3 The carry problem in adders

The adder section in Figure 7.3 has been optimised to the criteria 'a minimum number of gates' and 'simple gates'. With regard to the criterion 'minimum propagation delay time' this setup is less successful (see the timing analysis in Figure 7.5). For convenience all gates are supposed to have the same limits for the propagation delay time, lying between $t_{P(min)}$ and $t_{P(max)}$.

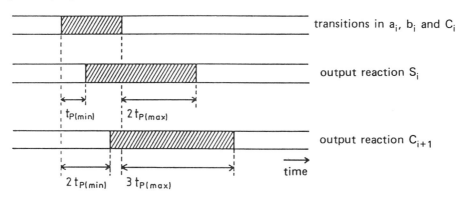

Figure 7.5 Analysis of propagation delay time of a full adder section.

The internal structure of the full adder section given in Figure 7.3 results in a delay of the sum signal of at most $2t_{P(max)}$ and of the carry signal of at most $3t_{P(max)}$. If a 4-bit full adder with a structure as in Figure 7.4.a is composed of these sections the total propagation delay time of the carry C_0 to C_4 is maximally $9t_{P(max)}$ ($3t_{P(max)}$ for the first section and $2t_{P(max)}$ for all following sections; it is assumed that the bits of A, B and C_0 are presented simultaneously). The maximum propagation delay time of a 32-bit full adder is then $65t_{P(max)}$. With a $t_{P(max)}$ of 10 ns per gate this would total 650 ns, which is unacceptably long for many applications.

An initial attempt was made to solve this problem by adapting the *internal structure* of each full adder section. See Figure 7.6. The circuit SN7482 (Texas Instruments) in this figure is composed of AND-OR-INVERT (AND-NOR) gates. (The variables are assigned in accordance with positive logic.) In TTL technology these gates have a propagation delay time of about one physical gate level.

With an internal organisation as in Figure 7.6 the carry in an n-bit adder is available about 50% sooner than according to Figure 7.3. However, as Figure 7.7 illustrates, the structure of the carry mechanism of this 4-bit full adder has remained, in principle, the same. This type of carry construction is called *ripple carry*.

In fact, with a carry circuit as in Figure 7.6, one has not been able to break away from the structure in which the addition of two binary numbers is done on paper. That it is also possible to do this in another way will be demonstrated below.

A 4-bit full adder with fast carry

An adder is a combinational circuit. As soon as the bits of the numbers to be added and C_0 are present, all the information needed to compute S_0 through S_3 and C_4 is available. The carry signal C_4 can therefore be directly expressed in the input signals:

C_4 = function(a_0 through a_3, b_0 through b_3, C_0).

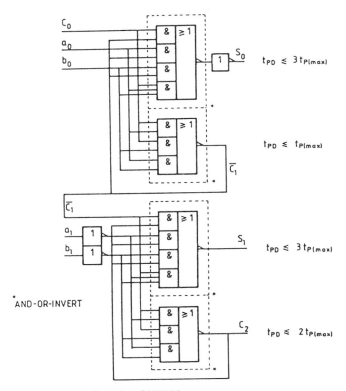

Figure 7.6 2-Bit binary full adder SN7482.

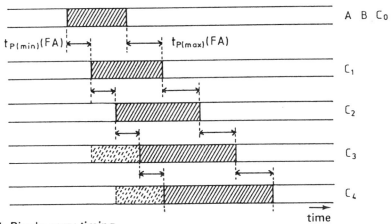

Figure 7.7 Ripple carry timing.

The following subfunctions *carry generate* and *carry propagate* are introduced because of *design rules*. These subfunctions are defined as follows:

The logic formula for the carry signal of a full adder section is

$$C_{i+1} = a_i b_i + (a_i + b_i)C_i \tag{7.4}$$

or
$$C_{i+1} = a_i b_i + (a_i \oplus b_i) C_i. \tag{7.6}$$

These formulas have been derived earlier. The term $a_i b_i$ indicates that there is an outgoing carry $C_{i+1} = 1$ if $a_i = b_i = 1$. This carry arises in the section itself, irrespective of C_i. So

$$G_i = a_i b_i \tag{7.7}$$

is called the *carry generate function*.

In the second term $(a_i + b_i)$ or $(a_i \oplus b_i)$ indicates the *condition* on which a possible incoming carry $C_i = 1$ results in an outgoing carry $C_{i+1} = 1$. This is called the *carry propagate function* P_i,

$$P_i = a_i + b_i \tag{7.8}$$
or
$$P_i = a_i \oplus b_i. \tag{7.9}$$

Equation (7.9) differs from (7.8) with regard to the term $a_i b_i = 1$. In that case, however, the carry generate function also has the value of 1, and as $1 + 1 = 1 + 0 = 1$ in switching algebra, it does not matter how the carry propagate function is defined.

For a 4-bit adder the formulas for the carry signals, expressed in P and G, become:

$$\begin{aligned} C_1 &= a_0 b_0 + (a_0 \oplus b_0) C_0 \\ &= G_0 + P_0 C_0. \end{aligned} \tag{7.10}$$

$$\begin{aligned} C_2 &= a_1 b_1 + (a_1 \oplus b_1) C_1 \\ &= G_1 + P_1 (G_0 + P_0 C_0) \\ &= G_1 + P_1 G_0 + P_1 P_0 C_0. \end{aligned} \tag{7.11}$$

$$\begin{aligned} C_3 &= a_2 b_2 + (a_2 \oplus b_2) C_2 \\ &= G_2 + P_2 (G_1 + P_1 P_0 C_0) \\ &= G_2 + P_2 G_1 + P_2 P_1 G_0 + P_2 P_1 P_0 C_0. \end{aligned} \tag{7.12}$$

$$\begin{aligned} C_4 &= a_3 b_3 + (a_3 \oplus b_3) C_3 \\ &= G_3 + P_3 (G_2 + P_2 G_1 + P_2 P_1 G_0 + P_2 P_1 P_0 C_0) \\ &= G_3 + P_3 G_2 + P_3 P_2 G_1 + P_3 P_2 P_1 G_0 + P_3 P_2 P_1 P_0 C_0. \end{aligned} \tag{7.13}$$

The equations (7.10) through (7.13) describe how the various carry signals are composed. They can easily be interpreted.

The carry in circuit SN7483A, a 4-bit full adder of Texas Instruments, has been realised on the basis of (7.13). See Figure 7.8. It is a TTL IC, characterised by a preference for the AND-NOR gate. The sum circuits have also been realised on this basis. (We shall not work out the conversion of the formulas to this realisation. For the enthusiasts: the variables are notated in accordance with the positive logic convention and the carry propagate function is based on (7.8)).

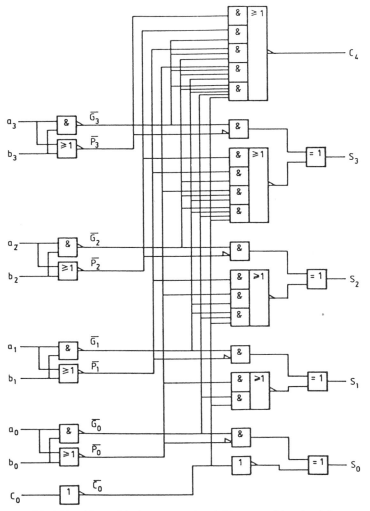

Figure 7.8 4-Bit full adder with fast carry (variables: positive logic).

In Figure 7.8 signal C_4 goes through a total of two gate levels. In the original solution, Figure 7.3, the logic depth is eight or nine.

This reduction in the propagation delay time was reached as soon as one realised that the structure of a logic circuit does not necessarily have to be the same as that in the original formulation of the problem.

7.4 The look-ahead carry generator

The number of sections in an adder over which a fast carry circuit like that in Figure 7.8 can be used is restricted because of the limited fanin of the gates and other design rules. However, the concept can be expanded, starting with (7.13):

$$C_4 = G_3 + P_3G_2 + P_3P_2G_1 + P_3P_2P_1G_0 + P_3P_2P_1P_0C_0.$$

Based on this we define

$$G_{0-3} = G_3 + P_3G_2 + P_3P_2G_1 + P_3P_2P_1G_0 \qquad (7.14)$$

$$P_{0-3} = P_3P_2P_1P_0. \qquad (7.15)$$

The *function* G_{0-3} indicates when a carry is generated in the four sections, independent of the incoming C_0. P_{0-3} indicates when an incoming carry C_0 is transferred to the output.

This system can now be expanded. For a 16-bit adder we find:

$$C_4 = G_{0-3} + P_{0-3}C_0. \qquad (7.16)$$

$$\begin{aligned} C_8 &= G_{4-7} + P_{4-7}C_4 \\ &= G_{4-7} + P_{4-7}(G_{0-3} + P_{0-3}C_0) \\ &= G_{4-7} + P_{4-7}G_{0-3} + P_{4-7}P_{0-3}C_0. \end{aligned} \qquad (7.17)$$

$$\begin{aligned} C_{12} &= G_{8-11} + P_{8-11}C_8 \\ &= G_{8-11} + P_{8-11}(G_{4-7} + P_{4-7}G_{0-3} + P_{4-7}P_{0-3}C_0) \\ &= G_{8-11} + P_{8-11}G_{4-7} + P_{8-11}P_{4-7}G_{0-3} + P_{8-11}P_{4-7}P_{0-3}C_0. \end{aligned} \qquad (7.18)$$

$$\begin{aligned} C_{16} &= G_{12-15} + P_{12-15}C_{12} \\ &= G_{12-15} + P_{12-15}(G_{8-11} + P_{8-11}G_{4-7} + \ldots + P_{8-11}P_{4-7}P_{0-3}C_0) \\ &= G_{12-15} + P_{12-15}G_{8-11} + P_{12-15}P_{8-11}G_{4-7} \\ &\quad + P_{12-15}P_{8-11}P_{4-7}G_{0-3} + P_{12-15}P_{8-11}P_{4-7}P_{0-3}C_0. \end{aligned} \qquad (7.19)$$

Within this system a 16-bit adder may be based on 4-bit full adders, with P and G outputs, and a so-called *look-ahead carry generator circuit*, which derives the signals C_4, C_8 and C_{12} from the signals P_{i-i+3} and G_{i-i+3} of the 4-bit sections. A look-ahead carry generator also supplies the signals G_{0-15} and P_{0-15}, in order to expand the system to, for example, a 64-bit adder. See Figure 7.9.

The signals G_{0-15} and P_{0-15} are defined as

$$G_{0-15} = G_{12-15} + P_{12-15}G_{8-11} + P_{12-15}P_{8-11}G_{4-7} + P_{12-15}P_{8-11}P_{4-7}G_{0-3} \quad (7.20)$$

and

$$P_{0-15} = P_{12-15}P_{8-11}P_{4-7}P_{0-3}. \qquad (7.21)$$

These definitions follow directly from (7.19) and (7.7) - (7.9), (7.14) - (7.15).

As a rule adders have either P and G outputs or a C_4 output. In a 16-bit adder C_{16} is required. There are two possibilities for generating C_{16}:

– the most significant 4-bit full adder is replaced by an ordinary 4-bit full adder and a '0' is set on the look-ahead carry generator G_{12-15} and P_{12-15} inputs, or

– an extra look-ahead carry generator is used.

The outgoing carry is worked out in Figure 7.10.

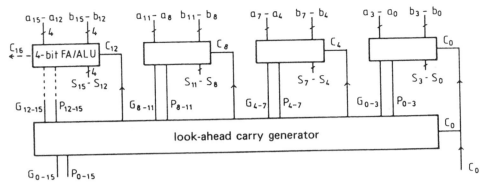

Figure 7.9 16-Bit adder with look-ahead carry generator.

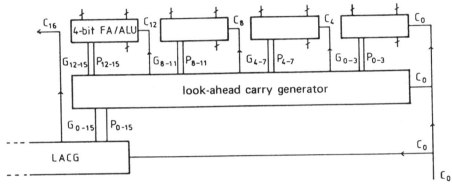

Figure 7.10 16-Bit adder with C_{16} output.

Figure 7.11 describes the detailed logic diagram of the look-ahead carry generator SN74182 (Texas Instruments). It is a relatively simple circuit, with a delay of about two gate levels. The assigned variables correspond to the positive logic interpretation. We will not derive this circuit here. (Note that inverted P and G signals have to be used!)

It is necessary to realise that all P and G signals are independent of any carry. Consequently, once it has the P and G signals of new numbers A and B, an LACG quickly has available the information necessary to determine an outgoing carry.

Very fast adders can be realised with look-ahead carry generators. Every multiplication of the number of bits by a factor of four causes an extra delay of about two gate levels in this structure. They are used in all computers, on the understanding that one has to read ALU for the 4-bit full adder mentioned here. An ALU (arithmetic logic unit) is a component which can perform other logic and arithmetic operations besides addition.

Note
The look-ahead carry generator SN74AS882 has been designed for the 32-bit adder/ALU structure, composed on the basis of the SN74AS881A. For a 32-bit adder/ALU a 1-level look-ahead carry generator is sufficient. □

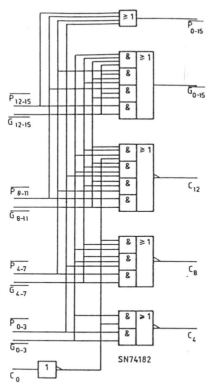

Figure 7.11 Look-ahead carry generator.

7.5 Comparators

Comparing binary encoded information is a common operation on data in digital equipment. Components that are able to do this, comparators, are divided into two types:

- *identity comparators*.
  These test whether two code combinations are equal/unequal.

- *magnitude comparators*.
  These test whether a code combination is within a certain range.

The former operation is found in, for example, addressing computer peripherals. A peripheral is activated after recognising an address that is unique for that peripheral. The latter type occurs in testing input data. An ASCII character that is fed into the computer then and only then represents a number c between $0 \leq c \leq 9$ if its ASCII code (Table 1.4) lies between

$$b_6b_5b_4b_3b_2b_1b_0 = 0110000 \leq c \leq 0111001.$$

A comparison may be performed in both hardware and software. We will restrict ourselves to the design of hardware comparators. In software a comparison is usually reduced to a subtraction, followed by a test to determine whether the result is zero or ≤ 0.

Identity comparators

Two bits are equal if $a = b = 0$ or if $a = b = 1$. With the propositions

a: 'Bit a has the value of 1'

b: 'Bit b has the value of 1'

S: 'Bit a is equal to bit b'

this property can be expressed in a logic formula,

$$S = \bar{a}\bar{b} + ab \qquad (7.22)$$
$$= \overline{a \oplus b}.$$

The inequality of bits a and b is specified by

$$\bar{S} = \bar{a}b + a\bar{b} \qquad (7.23)$$
$$= a \oplus b,$$

with the help of the above propositions.

In hardware the equality or inequality of two bits can be determined by means of an EXOR gate.

Figure 7.12 describes the logic design of two 8-bit comparators. This design enables two 8-bit words to be compared bit by bit. The design is based on (7.22) and (7.23).

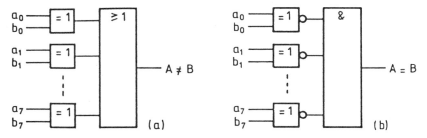

Figure 7.12 8-Bit comparators (logic diagrams).

Example

Figure 7.12.b is the basis on which the circuit SN74ALS518/519 (Texas Instruments) has been realised. The differences in Figure 7.13 with regard to the *logic design* in Figure 7.12 are:

- The detailed logic diagram contains *input inverters* to adapt the *fanin* to a fanin of 1. The EXOR gates have a fanin of 2.
- There is an extra *enable input* $\overline{G}$, used to expand the circuit and to fit it in as a component for constructing a 16-bit comparator. (It would have been more convenient if the circuit had had a G-input, in light of the output polarity.)

The AND operation at the output has been realised by the open-collector output construction (Section 6.3, Figure 6.18). In positive logic this *active-low wiring* realises the logic AND.

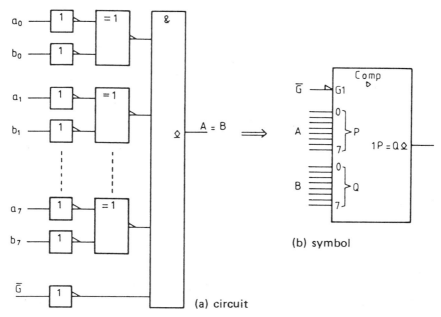

Figure 7.13 SN74ALS518/519 8-bit comparator (detailed logic diagram).

In Figure 7.13 two addresses are compared. One of these is often fixed. Comparators having one address which can be programmed, by means of fuses that can be blown or by putting the address in an internal memory, do exist. This saves half the number of input pins. □

Example

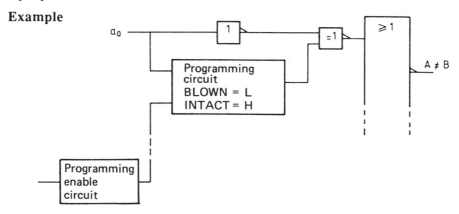

Figure 7.14 Fuse programmable identity comparator (Texas Instruments).

Magnitude comparators

Comparators for comparing/checking numbers within a predefined range are based on the *order relation*, as defined in the set of natural numbers or in the set of integers. These types of comparators are usually composed of 4-bit modules. In between there is an information transfer about whether $a_{i+3}a_{i+2}a_{i+1}a_i$ is larger, equal to or smaller than $b_{i+3}b_{i+2}b_{i+1}b_i$. See Figure 7.15 for the principle.

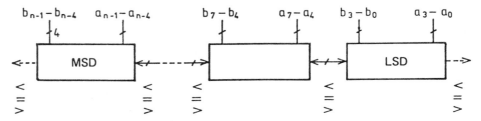

Figure 7.15 Principle of a magnitude comparator.

In the functional design of a comparator and the specification of one of its 4-bit modules, the designer is faced with some interesting design problems, for example:

- In which direction should the information '> = <' be transferred, or is data transfer in two directions possible/necessary?

A short examination shows that one direction is sufficient, but that this direction can, in principle, be chosen arbitrarily. However, the direction of information transfer does have consequences for the *specification* of the 4-bit module. When the direction and the kind of data have been determined, the next question arises:

- With how many bits should the data to be transferred be encoded and what is a convenient encoding?

A quick glance in a 'random' IC data book shows that there are many types of comparators, in which the information to be transferred is encoded differently. Table 7.3 shows some examples.

type: signals:	SN7485 > = <			SN74LS685 ≤ ≠		SN74AS885 ≥ ≤	
>	1	0	0	0	1	1	0
=	0	1	0	1	0	1	1
<	0	0	1	1	1	0	1

Table 7.3 Encoding of information between 4-bit comparators.

Given the scope of this book, this is not the place to discuss all of the motives that have led a manufacturer to a certain choice. Unfortunately, data books are also extremely brief on this point!

A less obvious, but nevertheless extremely important question the designer should ask is:

- Should a comparator be constructed according to the structure in Figure 7.15, or are other structures possible? If so, what are these other structures and what are the pros and cons of each structure?

The structure of Figure 7.15 contains the same effect as the adder for binary encoded numbers discussed previously, the so-called *ripple effect*. The more sections that are in series, the longer it takes before the result is available. Figure 7.16 shows another structure, which does not have this disadvantage. After having discussed the look-ahead carry generator this solution is actually rather obvious.

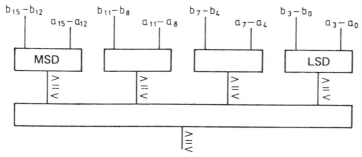

Figure 7.16 Comparator constructed in parallel.

Another important facet is suddenly added to the encoding problem of the data '> = <', for it turns out to be possible to apply a comparator at the second level as well, provided that the information is suitably encoded. The encoding used in type SN74AS885 complies with this prerequisite, as does that of type SN7485, using the > and the < outputs. The various data books supply details with regard to the wiring diagrams.

At this point we close our discussion of comparators. The reader has sufficient information to interpret the data sheets and, depending on the problem to be solved, to choose the type that best fits the selected structure.

7.6 Data transfer

The way in which data transfer in digital circuits is accomplished depends on many factors. With permanent data transfer a permanent link between transmitter and receiver is preferred. If data transfer only occurs incidentally, a common *interconnection network* is often used; a number of users are able to switch on and off this network. Figure 7.17 roughly describes how data transfer can be organised. With a selection mechanism one is able to choose from various sources and/or destinations. A certain choice of a source and/or a destination often blocks the others, unless the system is implemented in multiple (Figure 7.17.d_1 and Figure 7.17.d_2).

In digital systems data is transferred in various ways. On the transmission side two systems are usually found:

– transmitters which always put their internal value at the output. Separate *selectors* choose one of the outputs and link it with the interconnection network. The selectors are set by the system's *control unit*.

– transmitters whose output can be disconnected from the interconnection network with an 'output enable'. The selection takes place with the help of a 1-out-of-n decoder, which enables one of the outputs. The open-collector and 3-state output structures are used for this.

Demultiplexers allow data transfer from an incoming line to 1-out-of-n other lines. This may be n outgoing interconnections, or inputs of subsystems (receivers). In digital circuits data is usually transferred to all receivers. An 'input enable', coming from a *1-out-of-n decoder*, enables one of the receivers.

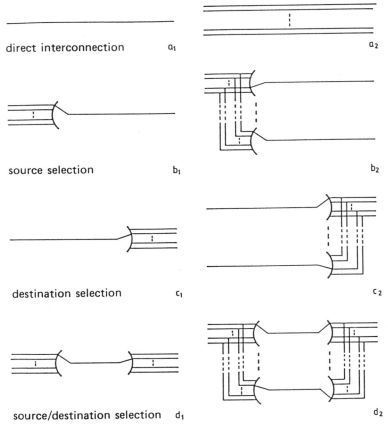

direct interconnection a_1 a_2

source selection b_1 b_2

destination selection c_1 c_2

source/destination selection d_1 d_2

Figure 7.17 Principles of data transfer systems.

Besides these hardware-oriented possibilities to construct interconnection networks there are many others at the system level. These will be discussed later.

Below we shall discuss some components which can be used to realise interconnection networks.

Selectors/multiplexers

The *logic design* of a *selector/multiplexer*, which can be used to select data from four sources, is described in Figure 7.18. Two *control signals* s_1 and s_0 determine which source is selected, according to

$$S = \bar{s}_1\bar{s}_0 \cdot a + \bar{s}_1 s_0 \cdot b + s_1 \bar{s}_0 \cdot c + s_1 s_0 \cdot d. \tag{7.24}$$

The control signals s_1 and s_0 originate in that part of the circuit called the *control unit*.

Detailed logic diagrams of two quadruple $2 \to 1$ selector/multiplexers are illustrated in Figure 7.19. In addition to a select input the IC SN74157 also has an 'enable input' (strobe) with which the selection mechanism can be switched off. The IC SN74S158 (Figure 7.19.b) is designed for a minimal propagation delay time. To this end AND-OR-

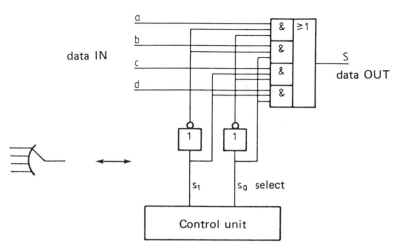

Figure 7.18 Logic design 4 → 1 selector/multiplexer.

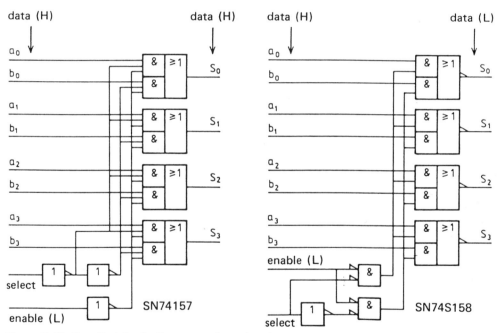

Figure 7.19 Detailed logic diagrams of quadruple 2 → 1 selector/multiplexers (SN74157 and SN74S158).

INVERT gates have been applied in the data path (TTL). A disadvantage is that inverted data is transferred. This usually poses no problems, provided that the data is interpreted correctly on the receiver side.

The manufacturer (Texas Instruments) does not explain the various realisations of the selection mechanism in the data books. We only note that, in gates, it can be done more simply. See Figure 7.20.

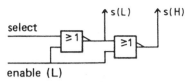

Figure 7.20 Realisation of a selection mechanism.

Larger selectors are often composed of smaller modules. Figure 7.21 illustrates schematically how $16 \to 1$ selectors/multiplexers can be composed of $4 \to 1$ selectors/multiplexers. (Note that the output enable only has to intervene at the second level now.)

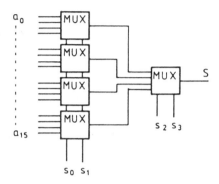

Figure 7.21 Modular structure of a $16 \to 1$ selector/multiplexer.

Decoders/demultiplexers

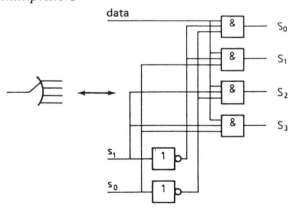

Figure 7.22 Logic design of a 2-line-to-4-line decoder/demultiplexer.

The receiver side of a data transfer system can be realised with decoders or demultiplexers. Figure 7.22 describes the *logic design* of a *2-line-to-4-line decoder/demultiplexer*. In data books the data input is often called an enable input, an instance of nomenclature that is clearly influenced by an application.

If the data line is omitted in Figure 7.22, a *1-out-of-4 decoder* is formed. The 1-out-of-n decoder is applied in data selection with open-collector or 3-state outputs.

Note
In detailed logic diagrams of decoders one often finds an AND gate in front of the data input, with 'data' and 'enable' as inputs. This gate serves to facilitate the extension, from two 1-out-of-4 decoders, for example, to a 1-out-of-8 decoder. □

The open-collector bus line

A galvanic interconnection of open-collector outputs, e.g. in TTL, is an active-low wiring (Section 6.3). This forms a *wired-AND* (in positive logic). Components with an open-collector output usually have an *output enable*. This input allows an output to be forced to 'High', or allows it to operate normally. In the latter case the output level is data-dependent. Based on this a system for data transfer can be constructed (Figure 7.23). With a 1-out-of-n decoder the output of one of the n modules is enabled. This module then controls the output line of the entire system. It is called an *open-collector bus line*. One or more modules on the receiver side can be activated by means of an (optional) decoder (Figure 7.23.b).

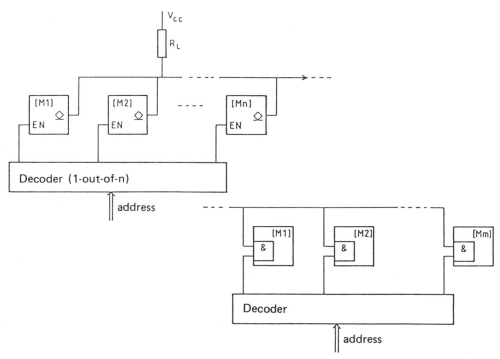

Figure 7.23 Open-collector bus line.

A TTL open-collector output switches to 'active Low'. When it is not active the output impedance of the gate is high. A load resistor R_L pulls the output to the 'High' level (active low, passive pull-up). Many components with an open-collector output already have an internal *pull-up resistor*, in which case an external resistor R_L is sometimes no longer necessary. Equation (7.25) gives the limits between which the load resistor has to lie (with TTL). In (7.25) n is the number of transmitters and m the number of receivers.

$$\frac{V_{cc} - V_{OH(min)}}{n \cdot I_{OH} + m \cdot I_{IH}} \geq R_L \geq \frac{V_{cc} - V_{OL(max)}}{n \cdot I_{OL} - m \cdot I_{IL}} \; \Omega, \qquad (7.25)$$

Characteristics

When switching from one source to another it is possible that two or more modules will transmit simultaneously. An open-collector bus line allows this, in principle. The signal on the bus is then the AND function of all activated outputs. The 3-state bus introduced below does not allow two modules to transmit simultaneously.

A disadvantage of the open-collector bus is its passive pull-up character. The rise times of the signals on the bus are much longer than the fall times. This restricts the maximum signal frequency on the bus. The 3-state bus has more or less identical rise and fall times.

Note

In principle it is possible to design an active high/passive low output circuit. In such a case, the OR function of all activated outputs is formed on the bus (positive logic). □

The 3-state bus line

A circuit with 3-state outputs can be switched into three output states:

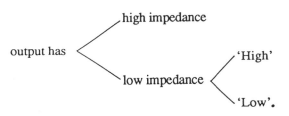

These outputs, too, may be interconnected galvanically, with the restriction that the 1-out-of-n decoder *must necessarily be used* on the transmission side. In this case it is not allowed to connect outputs that are 'High' and 'Low' (both in the low impedance state). This would result in a short circuiting of the power supply to ground, which may blow the outputs.

In principle it is possible for all outputs on a 3-state bus to be in the high impedance state. None of the outputs then controls the bus line. Practice shows that a 3-state bus is *extremely susceptible to noise* in that case, so that a certain (high-ohmage) load resistor is desirable here as well.

Figure 7.24 shows some pictures of signals on an open-collector bus and on a 3-state bus. The difference is obvious. In principle the 3-state bus allows a higher *bus frequency* because of its shorter rise and fall times.

Bus drivers

Because of their length bus lines usually have a relatively large capacitance, which negatively influences the rise and fall times. In long interconnections, therefore, special *line drivers* and *receivers* are used by way of compensation. In addition, bus receivers usually have a Schmitt-trigger input, for noise reduction purposes.

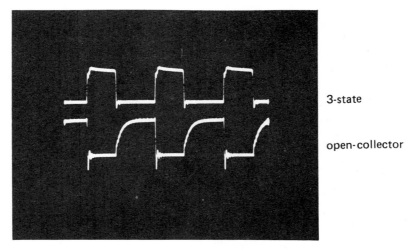

Figure 7.24 Oscillograms of signals on open-collector and 3-state bus.

In the above we have restricted ourselves to the discussion of *components* which can be used to construct bus systems. In practice, a so-called bus specification, such as the IEEE 488 bus, entails much more. This is not the place to discuss this in detail.

7.7 Error detection and error correction

During the transmission or storage of data errors may occur. Errors result, for example, from stuck-at 0/1 memory cells, but also as a result of cross talk between memory cells or between transmission lines. Many methods have been tried to protect data against 'noise'. A well-known example is the use of the *two-out-of-five code* or the *three-out-of-seven code*. In these codes every code word has a fixed number of ones. If not, the code word received is wrong. These codes have *error detecting properties*.

With an m-out-of-n code a number of code combinations, with exactly m ones, is chosen from the set of all possible 2^n code combinations. If one error occurs, the code word received contains m + 1 or m − 1 ones. One error can always be detected. With two errors m + 2 or m − 2 ones may be found, or m ones again (when there is a 0 → 1 and a 1 → 0 transition). So the code cannot protect against all two-bit errors. The *error detecting capacity* is limited.

Definition
The *Hamming distance* d of a pair of code combinations is understood to be the number of bits in which these code combinations differ.

Example

$$\left. \begin{array}{c} 0110110 \\ \updownarrow\updownarrow\updownarrow \\ 0101010 \end{array} \right\} \; d = 3.$$

Definition

The *Hamming distance* of a code is understood to be the *smallest Hamming distance* existing between any pair of code combinations, chosen from the set of all code combinations.

For the m-out-of-n code discussed previously the smallest Hamming distance is d = 2. It can easily be seen that a code (set of code combinations) has to have a Hamming distance of k + 1 in order to be able to *detect* any k errors in a code word.

Sometimes *error correction* is also possible. When the minimum Hamming distance in a code is d = 3, it can always be ascertained with one error what the original code word was. One may then decide that the word transmitted/stored was equal to the code word that has a distance d = 1 with regard to the word received/read. But one can never be sure of this. After all, the incorrect code word may also have been the result of two errors in another code word. With a small error rate the likelihood of this is extremely small and also much smaller than the likelihood of one error. This method of decoding is called *maximum likelihood decoding*.

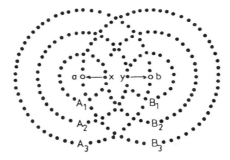

Figure 7.25 Maximum likelihood decoding.

In Figure 7.25 A_i is the set of combinations that may arise from the code word a by exactly i errors. B_j arises from code word b by j errors. We see that with a minimum Hamming distance of 3 between a and b the erroneous code word x is in A_1 and in B_2. We then decode x to a.

Error protection

Many codes do not have inherent error detecting or error correcting properties. The Hamming distance of the ASCII code, for example, is d = 1. One error changes an ASCII character into another ASCII character. See Table 1.4. In order to give the ASCII code some degree of error detecting properties this code has to be *extended* with one or more bits, such that the desired minimum Hamming distance is obtained. A well-known example is the *parity check*, which makes the number of ones in a code word odd or even.

The parity check

With even parity an eighth bit is added to the 7-bit ASCII character, such that the sum of all bits is even, i.e. the number of bits that are 1 is even. This can be checked at the

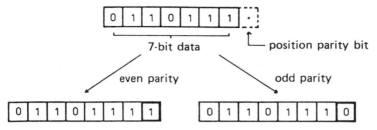

Figure 7.26 Addition of a parity check.

receiver side. If the sum of all bits is odd there has certainly been an error. However, the observation that the sum of bits received is even does not result in a 100 % certainty that no errors have occurred. Still, in a normally functioning system the likelihood that errors have occurred anyway is very small. This also holds for the case in which odd parity has been applied.

Example
In the ASCII code the letters D and d are encoded as

$$D: b_6b_5b_4b_3b_2b_1b_0 = 1000100$$
$$d: b_6b_5b_4b_3b_2b_1b_0 = 1100100$$

Hamming distance $d = 1$.

One error may change d into D and vice versa. With an extra parity check p these code combinations (even parity) become

$$D: b_6b_5b_4b_3b_2b_1b_0p = 10001000$$
$$d: b_6b_5b_4b_3b_2b_1b_0p = 11001001$$

Hamming distance $d = 2$.

In this case d can never directly change into D through one error. □

Further protection against undetected errors is provided by a *block* or *frame check*. A block check consists of one extra word, with which a vertical parity check is formed. See Figure 7.27. The errors marked x are not detected with the word check, but are detected with the block check.

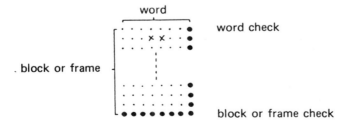

Figure 7.27 Block or frame check.

Adding bits to a word to allow error detection and/or correction takes place at the expense of the *channel capacity* of a transmission channel or the *storage capacity* of a memory. A comparison between the three-out-of-seven code and the parity code on this point is to the disadvantage of the former. The three-out-of-seven code, however, detects more errors.

Parity generators and parity checkers

The truth table of a circuit which determines whether a pair of bits is even/odd turns out to be the truth table of the EXOR operation (Table 2.5). We then have a circuit which can be used to carry out this check. Three bits being even/odd follows from the formula

$$S = (a_0 \oplus a_1) \oplus a_2 \qquad (7.26)$$
$$= a_0 \oplus (a_1 \oplus a_2),$$

as is shown when all values of a_0 through a_2 are filled in. Continuing in this way we find that the parity of a 7-bit ASCII character can be determined with the circuit of Figure 7.28.

Note

The 'exclusive OR', defined over an n-tuple of logic variables, is unequal to the function that determines whether the number of bits out of n bits that are 1 is even/odd. □

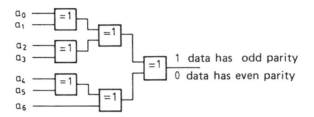

Figure 7.28 Parity determination over seven bits.

When the output of this circuit is added to the data as bit a_7, bits a_0 through a_7 automatically have *even parity*. If one still wants to be able to set the parity, even or odd, the circuit of Figure 7.29 should be used.

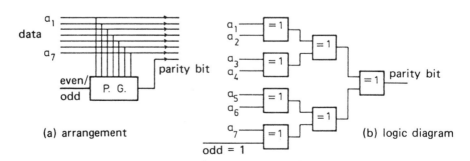

Figure 7.29 Setup and logic diagram of a parity generator.

The same circuit can be applied on the receiver side. The circuit then determines whether the number of ones in the word received is odd or even. When a *programmable output level* is desired for good/false detection, this level can be set with an extra EXOR gate. See Figure 7.30.

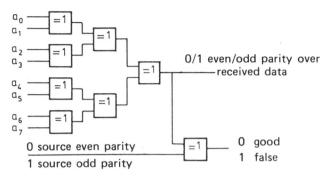

Figure 7.30 Parity checker, programmable.

Figure 7.31 provides a nice example of how the applied technology has an influence on the detailed logic diagram of a circuit (here the AND-OR-INVERT in TTL technology is preferred). This implementation of the 9-bit parity checker consists of five separate 3-input parity checkers. The outputs for even/odd parity have been implemented separately here, so that one does not need a separate mode input to switch over. Another type (SN74180) has two programming inputs and two outputs for that purpose, one for even and one for odd parity.

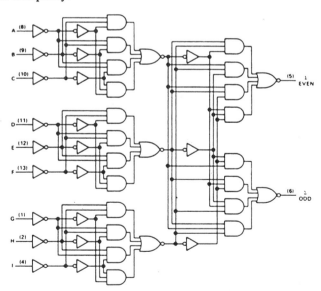

Figure 7.31 Implementation of the 9-bit parity checker SN74LS280 (Texas Instr.).

Note

The EVEN and ODD outputs in Figure 7.31 are complementary. The ODD output can be realised with only one negator, starting with the even output. The manufacturer has not done this because of the extra gate delay the negator would introduce. This is an example of interchanging components versus speed in a design. □

7.8 Summary

Section 7.1 gives an outline of the *design procedure* of logic circuits. The subsequent examples of designs of MSI combinational circuits have each elucidated one or more aspects. The major conclusions from these examples are:

- There is certainly no uniform path leading from problem to circuit. Depending on the criteria to be met the design can be *optimised* to the cost price (number of gate equivalents, for example), to the propagation delay time (number of gate levels), to the dissipation (choice of technology) or to the IC area. In addition, there is often a preference for certain logic operators (NOR/NAND, technology-dependent).

- *Optimisation criteria* are often *dependent* (fast = more parallel = more components/area = more dissipation, for example). So it is necessary to determine a *priority order* and to assign weight factors. This is the reason one often encounters different implementations of a component in data books.

- A specification at the logic level enables one to weigh all this even more. At this level logic and/or switching algebra may be used. These tools are problem- and realisation-independent and allow certain parts of the design process to be *systematised* and possibly to be *automated* (reduction of formulas, for example).

- The concept *structure* is given another interpretation, depending on the stage a design is at. Good designers will recognise this. Examples have been given in the discussion of the look-ahead carry generator and of the comparator.

- Some, but not all aspects of a *next design level* can already be *anticipated* at higher levels (in terms of top-down designs). In practice, designing is done via various iterations, with *top-down* as the *main direction*. This book concentrates on the logic part of the design procedure. Below the logic level there are still other levels, e.g. layout and/or assembly.

Besides the points mentioned above there are still other factors that determine the *applicability* of a piece of equipment for a certain application, for example the possibility of *error detection* and/or *correction*. We have not yet mentioned the *testability* of a design. A 100 % yield never occurs in IC fabrication. When the testability of a design is low, testing problems during assembly will be great. Maintenance must also be mentioned. Many computer systems cost more on maintenance during their life cycle than their initial cost. It is too early to go deeper into these aspects here.

Let us return to the subject of this chapter, standard MSI components. On many circuits we often find two kinds of signals, *data signals* and *control signals*. In the *data path/ control design model* these two categories of signals are distinguished separately. A separate part of the circuit, called the control unit, generates the control signals. The control unit determines which operation is executed on the data and when. The data itself is stored and/or processed in the data path. This part of the circuit performs the operations; it executes what the control unit sets.

The separation into data path and control unit greatly facilitates designing. It also promotes the surveyability of a design. The IEC symbol system has turned out to be crucial in this process.

References

1. E.R. Berlekamp, *Algebraic Coding Theory*, McGraw-Hill, New York, 1968.
2. B.E. Briley, *Some New Results on Average Worst Case Carry*, IEEE Trans. on Computers, Vol. C-22, 1973, pp. 459–463.
3. G.A. Blaauw, *Digital System Implementation*, Prentice-Hall, 1976.
4. C.Y. Chow and J.E. Robertson, *Logical Design of a Redundant Binary Adder*, Proc. 4th Symp. Comput. Arithmetic, 1978, pp. 109–115.
5. H.L. Garner, *Generalized Parity Checking*, IRE Trans. on Electronic Computers, Vol. EC-7, 1958, pp. 207–213.
6. E.L. Gilbert, *How Good is Morse Code*, Information and Control, Vol. 14, 1969, pp. 559–565.
7. R.W. Hamming, *Error Detecting and Error Correcting Codes*, Bell Syst. Tech. J., Vol. 29, 1950, pp. 147–160.
8. K. Hwang, *Computer Arithmetic, Principles, Architecture, and Design*, Wiley, New York, 1979.
9. Y. Kambayashi and S. Muroga, *Properties of Wired Logic*, IEEE Trans. on Computers, Vol. C-35, 1983, pp. 550-563.
10. T. Kilburn, D.B.G. Edwards and D. Aspinall, *Parallel Addition in Digital Computers: A New Fast Carry Circuit*, Proc. IEEE, Vol. 106, 1959, pp. 464–466.
11. D. Lee, *Comparator with Completion Signal*, IEEE Trans. on Computers, Vol. C-34, 1985. pp. 855–857.
12. M. Lehman, *A Comparative Study of Propagation Speed-up Circuits in Binary Arithmetic Units*, International Federation of Information Processing Societies, 1962, North Holland, Amsterdam, 1963.
13. T.K. Liu, K.R. Hohulin, L.E. Shiau and S. Muroga, *Optimal One-Bit Full Adders with Different Types of Gates*, IEEE Trans. on Computers, Vol. C-23, 1974, pp. 63–70.
14. E.J. McCluskey, *Logic Design Principles*, Prentice-Hall, Englewood Cliffs, New Jersey, 1986.
15. R.L. Morris and J.R. Miller, *Designing with TTL Integrated Circuits*, McGraw-Hill, New York, 1971.
16. R.M.M. Oberman, *Digital Circuits for Binary Arithmetic*, McMillan, London, 1979.
17. C.A. Papachristou, *Parallel Implementation of Binary Comparison Circuits*, Int. J. Electron., Vol. 47, 1979, pp.187–192.
18. W.W. Peterson and E.J. Weldon, *Error-Correcting Codes*, 2nd ed., John Wiley & Sons, New York, 1972.
19. T. Rhyne, *Limitations on Carry Lookahead Networks*, IEEE Trans. on Computers, Vol. C-33, 1984, pp. 373-374.
20. A. Sakurai and S. Muroga, *Parallel Binary Adder with a Minimum Number of Connections*, IEEE Trans. on Computers, Vol. C-32, 1983, pp. 969-976.
21. J. Sklansky, *Conditional Sum Addition Logic*, IRE Trans. on Electronic Computers, Vol. EC-9, 1960, pp. 226–231.
22. M.V. Subba Rao and S.C. Mittal, *SN7485 gives carry lookahead digital comparison*, Electron. Eng., Vol. 50, 1978, p. 21.
23. N. Takagi, H. Yasuura and S. Yajima, *High-Speed VLSI Multiplication Algorithm with a Redundant Binary Addition Tree*, IEEE Trans. on Computers, Vol. C-34, 1985, pp. 789–796.

24. Y. Tamir and C.H. Séquin, *Design and Application of Self-Testing Comparators Implemented with MOS PLA's*, IEEE Trans. on Computers, Vol. C-33, 1984, pp. 493–506.
25. Texas Instruments, *The TTL Data Book*, Texas Instruments, Dallas, 1984.
26. J. Verhoeff, *Error Detecting Decimal Codes*, Mathematical Centre, Amsterdam, 1969.
27. J.F. Wakerly, *Logic Design Projects Using Standard Integrated Circuits*, Wiley, New York, 1976.
28. C.S. Wallace, *A Suggestion for a Fast Multiplier*, IEEE Trans. on Electronic Computers, Vol. EC-13, 1964, pp 14–17.
29. N. Weste and K. Eshragian, *Principles of CMOS VLSI Design*, Addison-Wesley, 1985, pp. 310–335.

8
Programmable logic

8.1 Read Only Memories

An extensive range of components recently entering the market has made it possible to *implement* the desired logic function via programming. This *programmable logic* allows a certain degree of *flexibility*, as long as the logic function is not restricted by the external wiring between the ICs. This flexibility has made shorter design times possible. In addition, the *packing density* on printed circuit boards, expressed in gate equivalents per board, is higher in programmable logic. Circuits composed of 7400 series ICs are not flexible. The realisation of logic with discrete gates is therefore increasingly dropping into disuse.

A circuit designed in programmable logic also has disadvantages, e.g. a somewhat *higher dissipation*. In many applications, however, the pros outweigh the cons.

Read Only Memories

Read only memories have been designed to realise *memories*, intended for the permanent storage of programs and data. A ROM is a component which, once programmed, reproduces the contents of the current address on the output pins after having received an address code. With an address generator it is possible to read consecutive addresses in succession, in the order determined by the generator. In this way a program or data can be read. Figure 8.1 illustrates this concept.

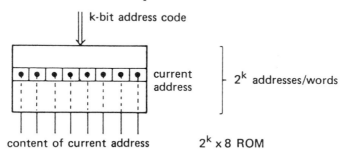

Figure 8.1 Concept of a ROM.

One soon realises that a ROM memory can replace combinational logic. After all, a combinational circuit also generates a *fixed* output combination after having received the input combination. It does not matter whether this output combination is computed every time, as with a gate circuit, or is read back from a ROM after having been stored once.

The gate circuit in Figure 8.2 has seven inputs and four outputs. With every input combination abcdefg the output field gets an appropriate value $S_0S_1S_2S_3$. If we compute the output field for every input combination and store it in a ROM at the corresponding address the ROM will subsequently react *logically identical*. Sometimes not all of the inputs of the ROM are needed because the number of input variables of the combinational circuit to be replaced is smaller. The superfluous address inputs are then set at a constant 0 or 1.

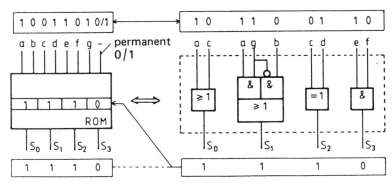

Figure 8.2 The implementation of combinational logic in ROMs.

Considered in this way a ROM and a combinational circuit are conceptually identical. The propagation delay time of a ROM is usually much longer, e.g. because of its complex address decoder.

The internal structure of a ROM

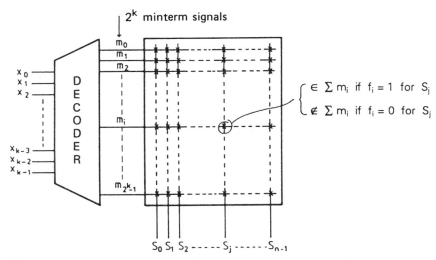

Figure 8.3 The internal structure of a ROM memory.

Figure 8.3 shows the internal structure of a ROM memory. An *address decoder* converts the k input signals (address lines) into 2^k internal signals. Each signal activates one address in the memory, the contents of which subsequently appear at the outputs.

According to the 'logic view' of a ROM each *internal* address line (behind the decoder) may be regarded as a *minterm signal*. If the corresponding minterm m_i belongs to the function S_j, $f_i = 1$ for m_i of S_j, then its m_i signal has to be included in the output summation. This is the OR function of all m_is of which $f_i = 1$ (for the concept 'minterm': see Section 2.5).

In this way n *random logic functions* can be realised in a ROM, one on each output line. Standard ROMs usually have eight outputs.

In a ROM the *decoder* is fixed and the so-called *OR array* is fully programmable. For practical reasons (layout at the IC level) the internal structure of large ROMs often differs from the theoretical structure as shown in Figure 8.3. In that case, address decoding is done in two steps, by *row decoding* and by *column decoding*. The memory array is composed of words with a width that is a multiple of the output field. Via row decoding a word is selected from the memory array. Column decoding subsequently selects the desired output field. This internal structure usually remains hidden from the user. To gain an insight into the *external* logic properties the model in Figure 8.1 can be used, i.e. a fixed AND array (address decoder) that supplies the minterm signals (address signals) to a programmable OR array in which the function is realised on the basis of the *minterm form*.

Figure 8.4 shows an internal structure with 128-bit words. Here column decoding selects an 8-bit output field out of 128 bits.

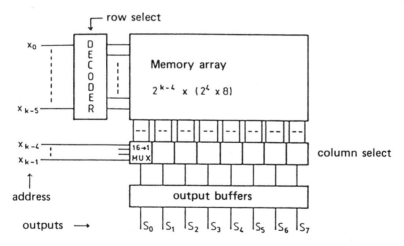

Figure 8.4 Internal structure of the selection mechanism in a ROM memory.

Some details

ROMs are implemented in various ways, varying from *mask programmable* during production to *electrically (re)programmable* by the user. Figure 8.5 gives some details of the OR array of an electrically programmable bipolar ROM. Each cell consists of a *bipolar transistor* and a *fuse*, the latter can be blown by a small current pulse. If the fuse is intact a voltage 'H' on the address line results in a voltage 'H' on the corresponding output line.

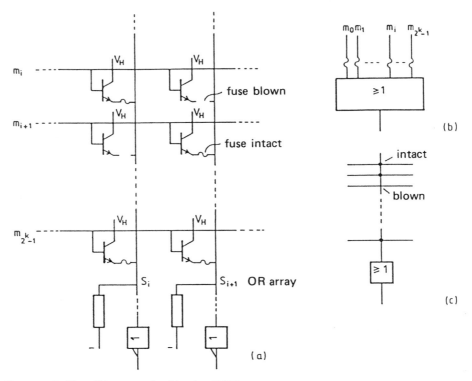

Figure 8.5 The OR array of a bipolar ROM.

Figure 8.5.b shows the logic equivalent of one output line in symbols. The usual shorthand drawing mode is shown in Figure 8.5.c. A dot indicates that the cell or minterm is included in the output sum.

Figure 8.6 describes some details of a *MOS memory array*. An output line realises the NOR function on all interconnected input lines (positive logic). Although MOS memories and programmable logic therefore have an internal *NOR-NOR array structure*, one usually talks in terms of the AND array and the OR array. After all, for the user the internal structure of the ROM memory remains hidden.

Programmable ROM memories exist in many varieties. Virtually all have eight outputs. The number of inputs is limited to about 16, with the present state of technology.

Note
During the production of memory ICs spare rows and columns are often added to the memory arrays. When it turns out during testing that certain parts are faulty these are replaced by good parts via a change of interconnections. This greatly increases the yield of the production process. □

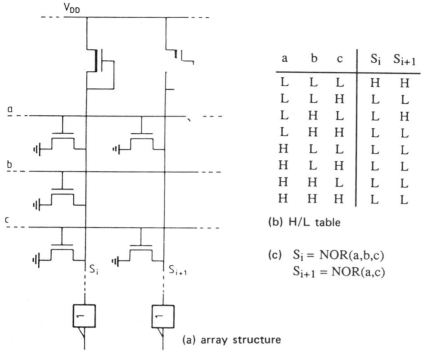

Figure 8.6 The MOS memory array.

8.2 Some applications of ROMs

In this section some examples in which logic is realised in ROMs are discussed. The first example shows a *BCD-to-7-segment decoder*, both in a ROM and with gate equivalents from the 7400 series.

BCD-to-7-Segment decoder

Decimal digits are often displayed on 7-segment displays. These consist of 7 bar-shaped lights. Figure 8.7 illustrates how the digits are displayed. When the digits are represented in the BCD code a code converter from the *BCD code* to the *7-segment code* is required to drive the display.

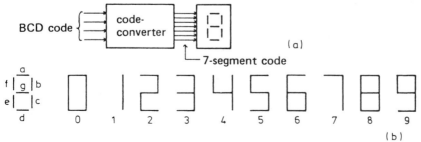

Figure 8.7 The representation of digits by 7-segment displays.

Table 8.1 specifies the code converter. The formulas of the signals S_a through S_g which drive the lights are specified next to the table.

BCD code	7-segment code
$b_3\ b_2\ b_1\ b_0$	a b c d e f g
0 0 0 0	1 1 1 1 1 1 0
0 0 0 1	0 1 1 0 0 0 0
0 0 1 0	1 1 0 1 1 0 1
0 0 1 1	1 1 1 1 0 0 1
0 1 0 0	0 1 1 0 0 1 1
0 1 0 1	1 0 1 1 0 1 1
0 1 1 0	1 0 1 1 1 1 1
0 1 1 1	1 1 1 0 0 0 0
1 0 0 0	1 1 1 1 1 1 1
1 0 0 1	1 1 1 1 0 1 1
1 0 1 0	- - - - - - -
⋮	⋮
1 1 1 1	- - - - - - -

$S_a = \overline{b_0}\overline{b_2} + b_0 b_2 + b_1 + b_3$
$S_b = \overline{b_0}\overline{b_1} + b_0 b_1 + \overline{b_2}$
$S_c = b_0 + \overline{b_1} + b_2$
$S_d = b_0 \overline{b_1} b_2 + \overline{b_0} b_1 + \overline{b_0}\overline{b_2} + b_1 \overline{b_2} + b_3$
$S_e = \overline{b_0} b_1 + \overline{b_0}\overline{b_2}$
$S_f = \overline{b_0} b_1 + \overline{b_0} b_2 + \overline{b_1} b_2 + b_3$
$S_g = \overline{b_0} b_2 + \overline{b_1} b_2 + b_1 \overline{b_2} + b_3$

Table 8.1 The relation between the BCD code and the 7-segment code.

The realisation of these formulas in 7400 series gates uses a lot of ICs. The entire code converter also fits into one ROM. Figure 8.8 shows the programming in a 16×8 ROM structure.

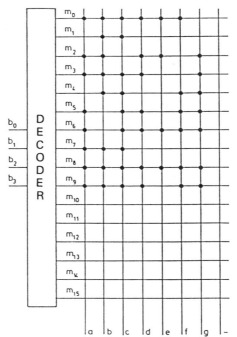

Figure 8.8 BCD-to-7-segment decoder in a ROM.

Don't cares

With realisations in gates it is important to convert design freedom into *don't cares*, in order to be able to use them when minimising the formulas. In Table 8.1 profitable use has been made of the don't cares for the binary input combinations 1010 through 1111 to minimise functions S_a through S_g.

When programming in a fixed ROM structure this kind of design freedom is of minor importance. The *number of cells* is determined by the number of functions (outputs) and the number of input variables (inputs). Don't cares are of no influence.

Sometimes, however, it is possible to fill in the don't cares in a useful way, for example, to extend the range of applications of the IC. In the above example it is possible to supplement the specification to a *hexadecimal-to-7-segment decoder*. In the hexadecimal code the bit combinations 1010 through 1111 do occur. The code converter is then suitable for two applications.

A form of *error detection* may also be useful. If a binary combination ≥ 1010 occurs the display can report an error. While both alternatives are useful, they do exclude each other. When a standard $2^4 \times 8$ ROM is used, the eighth output can be used to report that the contents are $\geq 10_{DEC}$. This signal may then be used as a test signal.

Replacing random logic

Many older designs are based on 7400 SSI/MSI integrated circuits. The number of *gate equivalents* per PC board is low; the devices use a lot of space. In the event of a redesign the step to programmable logic is an easy one. One ROM with k inputs/n outputs can replace *any* combinational circuit that does not have more *external* inputs/outputs, on condition that the somewhat longer propagation delay time of a ROM does not conflict with specified timing constraints. See Figure 8.9. The procedure for programming the ROM has already been discussed in Section 8.1.

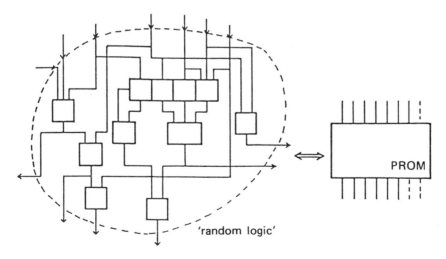

Figure 8.9 Replacing combinational logic by a PROM.

Programmable ROMs are available in two versions, so-called PROMs (programmable once and only once) and RePROMs (reprogrammable after erasing with UV light, for example). Prototypes are usually realised in RePROMs. In the final design PROMs are more economical, faster and possibly somewhat more reliable (also see below, PLAs and PALs).

The programming of a set of logic functions often does not take up the entire contents of a ROM. The following example illustrates some of the resulting possibilities.

Example

For the realisation of eight functions, defined on a maximum of nine input variables, a standard $2^{10} \times 8$ PROM is available. The programming takes up half of the PROM in Figure 8.10.a. The input x_9 is set to 0.

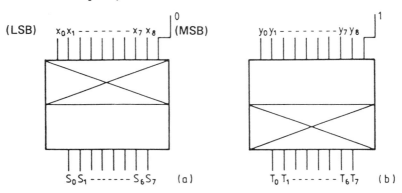

Figure 8.10 Logic functions in a PROM.

When the input x_9 is set to 1 another eight functions, defined on at most nine input variables, can be programmed. Of course, only one set at a time can be used. The selection is done by x_9. From the point of view of stock-piling this is an interesting solution. This convertibility can sometimes be used to adapt the function of a module in a network. □

Correcting a transfer characteristic

When measuring certain quantities one often has trouble with the non-linearity of the converters. An example is measuring a temperature with a thermocouple. The DC current delivered by the thermocouple, which is not completely proportional to the temperature, is converted into a binary number by an analog-to-digital converter (A-D converter). See the setup in Figure 8.11. For an 8-bit binary number the result of the measurement can be easily corrected with a 256×8 PROM. For each 8-bit combination, representing a measured voltage, the correct value of the temperature is stored in the memory.

For the realisation of *random logic*, ROMs and PROMs offer a multitude of possibilities. In the above example it is most impractical to express the corrections of the measured values in logic functions and subsequently realise them in gates. The circuit would become too large and too expensive. With PROMs, however, it is even possible to take the characteristic of each individual thermocouple into account.

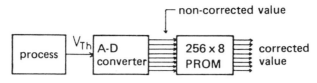

Figure 8.11 Application of a PROM for measurement correction.

Instruction decoding

Computers have an instruction set. An instruction is usually carried out by executing a series of micro-instructions. These program the hardware. With every micro-instruction a number of signals has to be set. The specific value of these signals is not incorporated in the operation code of the micro-instructions, as it would become too wide. The setting of the control field is linked up to the operation code and it is decoded from it by, for example, ROMs. See Figure 8.12. PLAs, to be introduced below, are also used for this purpose.

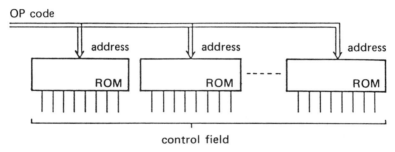

Figure 8.12 Instruction decoding.

With non-time-critical applications the corresponding control field is often stored at *successive addresses* in one ROM. The operation code is then presented to an *address generator*, which generates the corresponding addresses. These are subsequently sent to the ROM memory. Registers (Chapter 12) at the output intercept the ROM contents and compose the control field step by step.

8.3 Programmable logic arrays

In conceptual terms a PROM consists of two arrays:

- a *fixed, fully decoded AND array*, in which signals corresponding to minterms are generated;

- a *programmable OR array*, in which the desired logic function can be programmed, based on the minterm form.

In addition, there are input and output buffers, used on the one hand to adjust the fanin and fanout, and on the other hand to adapt the internal signal levels. This internal structure is not always necessary for the implementation of logic, as is shown below.

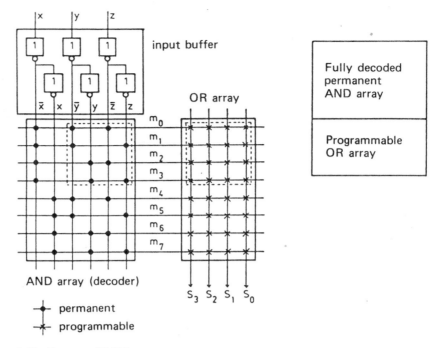

Figure 8.13 Concept PROM.

The dotted line in Figure 8.13 denotes how large the AND and the OR array would be if the PROM had one input variable less. Each extra input variable results in more than double the necessary number of cells. For a function in many variables this effect is unacceptable. Therefore, PROMs are less suitable for this purpose.

An advantage of realising logic functions with PROMs is that the *number of product terms* is irrelevant. The function is realised at the *minterm level*. The truth table can be read in directly, without minimisation of the formula. At the same time this is also one of its limitations. It is preferable not to implement an OR gate with eight inputs in a PROM.

The 'silicon efficiency' is improved by also making the AND array programmable. This opens up perspectives for two reasons. For every product term in a logic formula only one row in the AND array has to be reserved. The resulting product terms can be shared by several functions. Besides, minterms that do not belong to any of the output functions do not take up any area. In the AND array of a ROM/PROM every minterm automatically does. The average number of product terms of a logic function is much smaller than the number of minterms.

A logic array with both a programmable AND array and a programmable OR array is called a PLA (*programmable logic array*). When it can be programmed by the user it is called FPLA (*field programmable logic array*). PLAs are not (yet) available in a reprogrammable form.

PLAs are used for the implementation of control logic in VLSI ICs. A PLA with 16 inputs and 64 product rows contains 2048 cells in the AND array. A PROM with 16 inputs would have 1,048,576 cells.

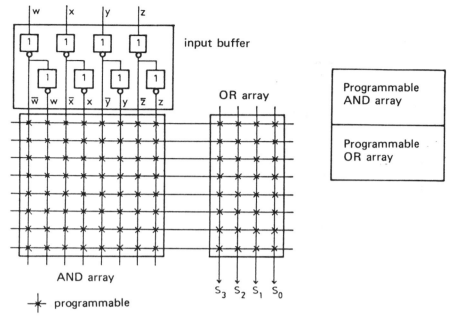

Figure 8.14 The PLA structure.

A *limitation of PLAs* is that not every logic function in k variables (k is the number of input variables of the PLA) can be programmed in them. After all, the number of product terms is limited and, as a rule, much smaller than 2^k. ROMs, which do not have this restriction, and PLAs therefore both have their own application field.

PALs

A more restricted form of programmable logic is found in PALs (*programmable array logic*, a trademark of Monolithic Memories, Inc.). In PALs the OR array is fixed. Each output function can only contain a few product terms. Although this makes a PAL cheaper, it cannot do as much. See Figure 8.15.

Example
Two functions are given,

$$S_0 = abc\bar{d} + a\bar{b}cd + \bar{a}\bar{b}c\bar{d} + \bar{a}\bar{b}cd$$

and

$$S_1 = ac\bar{d} + \bar{b}c\bar{d} + \bar{a}cd + \bar{b}cd,$$

both in their minimal sum-of-products forms. Programming directly in a PLA in this form requires eight product rows.
The function S_1 can also be written as

$$S_1 = abc\bar{d} + a\bar{b}c\bar{d} + \bar{a}\bar{b}c\bar{d} + \bar{a}\bar{b}cd + \bar{a}bcd + a\bar{b}cd.$$

S_0 and S_1 then have four product terms in common and two which are not. In the PLA six product rows are now occupied. See Figure 8.16. □

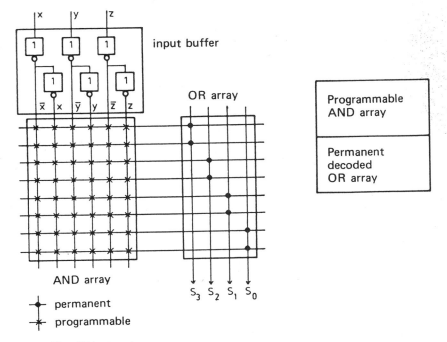

Figure 8.15 The PAL structure.

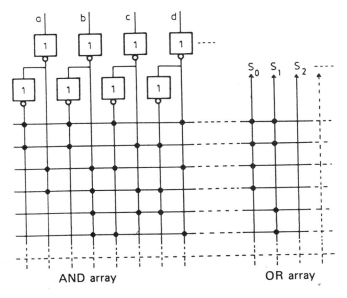

Figure 8.16 Realisation of S_0 and S_1 in a PLA.

Programming the same functions in a PAL, with at least four product rows for each output function, always requires eight product rows. (If each output function in a PAL contains more than four product terms, some of them remain unused in this example.)

Note

The negation $\overline{S}$ of a function S sometimes has fewer product terms than S itself. Some versions of PALs and PLAs take advantage of this. Owing to their programmable output polarity, one may choose to program on the basis of S or $\overline{S}$. □

A considerable advantage of PROMs/ROMs is that the logic functions do not have to be minimised. The number of input variables is the criterion used for fitting. For PALs and PLAs a reduction of the functions is required. There exist programs for these components which can do the necessary reduction of a logic function and/or the examination of fitting in a certain type of IC. At the moment these programs are still far from ideal. They sometimes claim to be able to process don't cares, whereas these are replaced by a fixed 0 or 1 once the function has been read in.

The following table gives a very rough indication of the pros and cons of each type. The table may in no way be regarded as an *absolute guideline* for component selection!

	PROM	PLA	PAL	GATES
Area PCB	+	++	+/–	– –
Area IC	–	+	+/–	+
Dissipation	–	–	–/+	++
Propagation delay time	–	–/+	–/+	++
Flexibility	++	+	+/–	–
Design procedure	++	+	+/–	–
Number of variables	+/–	++	+/–	–
Number of product terms	++	–	–	–

Table 8.2 Properties of programmable logic versus gates.

Examples of the application of programmable logic in the realisation of logic circuits will be given in Chapter 10. Those types having an internal memory are introduced in that chapter as well.

Besides PROMs and PLA/PALs there is a third group of customisable components, the *uncommitted logic arrays* (ULAs). These are integrated circuits containing a number of gates, other standard cells and input/output buffers. On all these 'cells' the supply and ground interconnections have been mounted. Only the wiring for the *logic signals* is missing. The logic wiring is done in some final processing steps, according to customer specifications. Gate arrays, a specific kind of ULA, allow a design to be realised fast, at some loss of silicon efficiency. In the production of large computers and other digital equipment gate arrays are used frequently. Designers of smaller companies (still) have poor access to these components, unless they are assisted by specialised *design houses*.

8.4 Optimisation in array logic

Sometimes a logic function cannot be mapped into an available array. In this situation the present generation of software tools often refers to a larger type, with more product rows per output. Many PALs and PLAs already have *internal facilities* for extensions.

An *open-collector output* allows a wired-AND construction (positive logic). With open-collector outputs functions such as

$$S = S_0 \cdot S_1 \cdot \ldots \cdot S_{n-1}$$

can be formed, of which the factors S_0 through S_{n-1} can be implemented directly. With an external OR gate functions of the form

$$S = S_0 + S_1 + \ldots + S_{n-1}$$

can be implemented. Both constructions increase the possibilities of PALs and PLAs considerably.

With many PALs and PLAs the OR construction can also be realised internally. Most types have the possibility to feed the output signals back to the AND array internally, although this is done at the cost of one external input signal per signal feedback. See Figure 8.17 for an example. Which signal is fed back to the AND array is determined by an internal selector.

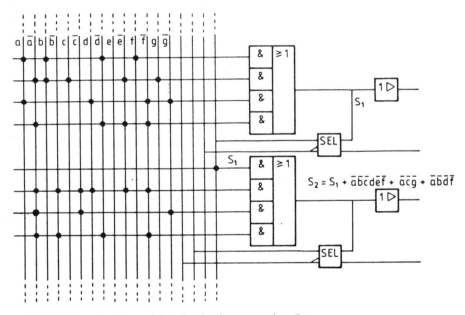

Figure 8.17 PAL with internal feedback of output signals.

Note
With this type of feedback one has to ensure that *no memory loops will be created, both static and dynamic*. See Chapter 9 for this problem. □

Bit partitioning

In a PLA with the structure of Figure 8.14 every external input on the AND array generates two internal inputs, one for the variable itself and one for its component. In the AND array each row forms a product term, based on the formula

$$\Pi_i = f_i(a)\cdot f_i(b)\cdot f_i(c)\cdot\ldots \qquad (8.1)$$
$$= (f_{i0} + a)\cdot(f_{i1} + \overline{a})\cdot(f_{i2} + b)\cdot(f_{i3} + \overline{b})\cdot(f_{i4} + c)\cdot\ldots$$

An interconnection corresponds to $f_{ij} = 0$. No interconnection corresponds to $f_{ij} = 1$, in which case the variable or its complement does not belong to the product term. Each product term Π_i contains one factor of the set

$$\{a, \overline{a}, b, \overline{b}, ab, \overline{a}b, a\overline{b}, \overline{ab}, 1(\text{not})\} \qquad (8.2)$$

for each pair of variables a and b. The set (8.2) is merely a subset of all of the functions in two variables. The efficiency of a PLA greatly increases if for every pair of variables one of its arbitrary functions participates in each product term Π_i. An arbitrary function of two variables a and b can be based on its *maxterm form* (see Section 2.6):

$$f(a, b) = (f_0 + a + b)\cdot(f_1 + a + \overline{b})\cdot(f_2 + \overline{a} + b)\cdot(f_3 + \overline{a} + \overline{b}) \qquad (8.3)$$
$$= (f_0 + \overline{\overline{ab}})\cdot(f_1 + \overline{\overline{a}b})\cdot(f_2 + \overline{a\overline{b}})\cdot(f_3 + \overline{ab}).$$

The structure of (8.3) corresponds to that of (8.1). When the presence/absence of an interconnection again corresponds to $f_i = 0$ or $f_i = 1$, a product term Π_i in a PLA corresponds to the following structure:

$$\Pi_i = f_i(a, b)\cdot f_i(c, d)\cdot f_i(e, f)\cdot\ldots \qquad (8.4)$$
$$= (f_{i0} + \overline{\overline{ab}})\cdot(f_{i1} + \overline{\overline{a}b})\cdot(f_{i2} + \overline{a\overline{b}})\cdot(f_{i3} + \overline{ab})\cdot(f_{i4} + \overline{\overline{cd}})\cdot\ldots$$

The PLAs based on (8.1) and (8.4) have the *same number of internal inputs* on the AND array. The implementation according to (8.4) requires 2-input NANDs as input buffers, compared to inverters for implementations based on (8.1). The next example shows the gain in efficiency.

Example
The realisation of the function

$$S = a\overline{b}c + \overline{a}bc + a\overline{b}d + \overline{a}bd \qquad (8.5)$$

in a standard PLA requires four rows of the AND array and one column of the OR array. The function (8.5) can also be written as

$$S = (a\overline{b} + \overline{a}b)\cdot(c + d) \qquad (8.6)$$
$$= f(a, b)\cdot f(c, d).$$

In this case the realisation of the function S only takes up one row of the AND array. Compared to (8.4) the fuses must be blown as $f_{i0} = f_{i3} = 0$ and $f_{i1} = f_{i2} = 1$ for the first two variables and $f_{i4} = 0$ and $f_{i5} = f_{i6} = f_{i7} = 1$ for variables c and d. □

This principle can, of course, be extended by partitioning the input sets into 3-bit or 4-bit groups. Three variables take up six input rows of the AND array, while the number of minterms/maxterms is eight. With four or more variables this ratio becomes even more disadvantageous. In practice therefore, one partitions the input set only into 2-bit groups. This is called *bit partitioning*.

Figure 8.18 describes the principle of a PLA with bit partitioning. Above the AND array are several input arrays, one for each pair of input variables. The maxterms of functions in two variables are formed in every input array. Products of these maxterms are formed in the AND array. In turn, these products are summed in the OR array, the output array.

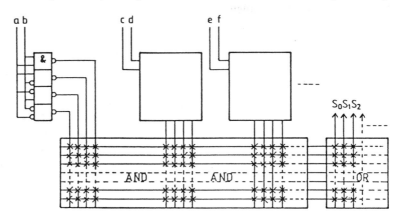

Figure 8.18 Principle of PLA with bit partitioning.

It is possible to realise a PLA with 2-bit partitioning in another and simpler way. The result of the i-th product row can be written as

$$\Pi_i = f_{i1}(a,b) \cdot f_{i2}(c,d) \cdot f_{i3}(e,f) \cdot \ldots \qquad (8.7)$$

$$= \overline{\overline{f_{i1}(a,b) \cdot f_{i2}(c,d) \cdot f_{i3}(e,f) \cdot \ldots}}$$

$$= \overline{\overline{f_{i1}(a,b)} + \overline{f_{i2}(c,d)} + \overline{f_{i3}(e,f)} \ldots}$$

The functions $\overline{f_{i1}(a,b)}$ etc. can in turn be expressed as a sum of the minterms:

$$\overline{f_{i1}(a,b)} = \overline{a}\,\overline{b}\overline{f_{j0}} + \overline{a}b\overline{f_{j1}} + a\overline{b}\overline{f_{j2}} + ab\overline{f_{j3}}, \qquad (8.8)$$

in which f_{j0} through f_{j3} are the negated function values of the original function $f_{i1}(a,b)$.

Because of this the formula for the i-th product Π_i can be written as

$$\Pi_i = \overline{\overline{a}\,\overline{b}\overline{f_{j0}} + \ldots + ab\overline{f_{j3}} + \overline{c}\,\overline{d}\overline{f_{k0}} + \ldots cd\overline{f_{k3}} + \overline{e}\,\overline{f}\overline{f_{l0}} + \ldots}\,. \qquad (8.9)$$

This is the NOR function. The OR and AND arrays, which together form the product terms Π_i in a PLA with bit partitioning, may be replaced by ANDs and a NOR array. Figure 8.19 shows the result.

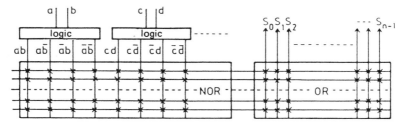

Figure 8.19 PLA with bit partitioning based on a NOR/OR-array.

Note

In comparing Figure 8.19 with Figure 8.14 we see that a PLA can be made more efficient by adding some extra gates. □

Folding

Bit partitioning is a method to effectively increase the number of possible product terms, defined on the same input set, thereby increasing the efficiency of a PLA. In general, not all of the input variables will occur in each product term. This means that there may be a great deal of open area in the AND array. A technique aimed at increasing the efficiency of this array is known as *folding*. It is explained in Figure 8.20.

The product Π_0 on the first product row is expressed in variables a_0 through a_m. The product Π_1 is expressed in b_0 and a_1 through a_m. The last product is expressed in variables b_0 through b_m. The output functions S_0 through S_n may contain the products formed in this way. The *folding* technique makes it possible to define the output functions on partly *overlapping input sets*.

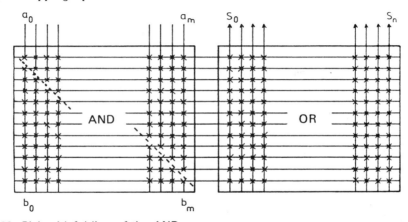

Figure 8.20 PLA with folding of the AND array.

In Figure 8.20 the 'fold' is drawn diagonally. It is also possible to work with a *variable fold*, in which the input rows of the AND array are split up at an arbitrary place. These splits are introduced during the production of the circuit.

Folding can also be applied on the OR array. In that case one finds separate output buffers on both sides of the OR array.

Bit partitioning and the various methods of folding allow the realisation of logic by PLAs to be optimised further. For larger PLAs specially developed computer programs are virtually indispensable. Problems such as the distribution of variables along a 'fold' can hardly be sorted out manually in a reasonable amount of time.

References

1. O.P. Agrawal and D. Laws, *The Role of Programmable Logic in System Design*, VLSI Design, 1984, pp. 44–50.
2. Z. Arvalo and J.G. Bredeson, *A Method to Simplify a Boolean Function into a Near Minimal Sum of Products for Programmable Logic Arrays*, IEEE Tr. on Computers, Vol. C-27, Nov. 1978, pp. 1028–1034.
3. J. Birkner, *PAL, Programmable Array Logic Handbook*, Monolithic Memories, 1978.
4. M.J.P. Bolton, A.C.G.I., D. Phil., *Designing with Programmable Logic*, IEEE Proc., Vol. 132, Pts E and I, 1985, pp. 73–85.
5. N. Cavlan and R. Cline, *Field Programmable PLA's Simplify Logic Designs*, Electronic Design 18, Sept. 1975, pp. 84–90.
6. H. Fleisher and L.I. Maissel, *An Introduction to Array Logic*, IBM J. Res. Develop., Vol. 19, March 1975, pp. 98–109.
7. W.I. Fletcher, *An Engineering Approach to Digital Design*, Prentice-Hall, Englewood Cliffs, N.J., 1980.
8. A. Hemel, *The PLA, a Different Kind of ROM*, Electronic Design 1, Jan. 1976, pp. 78–84.
9. M. Wild, *Converting PALs to CMOS Gate Arrays*, VLSI Design, 1984, pp. 58–64.
10. T.L. Larson and C. Downey, *Field Programmable Logic Devices*, Electronic Engineering, Vol. 52, Jan. 1980, pp. 37–54.
11. H.F. Li, *Programmable Logic Array Optimization Techniques*, Int. J. Electronics, Vol. 49, 1980, pp. 287–299.
12. G. de Micheli and A. Sangiovanni-Vincentelli, *Multiple Constrained Folding of Programmable Logic Arrays: Theory and Applications*, IEEE Tr. on CAD, Vol. CAD-2, 1983, pp. 151–167.
13. Monolithic Memories Inc, *Programmable Logic Handbook*, MMI, Santa Clara, 1985.
14. T. Sasao, *Input Variable Assignment and Output Phase Optimization of PLAs*, IEEE Tr. on Comp., Vol. C-33, 1984, pp. 879–894.
15. *Signetics Fiels Programmable Logic Array: An Application Manual*, Signetics Corp., Sunnyvale, Ca., 1977.
16. Toshiba Corporation, *MOS Memory Products*, Toshiba, Tokyo, 1985.
17. R.A. Wood, *A High Density Programmable Logic Array Chip*, IEEE Tr. on Comp., Vol. C-28, Sept. 1979, pp. 602–608.
18. Xicor Inc., *Xicor Data Book*, Milpitas, Ca., 1985.
19. Xilinx, *The Programmable Gate Array Design Handbook*, Xilinx, San Jose, Ca., 1987.

9
Memory elements I: latches

9.1 Memory elements

The output of a combinational circuit in the static state is completely determined by the present input combination. A characteristic of other circuits – sequential circuits – is that at least for one input combination the output signal in the static state (i.e. no signal transitions in the circuit) can have more than one value. In that case the present input combination does not determine the output signal completely. Some of the previous input combinations, as well as their order, also influence the present output signal. A sequential circuit therefore has to have memory. The ability to store data and circuits in which this property can be realised are the subjects of this chapter. To get some insight into the memory property and its realisation in a circuit we will first work our way through an example.

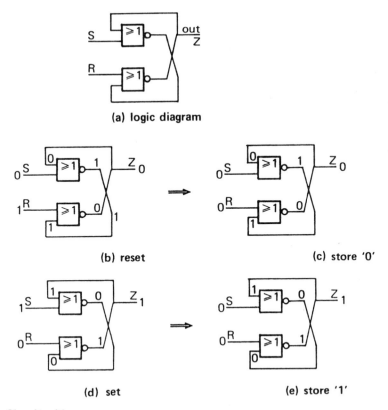

Figure 9.1 Circuit with memory.

The design described in Figure 9.1 consists of two NORs. This circuit is capable of storing data for the input combination SR = 00.

For the input combinations SR = 01 and SR = 10 all signal values in the circuit are completely determined (see Figures 9.1.b and 9.1.c.). If the input combination SR = 01 (Figure 9.1.b) changes into SR = 00 the circuit enters the state shown in Figure 9.1.d. This situation is stable and will continue to exist under the input combination SR = 00, as can easily be verified in the circuit. The same holds for the situation in Figure 9.1.e, which arises if, starting from SR = 10, the next input combination is SR = 00.

A comparison of Figure 9.1.d and Figure 9.1.e shows that two different output values $Z = 0$ and $Z = 1$ can exist for the input combination SR = 00. The output value depends on the previous input combination, SR = 01 or SR = 10. The circuit is capable of remembering this.

The memory function is realised by the *closed loop* in the circuit. This necessitates the circulating signal along the loop not being weakened, which is generally the case in modern ICs. Their gates regenerate the signals, and the output level of a gate always lies within the defined logic levels.

The state of a memory element.

It is customary to speak of the (internal) logic state of a memory element. *In the case of Figure 9.1 it is assumed that the (internal) state of the memory element resembles the value of the output signal Z.* The circuit can be in two stable states, which are encoded as 0 (Figure 9.1.d) or as 1 (Figure 9.1.e). Later we will introduce the concept 'state' in a more abstract way.

If we refer to the (internal) logic state of the memory element in Figure 9.1 as Z, we may conclude:

– With the input combination SR = 01 the memory element is put into state $Z = 0$; SR = 01 is the *reset combination*.
– With the input combination SR = 10 the memory element is put into state $Z = 1$; SR = 10 is the *set combination*.

With the input combination SR = 00 the circuit stores data. The state is then $Z = 0$ or $Z = 1$, depending on the previous input combination SR = 01 or SR = 10.

Note

A fourth input combination SR = 11 is possible. This input combination is superfluous for the control of a memory element. Besides, the use of SR = 11 entails great problems (see Section 9.4). We will therefore exclude SR = 11 for the time being. □

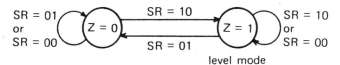

Figure 9.2 State diagram of a memory element.

In Figure 9.2 the function of the circuit of Figure 9.1 is described in a *state diagram*, input combination SR = 11 being excluded. The states Z = 0 and Z = 1 are indicated by two circles. Arrows denote the transition from one state into another. The supplementary input combination specifies under which condition a transition occurs. The state transition is effected as soon as the input combination changes.

This way of looking at a circuit is called the *level mode* view. This is in contrast to the *clock mode* view, in which the state transitions are assumed to take place when a so-called clock pulse is given (see Chapter 10).

The truth table and the formula of a memory element

If we call the old or present state of a memory element Z_{old} and the new state Z_{new}, we can directly generate Table 9.1 on the basis of the above reasoning. If we assume that input combination SR = 11 is not used we may specify for Z_{new} a don't care.

S	R	Z_{old}	Z_{new}	
0	0	0	0	} memory element stores data
0	0	1	1	
0	1	0	0	} memory element is being reset
0	1	1	0	
1	0	0	1	} memory element is being set
1	0	1	1	
1	1	0	–	} function not specified (SR ≠ 11)
1	1	1	–	

(a)

S	R	Z_{new}
0	0	Z_{old}
0	1	0
1	0	1
1	1	–

(b)

Table 9.1 Operation of a memory element.

Table 9.1.b is a concise form of table 9.1.a. and specifies the same information.

With a Karnaugh diagram a formula for Z_{new} can be formed, expressed in S, R and Z_{old} (see Figure 9.3).

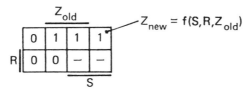

Figure 9.3 Karnaugh diagram.

From Figure 9.3 two simple forms of the formula for Z_{new} follow, the sum form

$$Z_{new} = S + \bar{R} \cdot Z_{old} \tag{9.1}$$

and the product form

$$Z_{new} = \bar{R} \cdot (S + Z_{old}). \tag{9.2}$$

These formulas are not equivalent algebraically. The don't cares have been chosen differently in each case.

Physical properties.

A memory element has not only logical properties, as expressed in (9.1) and (9.2), but also physical properties. One of them is the propagation delay time of the circuit. Figure 9.4 gives a survey of the timing model.

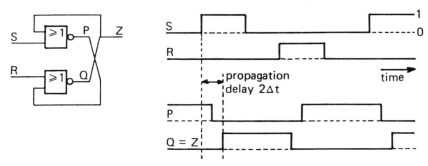

Figure 9.4 Circuit and timing model.

When starting from SR = 00 and Z_{old} = 0 the input combination becomes SR = 10, the output P of the upper gate will react after a propagation delay time Δt. This change is transferred to the input of the lower gate. The output Q of the lower gate will in turn react after another propagation delay time Δt. The new output value Q = 1 is then fed back to the upper gate. After this the circuit is again at rest.

The total propagation delay of the circuit is $2\Delta t$, assuming that each gate has a fixed delay Δt. After a time t = $2\Delta t$ the input combination may return to SR = 00. In the timing diagram of Figure 9.4 this happens somewhat later. It will be evident that a set command SR = 10 or a reset command SR = 01 must never be given a shorter time than $2\Delta t$. In the remainder of this chapter we will assume that this requirement, which follows from the physical characteristics of the components and from the structure of the network, has always been fulfilled. In practical cases this will have to be verified.

9.2 The $\bar{S}$-$\bar{R}$ and the S-R latches

We will now trace, starting from (9.1) and (9.2), how circuits that realise the above-described memory property can be designed. The formulas describe how the state Z_{new} depends on the present state Z_{old} and the present input combination. Figure 9.5 illustrates this schematically.

When a state $Z_{new} \neq Z_{old}$ is set it will appear at the output after some propagation delay time. In the model this has been schematically indicated by a delay element. Whether this delay in the feedback loop is necessary or not is still open to argument. When one observes that gates by nature always have a certain propagation delay, however small,

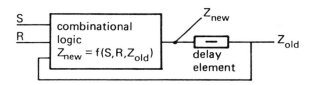

Figure 9.5 Model memory element.

this discussion is meaningless. We therefore agree to omit explicitly drawn delay elements in all of the following schemes.

Equation (9.1) then leads, based on the model in Figure 9.5, to the circuit in Figure 9.6. In this figure the delay element in the feedback loop has been omitted. We will further denote the (external) output of the circuit by Z.

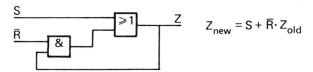

Figure 9.6 Realisation of a memory element.

Realisation with NANDs

A realisation with NANDs (sum form!) can be deduced from (9.1) as follows:

$$\begin{aligned} Z_{new} &= S + \bar{R} \cdot Z_{old} \\ &= \overline{\overline{S + \bar{R} \cdot Z_{old}}} \\ &= \overline{\bar{S} \cdot (\overline{\bar{R} \cdot Z_{old}})} \\ &= \text{NAND}(\bar{S}, \text{NAND}(\bar{R}, Z_{old})). \end{aligned} \qquad (9.3)$$

Figure 9.7 describes the circuit. The layout of Figure 9.7.b is the one most commonly found in practice.

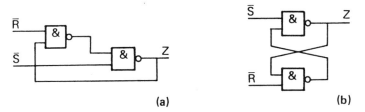

Figure 9.7 Memory element with NANDs.

Generally the realisation of a memory element is based on Figure 9.7, especially in the TTL technology in which the NAND is a preferred component. A disadvantage of this construction is that instead of the signals S and R their complements $\bar{S}$ and $\bar{R}$ have to be used. Most of the time this poses no problem. The circuit is called an $\bar{S}$-$\bar{R}$ latch.

Realisation with NORs

A realisation with NORs (product form!) can be deduced from Equation (9.2):

$$\begin{aligned} Z_{new} &= \overline{R} \cdot (S + Z_{old}) \\ &= \overline{\overline{\overline{R} \cdot (S + Z_{old})}} \\ &= \overline{R + \overline{(S + Z_{old})}} \\ &= NOR(R, NOR(S, Z_{old})) \end{aligned} \qquad (9.4)$$

Figure 9.8 describes the circuit.

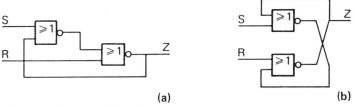

(a) (b)

Figure 9.8 Memory element with NORs.

With the circuit of Figure 9.8 we are back at our starting point, the circuit of Figure 9.1. This circuit is called an *S-R latch*.

Symbols for $\overline{S}\text{-}\overline{R}$ and S-R latches.

The memory elements in Figures 9.7 and 9.8 have been designed under the condition that the input combination SR = 11 not be used. This has been utilised in deducing (9.1) and (9.2), especially with respect to the choice of the don't cares. If SR = 11 is not applied, the logical values of the outputs of the circuits in Figures 9.7 and 9.8 are each other's complements (in the static state, i.e. when the memory element has stabilised). This can easily be verified. We then indicate the lower output of the memory elements by $\overline{Z}$.

We explicitly wish to point out that the outputs of the circuits when SR = 11 is used anyhow are not complementary!

Figure 9.9 shows the IEC symbols of the memory elements designed above. The circuits themselves are usually called $\overline{S}\text{-}\overline{R}$ latch and S-R latch. These symbols may only be used if the input combination SR = 11 is not applied. If this cannot be guaranteed more detailed symbols have to be used.

Note

In Figures 9.9, 9.11 and further the designations 'S' and 'R' occur in various places:
- outside the symbol outlines as the *name* of the signal lines;
- inside the symbol outlines.

In the latter case S and R have a meaning as defined in the IEC symbol system. They indicate the *nature of the influence* the signal has on the 'state' of the memory element. Inside the symbol outlines the designations are standardised, and outside the symbol outlines they are not.

Double use of the designations S and R (also compare Figure 9.11.a) is therefore not confusing. □

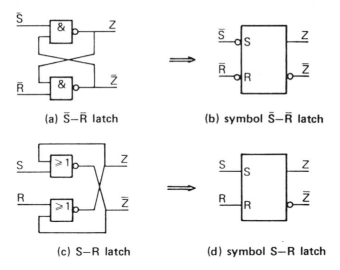

Figure 9.9 IEC symbols for latches.

9.3 The gated latch

The control circuits for the S and R input are usually combinational logic circuits. These circuits deduce the desired values of S and R from external input signals. There is usually a transient or uncertainty interval in the external signals. Such an interval is also, of course, found at the output of the control circuits. This uncertainty may result in short transient phenomena, called *spikes* or *glitches*. An input transition may, unintentionally, change the state of a latch, regardless of when it occurs. Therefore latches are said to be *transparent for their input signals*. Figure 9.10 gives an explanation.

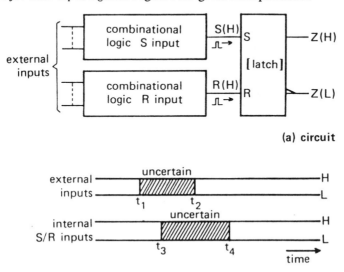

Figure 9.10 Latches and their control circuits.

There is usually no problem when the SR = 01 or the SR = 10 combination is briefly interrupted by SR = 00. When SR subsequently becomes SR = 01 or SR = 10 again the latch will still be set correctly. It is more serious when the SR = 00 combination is interrupted by SR = 01 or SR = 10. The wrong command, set or reset, is not cancelled when the input combination returns to SR = 00 again (see also Chapter 11).

To prevent an unintentional reaction of the latch because of such input transitions two gates are often placed before the S and R input. These gates are controlled by a so-called external *enable signal*. If these gates are disabled during the period in which the external signals are being set the transient phenomena in S and R have no influence on the state of the latch (see Figure 9.11). Such a circuit is called a *gated S-R latch*. The circuit realises the formula

$$Z_{new} = \overline{EN} \cdot Z_{old} + EN \cdot \overline{R} \cdot (S + Z_{old}) \qquad (9.5.a)$$

$$= \overline{EN} \cdot Z_{old} + EN \cdot S + \overline{R} \cdot Z_{old}. \qquad (9.5.b)$$

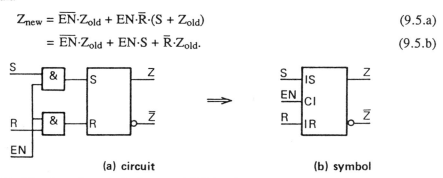

Figure 9.11 Circuit and symbol of a gated S-R latch.

Figure 9.11 gives the symbol for the gated S-R latch. If the command input C1 has the internal logic value 1, i.e. external EN = 1, the inputs 1S and 1R are enabled. The memory element follows and adapts to the value of the external signals S and R. If C1 = 0, however, S and R have no influence on the state of the memory element. The internal state remains unchanged.

In the circuit of Figure 9.11 the problem of transient phenomena at the S and R inputs has been made controllable. A problem that has arisen instead is that of transient phenomena at the enable input EN. An enable signal often controls several latches in parallel, e.g. eight data lines are either put through or not. The problems will then concentrate on one signal line instead of eight. In Chapter 10 we will discuss how the use of enable signals is related to the *internal organisation* of a circuit.

Example

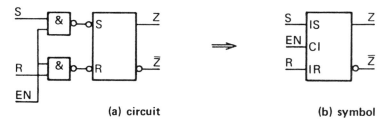

Figure 9.12 Gated S-R latch based on NANDs.

Figure 9.12 shows how a gated S-R latch can be realised, based on an $\overline{S}$-$\overline{R}$ latch and two NANDs. □

The gated D latch

Figure 9.13 describes a so-called *gated D latch*. The input signals S and R are linked by a negator. When EN = 0 the circuit holds the stored data. While EN = 1 the internal state of the circuit follows the D signal. The circuit realises the formula

$$Z_{new} = \overline{EN} \cdot Z_{old} + EN \cdot D. \qquad (9.6)$$

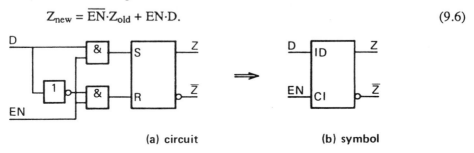

(a) circuit (b) symbol

Figure 9.13 Gated D latch.

The characteristics of the gated D latch obviously differ from those of the gated S-R latch. For EN = 1 the circuit follows the D signal. There is *no memory function*. During EN = 0 the circuit stores data. Unlike the gated S-R latch, this means that no problems arise when the input signal D of a gated D latch is still changing during EN = 1. The value of D just before the 1 → 0 transition of EN determines the internal state during EN = 0.

This is not the case for the S-R latch when SR = 00. If the command 'INHIBIT' is given, SR = 00, then transient phenomena are forbidden during the time that EN = 1. They may cause an unwanted set or reset command that cannot be cancelled. With an S-R latch *higher demands* are made upon the input signals.

The fact that a gated D latch has only one data input is another reason to prefer it to the gated S-R latch (see also Section 9.4).

Example

Figure 9.14 describes a design of the gated D latch based on the AND-OR-INVERT gate, as it is found in Texas Instruments IC SN74100. It is a typical TTL solution. The IC itself is an 8-bit *latch register*, consisting of two groups of four D latches with a *common enable input*. □

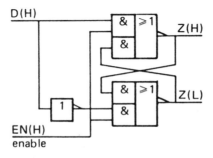

Figure 9.14 Gated D latch in TTL.

Example

Figure 9.15.b illustrates how an S-R latch can be realised in NMOS. It can easily be converted to a gated S-R latch by simply placing two transistors in series at the spot marked *. On one of these either S or R is fed at the gate, and on the other one the enable signal EN. Figure 9.15.c shows the transistor diagram of an S-R latch in CMOS (a smaller design is possible). □

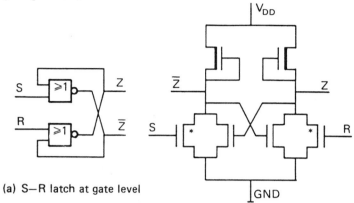

(a) S—R latch at gate level

(b) S—R latch (transistor level/NMOS)

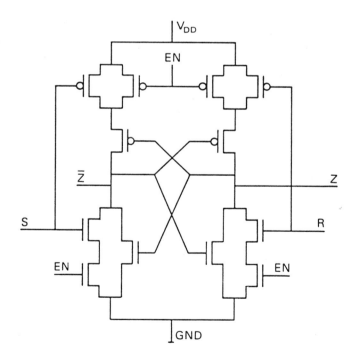

(c) gated S—R latch (CMOS)

Figure 9.15 S-R latch in NMOS and gated S-R latch in CMOS (positive logic).

9.4 Encoding problems and race conditions

The S-R memory elements are given three commands: SET, RESET and INHIBIT. These commands have been encoded for the S-R latch with the input signals S and R (Figure 9.8) as

    SET   : SR = 10
    RESET   : SR = 01
    INHIBIT : SR = 00.

For an $\bar{S}$-$\bar{R}$ latch (Figure 9.7) these commands are encoded with the input signals $\bar{S}$ and $\bar{R}$ as

    SET   : $\bar{S}\bar{R}$ = 01
    RESET   : $\bar{S}\bar{R}$ = 10
    INHIBIT : $\bar{S}\bar{R}$ = 11.

These commands cannot be encoded at random. The INHIBIT command may be given after a SET as well as a RESET command. This means that an assignment

    SET   : SR = 11
    RESET   : SR = 10
    INHIBIT : SR = 00

is not possible. This can be seen as follows.

Suppose that a SET command is given: SR = 11. Subsequently the memory element must go to the INHIBIT mode, SR = 00. Both S and R change from 1 to 0 in this sequence of commands. If R changes just a little earlier than S, the transient combination SR = 10 occurs during this sequence. This means that a RESET command is given, to which the element might react. The desired state, the 1-state, will not be reached.

The transient combination SR = 01 might also occur during this sequence. Now it is not specified what the circuit has to do. For SR = 01 the command SET or INHIBIT can be specified. When arriving at SR = 00 the memory element will be in the desired state.

Conclusion

Commands to circuits based on latches cannot be encoded at random; sometimes *race conditions* occur. A race condition occurs when different input signals change at approximately the same time. When it is possible for the circuit to enter a wrong internal state after a race, the race condition is called *critical*.

The SR = 11 input combination

Race conditions are also at the heart of the exclusion of the SR = 11 input combination for the S-R latch (see Section 9.1). (For the $\bar{S}$-$\bar{R}$ latch the $\bar{S}\bar{R}$ = 00 input combination is excluded.) In principle one might extend Table 9.1 to Table 9.2.

The state $Z_{(0)new}$ = 0 is chosen for the so-called $Z_{(0)}$ memory element, when the input combination SR = 11 is used. The formula for the $Z_{(0)}$ memory element is

$$Z_{(0)new} = \bar{R} \cdot (S + Z_{(0)old}). \tag{9.7}$$

S	R	Z_{old}	$Z_{(0)new}$	$Z_{(1)new}$	$Z_{(z)new}$
0	0	0	0	0	0
0	0	1	1	1	1
0	1	0	0	0	0
0	1	1	0	0	0
1	0	0	1	1	1
1	0	1	1	1	1
1	1	0	0	1	0
1	1	1	0	1	1

Table 9.2 Additional specification of a memory element.

The $Z_{(1)}$ memory element has as formula

$$Z_{(1)new} = S + \bar{R} \cdot Z_{(1)old}. \tag{9.8}$$

For the $Z_{(0)}$ memory element the reset input dominates, as it were, just as the set input does for the $Z_{(1)}$ memory element. It is also possible for SR = 11 to express a preference for maintaining the present state. This results in the so-called $Z_{(z)}$ memory element, with formula

$$Z_{(z)new} = S \cdot \bar{R} + (S + \bar{R}) \cdot Z_{(z)old}. \tag{9.9}$$

The $Z_{(0)}$ memory element has, in fact, been realised in the S-R latch of Figure 9.8 and the $Z_{(1)}$ memory element in the $\bar{S}$-$\bar{R}$ latch of Figure 9.7. In this case, however, $\bar{S}$ and $\bar{R}$ will have to be used instead of S and R. The $Z_{(z)}$ memory element can also be realised easily with some additional gates.

The above might suggest that the SR = 11 combination can be useful. Nevertheless, the SR = 11 combination is hardly ever used. If SR = 11 is used in accordance with one of the above interpretations *transient phenomena* will occur at the transition from SR = 11 to SR = 00. Depending on whether the signals S and R do or do not change simultaneously the following may happen:

SR = 11
- SR = 01 → SR = 00 → Reset; state $Z_{new} = 0$.
- SR = 00 → State Z_{new} is the state which holds for SR = 11.
- SR = 10 → SR = 00 → Set; State $Z_{new} = 1$.

One might try to find a solution to this problem by giving one of the inputs, the set or reset input, another threshold. This will secure the result of the race. However, it does not solve the problem! *The cause of the problem is not related to the internal structure of the memory element, but to the fact that S and R are assumed to change simultaneously, but do not always do so. This depends on the propagation delay time and other characteristics of their control circuits.*

Conclusion.
In practice the use of the SR = 11 input combination must be avoided as much as possible, because of the uncertainty about which state Z_{new} the memory element will reach after an SR = 11 to SR = 00 transition.

The three other input combinations of S and R are sufficient to control the memory element. We therefore agree to avoid, if possible, the SR = 11 input combination. □

With this conclusion the distinction drawn between the $Z_{(0)}$, the $Z_{(1)}$ and the $Z_{(z)}$ memory elements has become of theoretical importance only.

Encoding problems in circuits with latches have certainly not been discussed at length above. There are also great problems related to the choice of the *state assignment* of circuits based upon latches. State transitions can be very critical if the wrong assignment has been chosen. We will discuss these problems in Chapter 10.

9.5 Function tables

Until now we have concentrated on the analysis and synthesis of latches and gated latches. In the circuit design process the question of how to use them is foremost. The designer is confronted with the question of how, starting from a given state Z_{old}, the inputs of a memory element must be set in order to reach the specified state Z_{new}. Of course, the designer does not check this every time. Instead he uses *function tables*. A function table is a table that specifies which command, encoded by the external input signals, must be given in order to reach the desired state Z_{new}, starting in state Z_{old}.

For the S-R latch of Figure 9.8, for example, Table 9.3.a describes the logic function. Table 9.3.b is the corresponding function table.

S	R	Z_{old}	Z_{new}		Z_{old}	Z_{new}	S	R
0	0	0	0		0	0	0	–
0	0	1	1		0	1	1	0
0	1	0	0		1	0	0	1
0	1	1	0		1	1	–	0
1	0	0	1					
1	0	1	1		–2)	0	0	1
1	1	0	–1)		–2)	1	1	0
1	1	1	–1) (a)					(b)

1) input combination forbidden, Z_{new} not specified
2) state Z_{old} not known or information not used

Table 9.3 Function table for the S-R latch.

Explanatory note

From Table 9.3.a it follows that the transition

$$Z_{old} = 0 \rightarrow Z_{new} = 0$$

can be obtained under two input combinations, namely SR = 00 and SR = 01. The signal at the S input must therefore be 0; the signal on the R input can be chosen to be either 0 or 1. In the function table this is specified by the input condition SR = 0–.

For the transition

$$Z_{old} = 0 \rightarrow Z_{new} = 1$$

only one input combination can be used, namely SR = 10. This explains the second row of Table 9.3.b. The next two rows of Table 9.3.b follow similarly.

In the last two lines of Table 9.3.b the state Z_{old} is not specified; the signal of Z_{old} may be difficult to reach because of layout reasons, for example. In order to force a state $Z_{new} = 0$ independent of Z_{old} only the input combination SR = 01 can be used. Similarly, $Z_{new} = 1$ can only be reached with SR = 10 when information about the state Z_{old} is not used.

$\overline{S}$	$\overline{R}$	Z_{old}	Z_{new}
0	0	0	–1)
0	0	1	–1)
0	1	0	1
0	1	1	1
1	0	0	0
1	0	1	0
1	1	0	0
1	1	1	1

(a)

Z_{old}	Z_{new}	$\overline{S}$	$\overline{R}$
0	0	1	–
0	1	0	1
1	0	1	0
1	1	–	1
–2)	0	1	0
–2)	1	0	1

(b)

1) input combination forbidden, Z_{new} not specified
2) state Z_{old} not known or information not used.

Table 9.4 Function table for the $\overline{S}$-$\overline{R}$ latch.

Table 9.4 describes the construction of the function table of the $\overline{S}$-$\overline{R}$ latch. In Table 9.4.b the logical values of the inputs marked $\overline{S}$ and $\overline{R}$ are specified, rather than the logical values of S and R themselves.

Function tables for gated latches

These tables can be specified in a similar way. The enable input may be considered in two ways:

– as an input with the same priority as S and R (or D);

– as an input that controls *when* the latch is enabled, while S and R (or D) determine which command, SET, RESET or INHIBIT, is given.

In the first view the state transition

$$Z_{old} = 0 \to Z_{new} = 0$$

may, for example, be reached with SREN = ––0 (compare Figure 9.11). In the second view this is SR = 0–, while it is implicitly assumed that the enable signal EN will in due time have the value EN = 1. *The latter interpretation is the most customary.*

Table 9.5 is, *in this view*, the function table of the gated D latch. The tables of the gated S-R and $\overline{S}$-$\overline{R}$ latches are the same as Tables 9.3 and 9.4. The *interpretation* of the tables is different, however.

Z_{old}	Z_{new}	D
0	0	0
0	1	1
1	0	0
1	1	1
–2)	0	0
–2)	1	1

2) State Z_{old} not known or information not used.

Table 9.5 Function table for the gated D latch.

9.6 Application: the anti-bounce circuit

When commands are given to electronic circuits using switches and/or pushbuttons, two problems will have to be solved. First of all the information as to whether the contact is either conducting or not must be converted into the logic levels HIGH and LOW. Secondly, transient phenomena resulting from bouncing contacts must be removed. When contacts control relays the *mechanical inertia* of the relay will usually intercept *short transient phenomena* in the signals. The relay is too slow to react to them. Electronic circuits, however, do react to transient phenomena because of their higher speed.

Figure 9.16 gives an explanation. With the two resistors in Figure 9.16.a the position of the pushbutton (transfer contact) is converted into voltages which lie *within the defined regions* for the levels HIGH and LOW. The value of the resistance is dependent on the technology of the electronic circuit used (for TTL ca. 1kΩ).

In Figure 9.16.b it is indicated that, if the contact springs do bounce while switching, transient phenomena will arise. They may easily pass the threshold between H and L. The electronic circuit will interpret such signal transitions as consisting of several pulses.

The *anti-bounce circuit* in Figure 9.16.c resolves all these problems. Immediately after the contact spring makes contact on the other side for the first time the $\overline{S}$-$\overline{R}$ latch is set or reset. During the time that the contact spring is bouncing back the latch gets neither a set command nor a reset command. The latch stores its last command. This solves all problems concerning bouncing contacts.

Note
The anti-bounce circuit in Figure 9.16 only operates correctly if the contact spring does not completely return to the other side during bouncing. □

When a processor samples signal lines coming from a keyboard, bouncing phenomena will also occur. These can be resolved by:

– the anti-bounce circuit;

– repeated sampling of the signal lines and checking whether the signal just read remains unchanged for a period longer than the bouncing period.

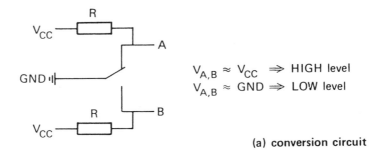

(a) conversion circuit

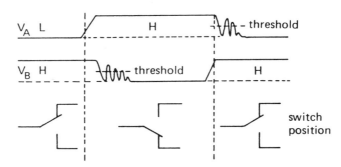

(b) output levels corresponding to switch positions

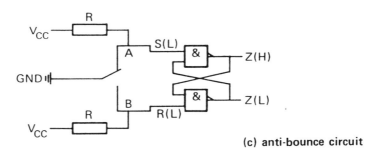

(c) anti-bounce circuit

Figure 9.16 Debouncing of contacts.

In the first case more hardware is needed, e.g. a latch and two resistors. In the second case the recognition of signal transitions takes longer. Besides, one must determine the length of the bouncing period by experimentation.

9.7 Application: the omnibus circuit

As a second example of the latch being applied as a memory element we will discuss the so-called *omnibus circuit*. If a passenger in a bus pushes the stop button a signal (usually a red lamp) is switched on in the front of the bus. For the driver this is a sign that a passenger wants to get off at the next stop. The stop signal remains on when the passenger releases the button. This requires a *memory*. After the stop the driver can press the reset button to reset the circuit, which switches off the signal.

The memory function of the circuit that operates the stop signal can be specified by two states:

> state q_0 : 'The lamp must be off';
>
> state q_1 : 'The lamp must be on'.

The state of the circuit, q_0 or q_1, indicates which button has been pushed last. The commands to the control circuit are described by two propositions A and B:

> A: 'The ON button is pushed' ⇔ 'Passenger wants to get off'
>
> B: 'The OFF button is pushed' ⇔ 'Passenger got off'

Figure 9.17 shows how the pushing of ON and OFF influences the state of the controller. If proposition A is true (ON has been pushed), the circuit enters state q_1 (lamp on) and/or remains in that state. If proposition B is true (OFF has been pushed), the circuit enters state q_0 (lamp off) and/or remains in q_0. When no one pushes the button, the circuit remembers its last command and remains in q_0 or in q_1, respectively.

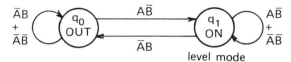

Figure 9.17 State diagram of the lamp controller.

In Figure 9.17 the case of both a passenger and the driver pushing simultaneously has not been specified. Clearly, it is useless to specify who must dominate in that case, the passenger or the driver. After all, the person who releases the button *last* determines the final state of the controller (see Section 9.4).

State assignment

When the two states in Figure 9.17 are encoded with one memory element Z as

> q_0: Z = 0
>
> q_1: Z = 1

and when the truth values of the propositions A and B are indicated by 0 (false) and 1 (true) the *encoded state diagram* of Figure 9.18 results.

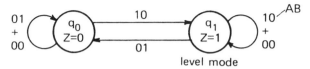

Figure 9.18 Encoded state diagram.

The encoded state diagram of Figure 9.18 corresponds to the diagram of Figure 9.2. This diagram describes an S-R latch with input signals S and R. It can therefore easily be concluded that the controller specified in Figures 9.17/18 can be realised, based on one S-R latch. In that case (compare Figure 9.18 with Figure 9.2):

> A = S and B = R.

The memory element Z must be SET for the input combination $A\bar{B}$, i.e. SR = 10, and RESET for $\bar{A}B$, i.e. SR = 01. With this conclusion we have obtained, with propositions A and B, the set and reset conditions of the memory element. Figure 9.19 describes the *logic diagram* of the controller. Its operation during the input combination AB = 11 is, in accordance with a previous note, not specified.

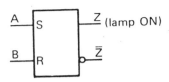

Figure 9.19 Logic diagram.

The realisation

Figure 9.20 describes the *detailed logic diagram* of the circuit. By means of a resistor to the positive voltage V_{cc} pushing a button is converted into a voltage level corresponding to the logic levels H and L. When a button is pushed its output level is Low (L). In other words:

$A(L) \Leftrightarrow$ proposition A is true;

$B(L) \Leftrightarrow$ proposition B is true.

The commands to the S-R latch correspond (in positive logic) to S(H) and R(H) respectively. The logic levels of A and B on the one hand and S and R on the other do not correspond and have to be adapted by an inverter.

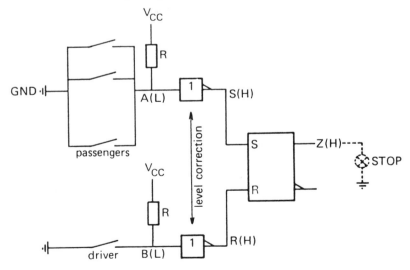

Figure 9.20 Detailed logic diagram.

If in Figure 9.20 an $\bar{S}$-$\bar{R}$ latch (positive logic) is chosen, S(L) and R(L) hold. In that case the logic levels correspond to the physical levels. The inverters are not necessary.

Note

During the design of the circuit in Figure 9.20 nothing was said about bouncing phenomena at the contacts of the pushbuttons. Bouncing results in repeated set or reset commands in such a case. Bouncing forms no problem! □

9.8 Mapping problems

A specification for a circuit design is usually drawn up in several stages. The result must be a design for a *realisable circuit*. Section 9.7 described an example of such a procedure. An initial specification is usually global and not complete. The details are filled in when it is worked out. In working out a specification one has to be careful. We will go into this. Figure 9.21 describes how the omnibus controller of Section 9.7 was specified in previous Dutch editions of this book for many years. This specification was clear to the reader.

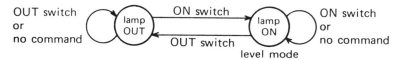

Figure 9.21 State diagram for omnibus controller.

If we 'translate' this specification directly into a logic specification with propositions A (ON button pushed) and B (OFF button pushed) Figure 9.22 results. This specification is incorrect.

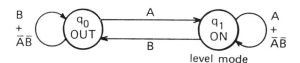

Figure 9.22 Logic specification.

In state q_0 two conditions are specified, condition

$$B + \overline{A}\overline{B}$$

under which the circuit must remain in state q_0, and condition

$$A$$

under which the circuit must go into state q_1. These conditions contain a common term AB:

$$B + \overline{A}\overline{B} = B(\overline{A} + A) + \overline{A}\overline{B} = \overline{A}B + AB + \overline{A}\overline{B}$$
$$A \quad = A(\overline{B} + B) \quad = A\overline{B} + AB.$$

The same holds for state q_1. According to one condition the circuit must remain in state q_1 for AB, and according to the other it must go into state q_0. This is contradictory.

Now condition AB = 11 corresponds to a situation which, in accordance with a previously made agreement, was to be left out of consideration because the behaviour of the circuit when both ON and OFF are pushed was problematic anyway. On the realisation level, however, it is necessary to choose one of the alternatives. The specification must then be *unambiguous*.

The overlapping of conditions as well as the incompleteness of a specification around a state can be simply examined with Karnaugh diagrams, by mapping all the conditions specified around a state in a diagram. This allows one to see directly whether a specification is incomplete and/or contradictory. Also see Section 3.3.

9.9 RAM Memories

Static RAMs

Chapter 8 contained a discussion of memories (ROMs) of which the programmable or non-programmable contents have been laid down in hardware. These memories are only designed to be *read*. In the previous chapter applications for realising combinational logic have been discussed. In addition, this type of memory is frequently used in computers, e.g. to store the system software.

Other memories are *read/write memories*. They are called a RAM (Random Access Memory), although this type of addressing also applies to ROMs. RAM memories are volatile, i.e. after the power is switched off the contents of the cells disappear after a short time. The memory cells of RAMs greatly resemble latches. Figure 9.23 gives some examples of memory cells in MOS technology.

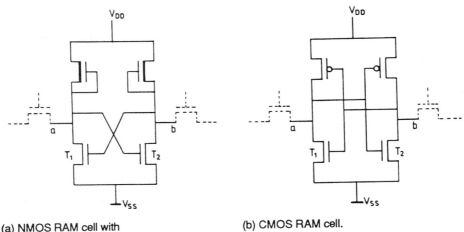

(a) NMOS RAM cell with depletion-type load.

(b) CMOS RAM cell.

Figure 9.23 Cell structures of static RAMs.

A memory cell is operated as follows. The logic voltages are fed (usually complementary) to points a and b. A voltage at point a with a value lower than V_t of transistor T_2 blocks this transistor. Consequently, the voltage in point b increases, as a result of which T_1 starts to conduct (if it was not doing so already). When points a and b are subsequently disconnected again the cell remains in the position taken previously.

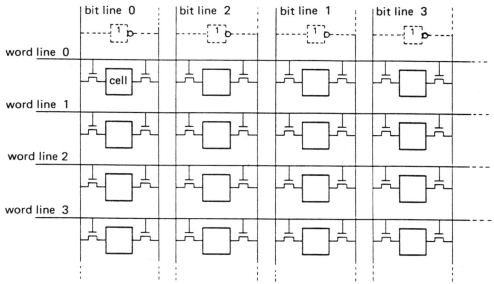

Figure 9.24 Layout of Random Access Memory (static).

Figure 9.24 illustrates how these cells are used as a basis for constructing the memory array of a RAM. Vertical bit lines running through the memory allow data to be written or read. When writing, the bit lines of each cell get a complementary value.

The reading of memory cells can be implemented in various ways. In one of them the bit lines are precharged (*precharge of (parasitic) capacitors*). When a certain word line is activated, the precharge either flows away via the cell or it does not, depending on the contents of the cell. A test for the potential of the bit line involved provides information about the contents of the cell. In another type the bit lines are connected to a *pull-up transistor*. A conducting T_1 or T_2 then pulls this bit line down.

RAMs usually also have an address selection by means of row decoding and column decoding. The internal structure greatly resembles that of Figure 8.4. Besides, RAMs have some additional facilities, including:

– a chip enable input. It directly controls the open-collector or 3-state output buffers. As a result it is possible to interconnect RAMs in parallel on an internal or external bus;

– the R/W line. It is set to reading (Read) or writing (Write); the R/W signal switches the RAM mode;

– a Write signal which activates the writing of the data to the address indicated.

All of these control signals have to be in a certain timing relation. In practice it is not easy to meet all of the timing requirements. One often comes across the most unlikely arrangements.

Dynamic RAMs

Figure 9.25 describes the principle of a cell for a so-called dynamic memory (DRAM). In its simplest form it is a capacitor which can be loaded with a certain charge.

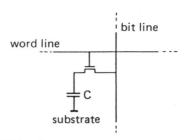

Figure 9.25 Dynamic RAM cell.

When *a dynamic RAM cell is read* its content is lost. This is because testing the voltage over C can only be done by activating the word line and the bit line. Dynamic RAMs therefore have a *read-modify-write* cycle when they are being read, during which time the data just read again can be written (in modified form or not).

Dynamic memories have to be refreshed from time to time. This is done by starting a so-called *refresh cycle*. During the refresh cycle the cell is not able to be read or written. Refresh controllers are available for dynamic RAM ICs. These are able to coordinate the read/write and refresh cycles of several RAM ICs at the same time.

Although *dynamic RAMs* have a number of disadvantages with respect to *static RAMs* they are used frequently. Firstly, the cell layout is much smaller. In addition, the power consumption is usually lower than that of static RAMs. As the technology is controlled better one is likely to see a shift from dynamic to static RAMs, because they are more convenient to use.

Note

Large RAM memories are usually realised in MOS technology. With fast memories one opts for bipolar technologies. See also Chapter 6. □

References

1. C. Mead and L. Conway, *Introduction to VLSI Systems*, Addison-Wesley, 1980.
2. R.M.M. Oberman, *Disciplines in Combinational and Sequential Circuit Design*, McGraw-Hill, New York, 1970.
3. Texas Instruments, The TTL Data Book, Texas Instruments, Dallas, 1984.
4. N. Weste and K Eshragian, *Principles of CMOS VLSI Design*, Addison-Wesley, 1985.

10
Memory elements II: flip-flops

10.1 Restrictions on the use of latches

The application of latches and gated latches is subject to restrictions; some of these have been discussed in Section 9.4. One of them is the poor noise immunity, because latches react directly to input transitions. We will introduce some other limitations on the use of latches by evaluating some applications. The discussion will lead to the *clock mode model* of sequential circuits.

Example *The toggle flip-flop*

Figure 10.1 specifies a 2-bit binary counter by its sequencing table and timing diagram. According to this specification the counter has to jump at each count pulse and returns to its initial state after every fourth count pulse. The timing diagram indicates how and when the counter is supposed to react.

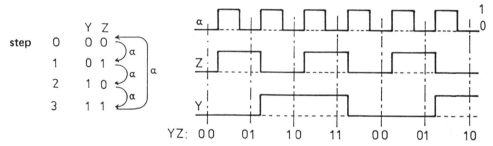

Figure 10.1 Design specification of a binary counter.

During its cycle the counter goes through four states. These are encoded with two memory elements Y and Z. The memory elements react upon or immediately after a $0 \rightarrow 1$ transition of the count pulse α (see the timing diagram). Looking at the output of the memory element Z, we see that the output signal of this memory element changes half the frequency when compared to the count pulse α. A circuit with such a relation between its input and output signal is called a *toggle flip-flop*. According to the timing diagram its function can be described as

$\alpha: 0 \rightarrow 1 \quad \Rightarrow Z_{new} = \overline{Z_{old}}$,

$\left.\begin{array}{l} 0 \rightarrow 0 \\ \alpha: 1 \rightarrow 1 \\ 1 \rightarrow 0 \end{array}\right\} \Rightarrow Z_{new} = Z_{old}.$

Such a circuit cannot be realised with a (gated) latch, as is shown in Figure 10.2.

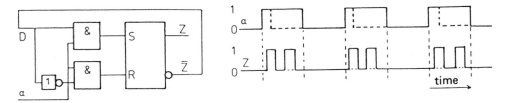

Figure 10.2 Toggle flip-flop with gated D latch (tryout).

The circuit in Figure 10.2 does not operate correctly. The circuit is in the hold mode when $\alpha = 0$. In that case $Z_{new} = Z_{old}$. At the $0 \rightarrow 1$ transition of α the input gates are opened and the latch enters a state that corresponds to the value of $\overline{Z}$. Then $Z_{new} = \overline{Z_{old}}$ holds. However, during the time that $\alpha = 1$, $Z_{new} = \overline{Z_{old}}$ continues to hold. As soon as the output has changed the input will start to adapt again. The circuit is said to oscillate. The oscillation between two states stops when α goes to 0. Depending on the moment when this happens the latch will sometimes remain in the state $Z = 0$ and sometimes in $Z = 1$. The oscillation frequency of Z is completely determined by the propagation delay time of the gates from which the D latch is composed. The desired toggle function is realised at the dotted width of the α pulse in Figure 10.2.b. This is not the case at the normal pulse width. The correct behaviour depends *critically* on the width of the enable pulse (also called the clock pulse or count pulse). In spite of this, some designers consider the circuit shown in Figure 10.2 to be a correct realisation of a toggle flip-flop. □

Digital circuits are manufactured with quite ample tolerances, which is one of the reasons their prices are low. This quickly leads to some variation in the timing properties of latches, which requires an *adapted width* of the count pulse in the example above.
On the other hand, if the required tolerances were to be maintained, the variable length of the wiring and the variable load of each line would cause a distortion of the pulse and the pulse width. Moreover, the propagation delay time of gates is temperature-dependent. This means that, in general, the toggle flip-flop as designed does not function correctly.

Conclusion

If the desired state Z_{new} of a (gated) latch also depends on the old state Z_{old} or $\overline{Z_{old}}$ of *the same* memory element this is, in principle, a source of problems. As soon as the new state is reached, this state replaces the state Z_{old}. Any information about it is lost directly and the new internal state immediately influences the external state of the circuit. See Figure 10.3 for an explanation.

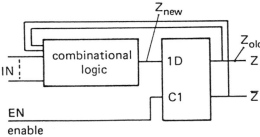

Figure 10.3 Input conditions of a gated latch.

Example *Data transfer between latches*

Data transfer between (gated) latches does not proceed without problems either. In many applications, for example, one needs to transmit data serially, i.e. bit by bit. The data is therefore loaded into a so-called shift register. Every time the register gets a pulse α, the bits have to shift one place. Such a circuit cannot be constructed as in Figure 10.4.

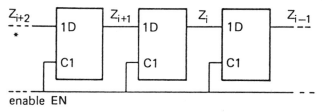

Figure 10.4 Shift register (tryout).

When the α pulse enables the inputs of the latches, every latch is initially set to the output state of its predecessor. However, if the α pulse lasts longer, all latches will finally be in the same state, which corresponds to the value of the signal on the input marked *. Data in the latches in between is lost.

In this case, too, an α pulse, with a width such that a latch only sets to the state of its predecessor in the chain, is usually not a reliable solution. □

10.2 The master-slave principle

The analysis of the above problems shows that they can be traced back to the same origin. When a memory element of the latch or gated latch type changes, all information about the previous state is lost as soon as the new state is set. However, the information about the previous state is necessary in the examples described until setting is ended. This is a consequence of the *transparent nature* of (gated) latches. The input signals of these memory elements directly influence the outputs via their internal state.

This chain of *cause* and *effect* has to be broken. Initially a solution was sought in doubling each memory element, with each section having a separate enable signal. For the first section, the input section, this is α. The second section has the enable signal β. See Figure 10.5.

On every α pulse the first section reads in data and enters the desired next state. On the subsequent β pulse the output section follows and takes over the state of the input section. The external output Q of the memory element then reacts as well.

Owing to this construction the information on the old state of a memory element is saved when its input section is set. When the new state is passed to the output section (β pulse), its input section is disabled and stores the data just loaded. Memory elements of this double type are *never transparent*.

It turns out that it is possible to deduce the β pulse from the α pulse with a negator: $\beta = \overline{\alpha}$. The propagation delay time of this negator must be less than that of the Y latch. After all, the Z latch has to be disabled before the Y latch is enabled again. In integrated circuits it is certainly possible to meet this requirement. See Figure 10.6. In a design with

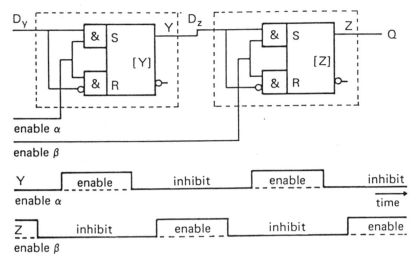

Figure 10.5 Memory element with double excitation.

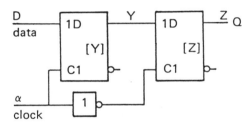

Figure 10.6 Master-slave D flip-flop.

distinct gates this delay requirement has to be verified accurately, taking into account the distribution of the propagation delay times of the gates. (See Section 11.9 as well.)

With Figure 10.6 we have arrived at the *principle* of the *master-slave memory element*. An MS memory element consists of two sections, one of which is enabled and reads in data while the other is disabled and stores data. A memory element of this type is *never transparent*. Memory elements based on the master-slave principle are called flip-flops. Their internal structure will be discussed below and in Chapter 11.

Example *Toggle flip-flop*
Figure 10.7 illustrates how a toggle flip-flip can be realised based on a so-called D flip-flop. When $\alpha = 1$ the Y section reads in data. The new state of Y corresponds to $\overline{Z}$. At the $1 \rightarrow 0$ transition of α the inputs of the Y latch are disabled and Z sets to the state of Y.
After one α pulse the Q output of the toggle flip-flop has changed once. This is *independent* of the width of the α pulse. A certain minimal pulse width is necessary. Each memory element must have sufficient time to be set. (See also Section 11.1.) □

Note
The output transition of the toggle flip-flop in Figure 10.7 occurs immediately after the

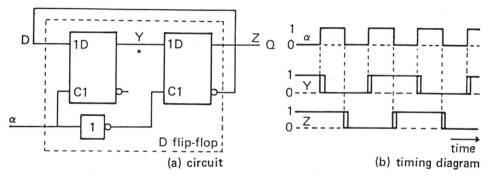

Figure 10.7 Toggle flip-flop based on a D flip-flop.

$1 \rightarrow 0$ transition of the α pulse. By using $\overline{\alpha}$, or by reading the output marked *, the $0 \rightarrow 1$ transition of α determines the moment of the output reaction. We will discuss these timing details in Chapter 11. □

10.3 The level mode and the clock mode views

The memory elements of digital circuits are usually based on the MS principle. All memory elements are generally controlled by the same signal, called the *clock signal* or *clock*. The clock signal is a sequence of pulses (usually periodical), which determine the time at which the flip-flops in a circuit are set and change their states and outputs.

In the following we assume that all data and input signals of the flip-flops are set when the input sections of the flip-flops are disabled. When the clock pulse arrives all input settings are ready. We will see later (Chapter 11) that this requirement will not always have to be met as strictly as formulated here.

The level mode view

Figure 10.8.a describes the internal structure of a D flip-flop. This flip-flop consists of two gated latches, Y and Z. When the clock signal $\alpha = 1$ the Y section is enabled and reads data. The data on the D input is then supposed to be stable. The Z latch sets to the input data at the $1 \rightarrow 0$ transition of the clock signal α.

The timing diagram in Figure 10.8.b describes in detail some examples of how the D flip-flop reacts to the signal at the D input. All of the reactions of the circuit are controlled by the clock.

The same information can also be represented in a state diagram. The *four combinations of states of the latches* Y and Z correspond to the *four states in the state diagram* 10.8.c. Each state is represented by a circle. The state encoding within the circles describes the relation with the states of the latches in the circuit.

The arrows indicate the state transitions. An arrow pointing to the same state indicates a *stable* state, as far as the conditions specified next to it are concerned. An arrow to another state indicates a *state transition* into that state, for the specified conditions. A state transition of Figure 10.8.c takes place *as soon as the corresponding input condition is true*. During a state transition the latches change their internal states. The timing diagram and the state diagram follow directly from an analysis of the circuit.

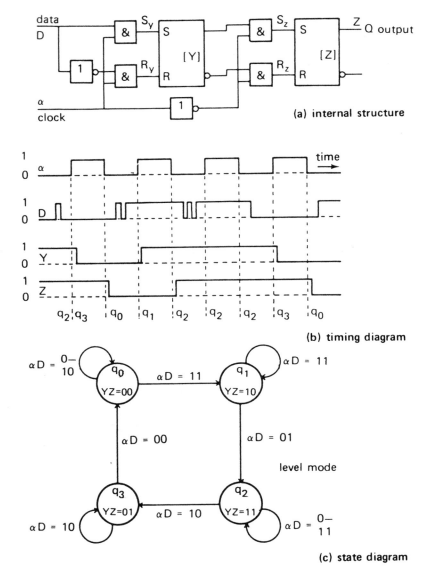

Figure 10.8 Level mode description of a D flip-flop.

In the descriptions of the D flip-flop in Figures 10.8.b and 10.8.c the fact that the clock signal has another function than the data signal on the D input has not been emphasised. The influences of both signals are looked at on the same level. The timing diagram and the state diagram describe how the given circuit reacts to each *signal transition*. Such a description, based on level changes and their immediate effects, is called a *level oriented* or *level mode description* or *view*. It is characteristic of this view that all input transitions are similarly involved in the discussions.

Note

In Figure 10.8.b/c it is assumed that D only changes when $\alpha = 0$. If D may also change when $\alpha = 1$, the state diagram has to be extended. □

The clock mode view

When *all* memory elements in a circuit operate according to the *MS principle* and all flip-flops are controlled by one (periodical) *clock signal* and react to it *simultaneously*, then the so-called *clock mode view* can be used. It is assumed, then, that data and control signals change only in predefined intervals. These intervals lie, to begin with, in between the clock pulses.

Definition

In the clock mode view the (internal) *state of a flip-flop* is understood to be the state of the latch that determines the output signal.

In the clock mode view the *description of the logic behaviour* of a flip-flop consists of a list of logic values of the control and data signals of that flip-flop and of its internal state, just before a clock pulse activates the flip-flop, in relation to the state of the flip-flop after this clock pulse. In addition to the description of the logic behaviour there is the *timing specification*. It specifies exactly when the data signals and/or control signals have to be set and when the output reaction of the flip-flop occurs. All timing parameters are usually related to a fixed voltage level V_{ref} of the corresponding signals: the clock pulse versus data signals, control signals and output signals. See Chapter 11.

In the clock mode view a distinction is made between the logic behaviour of a circuit and its timing. This is possible under the above-mentioned conditions. A great advantage is that the description of the logic behaviour is simplified, compared to the level mode.

Figure 10.9 illustrates how this works out for the clock mode description of a D flip-flop. See Figure 10.8.a for its internal structure.

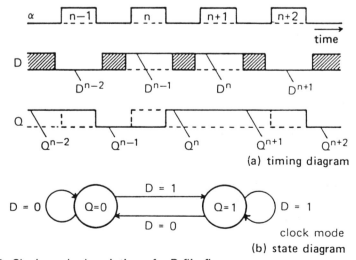

Figure 10.9 Clock mode description of a D flip-flop.

Notation

The state Q^n of a flip-flop is the state the flip-flop has reached after the n-th clock pulse. (See also the above definition.)

Notation

The *logic value* D^n of the data signal on the D input is the value of the D signal *at a time which determines its effect on the state* Q^{n+1} the flip-flop enters at the (n+1)-th clock pulse. It is the value of D resulting from the n-th clock pulse.

According to previous agreements D^n is the value of the signal on the D input during the (n+1)-th clock pulse. The signal on the D input has to maintain this value until after the (n+1)-th clock pulse. This is explained in more detail in Chapter 11.

Figure 10.9.a indicates how the output Q of the flip-flop in Figure 10.8.a depends on the value of the signal on the D input. We see that when D = 0 during the clock pulse, the Q output after the clock pulse is or becomes 0 as well. Similarly D = 1 sets the flip-flop to the state Q = 1. This behaviour is expressed by the formula:

$$Q^{n+1} = D^n.$$

The output reactions of this flip-flop occur around the $1 \rightarrow 0$ transition of the clock pulses. Many flip-flops react around the $0 \rightarrow 1$ transition. Q then changes in accordance with the dashed lines in Figure 10.9.a.

The state diagram in Figure 10.9.b gives the clock mode description of the D flip-flop. When, for example, the present state is $Q^n = 0$ and the D signal has the value of 1 during the next clock pulse, the new state after this clock pulse becomes the state $Q^{n+1} = 1$. The designation *clock mode* next to the diagram indicates that an additional timing specification is required to specify exactly when D has to be set and when the Q output changes.

Symbols

In Figure 10.10 two symbols are drawn for flip-flops. The symbol in Figure 10.10.a is used for D flip-flops with an output reaction at the $0 \rightarrow 1$ transition of the clock pulse. Flip-flops reacting around the $1 \rightarrow 0$ transition are represented by the symbol in Figure 10.10.b. See also Chapter 11.

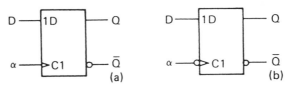

Figure 10.10 Symbols for a D flip-flop.

A clock mode description of a circuit is a description at a *higher level of abstraction* than a level mode description. Thus in a clock mode description the concept 'state' means more than in a level mode description. This is better suited to a *top-down design process*. In a top-down design process the degree of detail increases at every subsequent design level, until a design is obtained that can be directly mapped onto hardware and/or software.

In order to be able to apply a clock mode description some requirements have to be met, for example:

- All flip-flops are controlled by the same clock signal.
- All remaining input signals of the flip-flops only change within predefined intervals, which are determined, for example, by the timing properties of the type of flip-flop and by the propagation delay times of the control circuits.

At first sight these restrictions appear to be quite drastic, but they do not turn out to be so in practice, due to the lack of a reasonable alternative. Nor does it appear to be difficult, in general, to meet the timing requirements. In the following sections we will evaluate the logic properties of some types of flip-flops, at the clock mode level. Subsequently, some examples of sequential circuits are worked out.

10.4 The D flip-flop

In practice various realisations of flip-flops, based on the master-slave principle, are found. Depending on their input structures they are called D, J-K, T and S-R flip-flops. Of these, only D and J-K flip-flops are available in separate ICs. The other types are sometimes used as components within larger ICs, such as in counters.

The logic diagram of a D flip-flop has already been described in Figure 10.8.a. The external set and reset inputs of the Y latch are controlled directly by D:

$$S = D \quad \text{and} \quad R = \overline{D}.$$

For the clock mode view the logic behaviour of the D flip-flop can be described by (10.1):

$$Q^{n+1} = D^n \text{ (clock mode).} \tag{10.1}$$

From (10.1) it follows that the state Q^{n+1} after the clock pulse is completely determined by the signal D^n on the D input before/during that clock pulse.

Tables 10.1.a and 10.1.b are the *function tables* of the D flip-flop. Table 10.1.a follows directly from (10.1). Table 10.1.b fits better into the design process, in which the question usually is how to control the flip-flop such that it reaches the desired new state at the next clock pulse. Having Tables 10.1 available relieves the designer from having to go into their internal structure each time D flip-flops are applied.

D^n	Q^n	Q^{n+1}
0	0	0
0	1	0
1	0	1
1	1	1

$Q^{n+1} = D^n$

Q^n	Q^{n+1}	D^n
−	0	0
−	1	1

(a)   (b)

Table 10.1 Clock mode specification of a D flip-flop.

Note

The state Q^{n+1} of a D flip-flop only depends on D^n. This does not mean that D^n is always independent of Q^n. See the design of the toggle flip-flop in Figure 10.7. □

10.5 The J-K flip-flop

The problems arising in a latch when SR = 11 have already been discussed in Section 9.4. In the D flip-flop these problems have been avoided by coupling S_y and R_y externally by a negator. They can also be avoided in another way, as the logic diagram of a J-K flip-flop in Figure 10.11 shows.

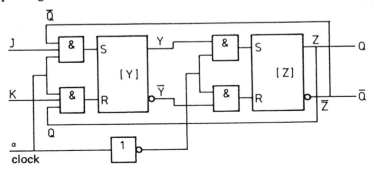

Figure 10.11 Logic diagram of a J-K flip-flop.

Two *internal feedback lines* feed $\overline{Q}$ to the set gate and Q to the reset gate of the Y latch. Depending on the state of the Z latch one of the two input gates is disabled. For $Q = 1$ the *set gate* is disabled and, for $Q = 0$ the *reset gate*. In the first case the signal on the J input has no effect on the state of the Y latch. In the second case, when $Q = 0$, the signal on the K input has no effect. Input combination $S_y R_y = 11$ and related problems have thus been eliminated.

The formula, which describes the logic behaviour of the J-K flip-flop, is derived from Table 10.2. See for example the fourth row of this table. Before the clock pulse the flip-flop is in the state $Q^n = 1$. The set gate is disabled and the reset gate is enabled. The J input now has no influence on the new state, but K does. Because K = 1, R_y will be 1. The Y latch gets a *reset command*. After the clock pulse, when Z is set to the state 0 of Y, the flip-flop is in state $Q^{n+1} = 0$. The other rows of Table 10.2 are similarly derived from Figure 10.11.

J^n	K^n	Q^n	Disabled gate	Command to flip-flop by J/K	Result	Q^{n+1}
0	0	0	reset gate	none	$Q^{n+1} = Q^n$	0
0	0	1	set gate	none		1
0	1	0	reset gate	none	$Q^{n+1} = 0$	0
0	1	1	set gate	reset command		0
1	0	0	reset gate	set command	$Q^{n+1} = 1$	1
1	0	1	set gate	none		1
1	1	0	reset gate	set command	$Q^{n+1} = \overline{Q^n}$	1
1	1	1	set gate	reset command		0

Table 10.2 Logic specification of the J-K flip-flop.

The logic formula of the J-K flip-flop is found with the help of the Karnaugh map in Figure 10.12. This map describes the same information as Table 10.2. The formula of the J-K flip-flop is:

$$Q^{n+1} = J^n\overline{Q}^n + \overline{K}^nQ^n \qquad (10.2)$$
$$= [J\overline{Q} + \overline{K}Q]^n.$$

The index n, which represents the clock pulse number, is often placed outside the square brackets. The index then refers to all logic variables within them.

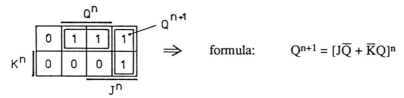

formula: $Q^{n+1} = [J\overline{Q} + \overline{K}Q]^n$

Figure 10.12 Karnaugh map.

When designing with J-K flip-flops, Table 10.3 is more convenient than Table 10.2. In a design one wants to know which input combination(s) set the flip-flop in the desired state Q^{n+1} after the clock pulse. Table 10.3.a contains the same information as Table 10.2. However, it now also indicates that for each input combination the value of J or K is a *don't care* because the corresponding input gate is disabled. Table 10.3.b follows directly from Table 10.3.a.

J^n	K^n	Q^n	Q^{n+1}		Q^n	Q^{n+1}	J^n	K^n
0	–	0	0		0	0	0	–
1	–	0	1		0	1	1	–
–	0	1	1		1	0	–	1
–	1	1	0		1	1	–	0
0	1	–	0		–	0	0	1
1	0	–	1 (a)		–	1	1	0 (b)

Table 10.3 Input conditions of a J-K flip-flop.

In Table 10.3 a distinction is made between situations in which the old or present state Q^n of the J-K flip-flop is and is not known. In the first case this information can be used to deduce the setting of J and K. In all cases either J or K turns out to be a don't care. If the information about the present state Q^n is not known or cannot (or is not allowed to) be used, then J and K must be set as is indicated below the dashed line. This is, for instance, the case with set and reset commands.

The clock mode state diagram of the J-K flip-flop has two states, just as the D flip-flop does. They are encoded with Q = 0 and Q = 1. Depending on J and K the circuit stays in state Q = 0, in state Q = 1, or switches between the states Q = 0 and Q = 1. See Figure 10.13.

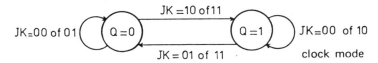

Figure 10.13 State diagram of the J-K flip-flop.

10.6 The T flip-flop

The logic diagram of the *T flip-flop* is shown in Figure 10.14. The circuit results from the J-K flip-flop by combining J and K into one external input signal T. In contrast with the D flip-flop, no negator is used.

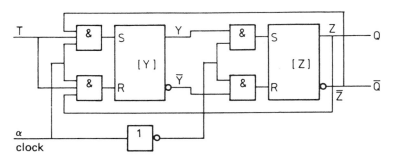

Figure 10.14 Logic diagram of a T flip-flop.

The logic formula of the T flip-flop can be deduced from that of the J-K flip-flop by filling in T for both J and K:

$$\left. \begin{array}{c} Q^{n+1} = [J\overline{Q} + \overline{K}Q]^n \\ J = K = T \end{array} \right\} Q^{n+1} = [T\overline{Q} + \overline{T}Q]^n = [T \oplus Q]^n. \qquad (10.3)$$

We distinguish two cases:

$T = 0$: $Q^{n+1} = [0 \oplus Q]^n = Q^n$, the flip-flop is in the inhibit mode;

$T = 1$: $Q^{n+1} = [1 \oplus Q]^n = \overline{Q}^n$, the flip-flop toggles.

A T flip-flop is a controllable toggle flip-flop. Depending on T the toggling can either be stopped (T = 0) or started (T = 1). This is the property to which the *toggle flip-flop* owes its name.

T^n	Q^n	Q^{n+1}		Q^n	Q^{n+1}	T^n
0	0	0		0	0	0
0	1	1		0	1	1
1	0	1	$Q^{n+1} = [T \oplus Q]^n$	1	0	1
1	1	0		1	1	0
0	–	undefined		–	0	impossible
1	–	undefined (a)		–	0	impossible (b)

Table 10.4 Logic specification of the T flip-flop.

The logic specification of the T flip-flop is shown in Table 10.4. Table 10.4.a indicates how Q^{n+1} depends on T^n and Q^n. Table 10.4.b is more suitable for use in the design process. The tables follow directly from the formula for Q^{n+1}.

Note

With a T flip-flop Q^{n+1} always depends on Q^n. The set or reset command cannot be effected without applying information about Q^n. □

A T flip-flop cannot be purchased. After all, a T flip-flop can be made from a J-K flip-flop by making J = K = T. However, the T flip-flop is sometimes found as a component in integrated circuits, as in counters.

Note

In the IEC system the *clock input* of a toggle flip-flop is called the T input. The input denoted by T in Figure 10.14 is then called *enable input*. See, for example, the description of the circuit SN7497. The name 'T flip-flop' is also used for an ordinary toggle flip-flop, i.e. without an enable input. □

10.7 S-R flip-flops

The logic diagram of an S-R flip-flop is shown in Figure 10.15. This diagram has the following characteristics:

- There is no feedback of Q and $\overline{Q}$ to the inputs.
- The signals S_y and R_y of the input latch Y are 1 when S and R are 1. There is no complementary exclusion as in the D flip-flop.

This memory element therefore has the unpleasant property that, when the external S and R are both 1 while the clock pulse goes from $1 \rightarrow 0$, the state of the Y-latch is *undefined*. As a result the states of Z and of the Q output after the clock pulse are undefined as well.

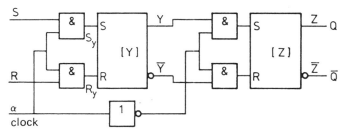

Figure 10.15 Logic diagram of an S-R flip-flop.

For S-R flip-flops it is up to the designer to see that the external S and R inputs are not 1 simultaneously when the Y latch is disabled. This requirement imposes restrictions upon the use of the S-R flip-flop. It may also be a source of errors. In the J-K flip-flop these problems are avoided by a feeding of Q and $\overline{Q}$ back to the inputs and in the D flip-flop by complementarily excluding S_y and R_y ($R_y = \overline{S_y}$). Table 10.5 summarises the other possibilities. The following types of S-R flip-flops result:

1. **The S-R flip-flop**
   The inputs are not safeguarded. The circuit that controls such a flip-flop has to exclude SR = 11. The problem is shifted to the user/designer.

2. **The S-R-Q flip-flop**
   When SR = 11 the old state Q^n must remain unchanged. The external input combination SR = 11 must internally be converted into $S_y R_y = 00$. Latch Y then does not get a set or reset command, and its internal state is left unchanged. Likewise, the state of the Z latch and of Q^{n+1} is left unchanged.

3. **The S-R-S flip-flop**
   A dominant set signal has been chosen in this case. The combination SR = 11 is internally converted into $S_y R_y = 10$.

4. **The S-R-R flip-flop**
   The reset input dominates in this flip-flop. The combination SR = 11 is internally converted into $S_y R_y = 01$.

5. **The S-R-$\overline{Q}$ flip-flop**
   Unlike the S-R latch this is a useful component. Oscillation is not possible in a flip-flop. The S-R-$\overline{Q}$ flip-flop is identical to the previously introduced J-K flip-flop.

S	R	S-R flip-flop $S_y\ R_y$	S-R-Q flip-flop $S_y\ R_y$	S-R-S flip-flop $S_y\ R_y$	S-R-R flip-flop $S_y\ R_y$
0	0	0 0	0 0	0 0	0 0
0	1	0 1	0 1	0 1	0 1
1	0	1 0	1 0	1 0	1 0
1	1	forbidden	0 0	1 0	0 1

Table 10.5 Input conversions in S-R flip-flops.

In the discussion about latches it was shown that a dominant (re)set did not solve the input problem. When the external S and R are simultaneously changed from 1 to 0 it depends on the *signal changed last* whether the resulting state of the latch is 0 or 1. This may result in an undefined state.

This race condition does not occur in S-R flip-flops, because our starting point in flip-flops is that all input signals are stable when the input latch is disabled by the clock signal.

If we leave the S-R flip-flop out of consideration (external safeguarding necessary), then three types of flip-flops which can be used in a design result. The internal S_y and R_y of these flip-flops depend on the input signals S and R as follows:

S-R-Q flip-flop: $S_y = S\overline{R};$ $R_y = \overline{S}R;$

S-R-S flip-flop: $S_y = S;$ $R_y = \overline{S}R;$

S-R-R flip-flop: $S_y = S\overline{R};$ $R_y = R.$

The S-R-Q flip-flop is shown in Figure 10.16.

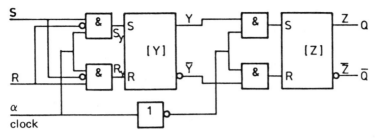

Figure 10.16 Realisation of the S-R-Q flip-flop.

In general, S-R flip-flops are not used as separate components. However, one often finds S-R flip-flops in logic diagrams of integrated circuits, especially in older types. Within an integrated circuit it is not difficult to couple the input signals S and R, so that SR = 11 does not occur. Usually R is derived from S by an negator/inverter. In fact a D flip-flop is then created.

Truth tables and formulas for S-R flip-flops

An S-R flip-flop is of the same complexity as, for example, a J-K flip-flop. The preference for the J-K flip-flop will be made clear from the following. The truth table and the formula of the S-R-Q flip-flop are shown in Table 10.6.

S^n	R^n	Q^n	Q^{n+1}
0	0	0	0
0	0	1	1
0	1	0	0
0	1	1	0
1	0	0	1
1	0	1	1
1	1	0	0
1	1	1	1

$$Q^{n+1} = [S\overline{R} + SQ + \overline{R}Q]^n \qquad (10.4)$$

Table 10.6 Specification of S-R-Q flip-flop.

Q^n	Q^{n+1}	S^n	R^n	
0	0	0	0	
		0	1	$\Rightarrow S \leq R$
		1	1	
0	1	1	0	$\Rightarrow S = 1, R = 0$
1	0	0	1	$\Rightarrow S = 0, R = 1$
1	1	0	0	
		1	0	$\Rightarrow S \geq R$
		1	1	

Table 10.7 Input combinations for S-R-Q flip-flop.

The presentation in Table 10.6 is inconvenient when the appropriate input combinations have to be determined for a desired transition $Q^n \rightarrow Q^{n+1}$. Table 10.7 is more suitable for this purpose.

To realise any desired state transition of $Q^n \rightarrow Q^{n+1}$ in a J-K flip-flop one can also choose from more than one input combination. However, for the J-K flip-flop it holds that for every state transition either J or K is a don't care. With S-R flip-flops S and R *cannot be chosen independently*. This dependence is inconvenient in the determination of simple input formulas, as the following example shows.

Example
In Figure 10.17 it is specified how Q^{n+1} of a memory element Q depends on the present state of the element and on two input signals A and B.

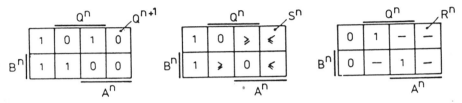

Figure 10.17 Next state and input conditions for an S-R-Q flip-flop.

The input conditions for the S and R inputs can be determined from the first diagram by using Table 10.7. It is not easy to determine the simplest formulas for S and R because both signals are *dependent*. Some possible solutions are, for example:

$$S^n = [\overline{AQ}]^n \quad \text{and} \quad R^n = [\overline{A}BQ + AB]^n,$$

or:

$$S^n = [\overline{Q} + A\overline{B}]^n \quad \text{and} \quad R^n = [A + \overline{B}Q]^n.$$

The fact that the input signals cannot be chosen independently of each other is the main reason that S-R-Q flip-flops are not used frequently. Therefore they are hardly ever encountered in practice. The same holds for the S-R flip-flops with a dominant set or reset, respectively. □

10.8 The derivation of input signals for flip-flops

In the previous part of this chapter we have evaluated how various types of flip-flops react to their input signals. In general, a circuit has *external input signals* as well, which also influence the mode of the flip-flop and its state Q^{n+1}. In this section we discuss how this relation has to be derived for circuits with one flip-flop. Section 10.9 deals with clock mode sequential circuits with more than one flip-flop.

According to the clock mode sequential circuit model the state Q^{n+1} of a D flip-flop is determined by (Figure 10.18):

– the present state Q^n;
– the present input combination $[a_1...a_k]^n$.

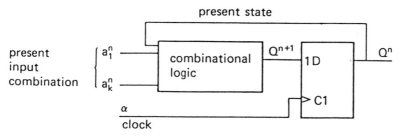

Figure 10.18 Model of a clock mode sequential circuit.

Written in formula, this becomes:

$$Q^{n+1} = \text{function}(Q^n, [a_1...a_k]^n). \tag{10.5}$$

In a circuit (10.5) is realised by a *combinational circuit*. Its logic function can be specified in a truth table, with the present input combination $[a_1...a_k]^n$ and the present state Q^n on the left-hand side and the next state Q^{n+1} on the right-hand side. After this table has been specified completely, the appropriate input conditions for each $Q^n \to Q^{n+1}$ transition are determined with the *function table* of the selected flip-flop. Subsequently, the formulas for the input signals can be derived.

Example *Conversion of a J-K flip-flop into a D flip-flop*
A J-K flip-flop is given, while a D flip-flop is required. Adapt the circuit as shown in Figure 10.19.

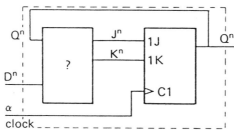

Figure 10.19 Conversion of a J-K flip-flop into a D flip-flop.

D^n	Q^n	Q^{n+1}		Q^n	Q^{n+1}	J^n	K^n	
0	0	0		0	0	0	−	
0	1	0		0	1	1	−	
1	0	1		1	0	−	1	
1	1	1	(a)	1	1	−	0	(b)

D^n	Q^n	Q^{n+1}	J^n	K^n	
0	0	0	0	−	
0	1	0	−	1	
1	0	1	1	−	
1	1	1	−	0	(c)

Table 10.8 Input conditions.

Table 10.8.a describes the *external behaviour* of the circuit to be designed (outside the dashed lines of Figure 10.19). This table is identical to that of the D flip-flop.

The circuit must be realised with a J-K flip-flop. Table 10.8.b gives its function table. Together these tables lead to Table 10.8.c. This table gives for each $Q^n \to Q^{n+1}$ transition the required value of J^n and K^n. Table 10.9 follows from Table 10.8.c, and describes directly how J^n and K^n depend on D^n and Q^n.

D^n Q^n	J^n K^n
0 0	0 –
0 1	– 1
1 0	1 –
1 1	– 0

Table 10.9 Input conditions.

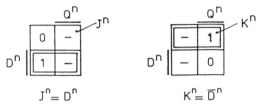

Figure 10.20 Karnaugh maps.

The (simplest) formulas for the signals on the J and K inputs follow from Figure 10.20. The resulting circuit is shown in Figure 10.21. □

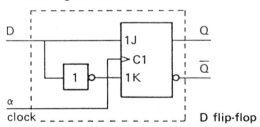

Figure 10.21 D flip-flop, realised with a J-K flip-flop.

J^n	K^n	Q^n	Q^{n+1}	D^n
0	0	0	0	0
0	0	1	1	1
0	1	0	0	0
0	1	1	0	0
1	0	0	1	1
1	0	1	1	1
1	1	0	1	1
1	1	1	0	0

Table 10.10 Input conditions of a D flip-flop

$D^n = [J\bar{Q} + \bar{K}Q]^n$

Figure 10.22 Karnaugh map.

Example *Conversion of a D flip-flop into a J-K flip-flop*

This conversion is based on the first four columns of Table 10.10. The fifth column indicates which signal on the D input is required to realise the desired $Q^n \rightarrow Q^{n+1}$ transition. Using the Karnaugh map in Figure 10.22 we then find

$$D^n = [J\overline{Q} + \overline{K}Q]^n.$$

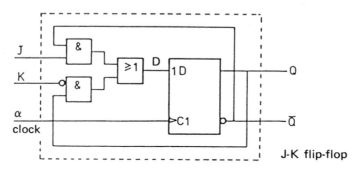

Figure 10.23 J-K flip-flop, based on a D flip-flop.

Figure 10.23 shows the circuit. In practice, the J-K flip-flop is frequently realised on the basis of a D flip-flop, for reasons having to do with the *timing properties*. The timing of a J-K flip-flop realised in this way is then based on the timing of the D flip-flop. See Section 11.3. □

Example *D flip-flop with enable input*

A D flip-flop reads in new data at every clock pulse. The old data is then lost. For many applications it is necessary for the data just read to be stored longer than one clock pulse period. In a D flip-flop this can be achieved in several ways:

– Maintain the signal on the D-input.

– Suppress clock pulses coming via the clock line to the input of the flip-flop.

– Extend the flip-flop with an extra enable input, with which it can be controlled whether the flip-flop is disabled ($Q^{n+1} = Q^n$) or is enabled ($Q^{n+1} = D^n$).

The last alternative leads to the design in Figure 10.24.a.

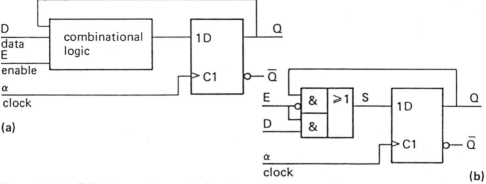

Figure 10.24 D flip-flop with enable input.

Table 10.11 specifies how Q^{n+1} of the D flip-flop depends on D^n, E^n and Q^n. The specification of the signal S on the D input follows from the third and fourth columns of this table. Via the Karnaugh map in Figure 10.25 the circuit diagram in Figure 10.24.b follows directly. □

E^n	D^n	Q^n	Q^{n+1}	S^n
0	0	0	0	0
0	0	1	1	1
0	1	0	0	0
0	1	1	1	1
1	0	0	0	0
1	0	1	0	1
1	1	0	1	1
1	1	1	1	1

Karnaugh map:

	Q^n			
D^n	0	1	0	0
	0	1	1	1

E^n

$S^n = [ED + \bar{E}Q]^n$

Table 10.11 Truth table.   Figure 10.25 Deduction of the formula of S^n.

Memory elements which have both *enable inputs* and data inputs are used frequently. Many registers, counters and larger circuits have enable inputs.

Note
A J-K flip-flop can store data without additional circuitry, i.e. with the input combination JK = 00. An enable input is then superfluous. □

10.9 Sequential circuit design

Circuits which can store data are called *sequential circuits*. For smaller circuits the behaviour can be specified with a *state diagram*. Its states describe the memory function. State transitions are indicated with arrows. We will demonstrate by means of examples how such a state diagram can be converted into a circuit.

Example *Cyclic counter of up to four pulses*
A cyclic counter of up to four pulses can be specified in a state diagram with four states. In the state diagram of Figure 10.26 the clock mode interpretation is used. No input combination is assigned to the arrows, for, in each state the next state is completely determined, making input signals unnecessary. Sequential circuits without inputs are called *autonomous sequential circuits*. (In the clock mode the clock signal is not considered to be an input signal.)

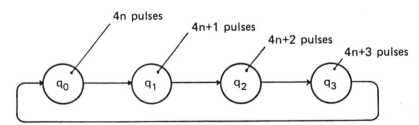

Figure 10.26 Specification of a cyclic counter (clock mode).

The realisation of the counter starts with the *state assignment*. With the state assignment it is decided how each state of the circuit is encoded, i.e. registered as a combination of states of memory elements. Each memory element (flip-flop) encodes the value of one *state variable* (this is one bit of the state assignment). The present state of the counter is then visible in the states of the corresponding flip-flops.

In clock mode circuits the state assignment may be freely chosen. Therefore one may, for example, indicate with the state assignment how many pulses have been counted in a counter. For the four states in Figure 10.26 two state variables are required; these are encoded in the states of two flip-flops A and B. Figure 10.27 shows a state assignment according to the normal *binary code*.

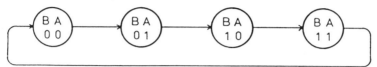

Figure 10.27 Encoded state diagram of a cyclic counter.

The next step in the realisation procedure starts by specifying a table, in which it is indicated how the next state $[Q_b Q_a]^{n+1}$ depends on the present state $[Q_b Q_a]^n$. This results in the first two columns of Table 10.12.

$[Q_b\ Q_a]^n$		$[Q_b\ Q_a]^{n+1}$		D_b^n	D_a^n	J_b^n	K_b^n	J_a^n	K_a^n
0	0	0	1	0	1	0	–	1	–
0	1	1	0	1	0	1	–	–	1
1	0	1	1	1	1	–	0	1	–
1	1	0	0	0	0	–	1	–	1

Table 10.12 Design table of a counter.

Using Tables 10.1.b and 10.3.b we can now fill in the columns for D^n or for J^n and K^n. Table 10.12 now specifies how both D_a^n and D_b^n and J_a^n through K_b^n depend on Q_a^n and Q_b^n. The formulas for these signals can easily be found with Karnaugh maps. See Figure 10.28.

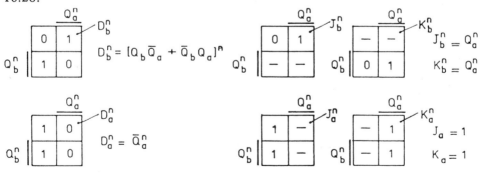

Figure 10.28 Karnaugh maps.

A realisation with two D flip-flops is shown in Figure 10.29. Three additional gates are necessary. With J-K flip-flops the circuit does not require extra gates. □

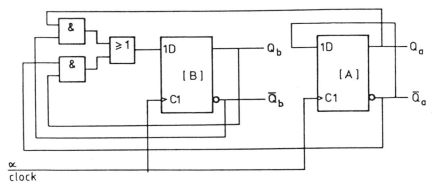

Figure 10.29 Detailed logic diagram with D flip-flops.

Example *Counter with enable and reset inputs*

When the count pulses in a counter are directly fed to the clock input of the flip-flops the counter stops automatically when no more clock pulses are received. Usually, however, the clock signal of the circuit, of which the counter is a subsystem, is directly supplied to the clock inputs of the flip-flops of the counter. With a *control signal* E (*enable*) the modes count/inhibit can be selected. An advantage of this method of counting is that such a counter better fits the data path-control design model, which will be introduced later.

As an example we show the design of a counter of up to four pulses, with *enable* and *reset*. See Figure 10.30 for the specification. The reset signal R directs the counter into a predefined initial state, independent of the present state of the counter. This is necessary, for example, after the power has been switched on. A reset command can also be applied after a certain operation cycle.

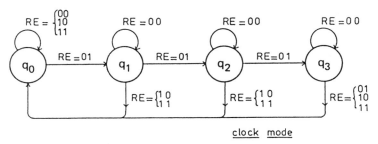

Figure 10.30 Specification of a counter with enable and reset inputs (clock mode).

The first step in the design process is again the state assignment, for example

$q_0 : Q_b Q_a = 0\ 0;$

$q_1 : Q_b Q_a = 0\ 1;$

$q_2 : Q_b Q_a = 1\ 0;$

$q_3 : Q_b Q_a = 1\ 1.$

Truth Table 10.13 is deduced subsequently.

[R E Q_b Q_a]^n	[Q_b Q_a]$^{n+1}$	[D_b D_a]^u	[J_b K_b]^n	[J_a K_a]^n
0 0 0 0	0 0	0 0	0 —	0 —
0 0 0 1	0 1	0 1	0 —	— 0
0 0 1 0	1 0	1 0	— 0	0 —
0 0 1 1	1 1	1 1	— 0	— 0
0 1 0 0	0 1	0 1	0 —	1 —
0 1 0 1	1 0	1 0	1 —	— 1
0 1 1 0	1 1	1 1	— 0	1 —
0 1 1 1	0 0	0 0	— 1	— 1
1 0 0 0	0 0	0 0	0 —	0 —
1 0 0 1	0 0	0 0	0 —	— 1
1 0 1 0	0 0	0 0	— 1	0 —
1 0 1 1	0 0	0 0	— 1	— 1
1 1 0 0	0 0	0 0	0 —	0 —
1 1 0 1	0 0	0 0	0 —	— 1
1 1 1 0	0 0	0 0	— 1	0 —
1 1 1 1	0 0	0 0	— 1	— 1

Table 10.13 Truth table.

The formulas for the D and the J/K signals are found with Karnaugh maps. For D_b and for D_a these maps are shown in Figure 10.31.

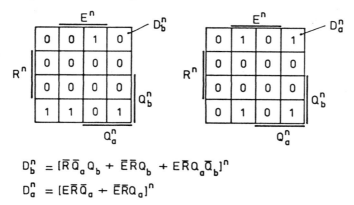

$$D_b^n = [\bar{R}\bar{Q}_a Q_b + \bar{E}R Q_b + ER Q_a \bar{Q}_b]^n$$
$$D_a^n = [E\bar{R}\bar{Q}_a + \bar{E}R Q_a]^n$$

Figure 10.31 Karnaugh maps.

Similarly, the formulas for J_b through K_a follow from Karnaugh maps. These formulas are

$$J_b^n = [E\bar{R}Q_a]^n \qquad J_a^n = [E\bar{R}]^n$$
$$K_b^n = [R + EQ_a]^n \qquad K_a^n = [R + E]^n \text{ or } K_a^n = [R + EQ_a]^n.$$

By not selecting the simplest variant for K_a, one gate can be saved. In that case $K_b = K_a$. The circuit with J-K flip-flops is shown in Figure 10.32. With D flip-flops the circuit requires more and larger gates. □

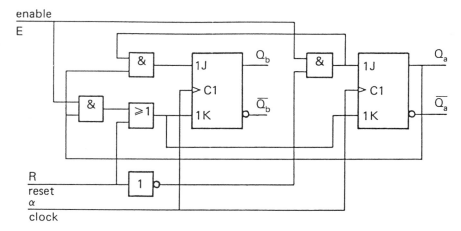

Figure 10.32 Counter circuit with enable and reset.

Note

In the previous examples formulas for the control signals of the flip-flops were deduced with the use of Karnaugh maps. These formulas are expressed in the form

$$Out^n = [function (in_1, ..., in_i)]^n,$$

etc. This formula specifies how, after clock pulse n, the signals Out and others depend on the signals $in_1, ..., in_i$. This relation can be realised in a combinational circuit, having the formula

$$Out = function (in_1, ..., in_i)$$

but in this case without indices. The indices n for the input signals are therefore usually left out of the formulas from the start. □

Selecting the type of flip-flop

In practice, it appears that, when designing *sequential circuits*, the application of J-K flip-flops generally leads to simpler gate circuits for the control signals than with D flip-flops or other types of flip-flops. This is caused by the large number of don't cares in the Karnaugh maps for J and for K. J or K can be chosen at random for any state transition, which is the reason that J-K flip flops are frequently used.

10.10 The state assignment

In the design process the concept 'state of a sequential circuit' is used at different levels of abstraction. A decimal counter section, for example, has ten states to describe its external function. At the realisation level the state of the circuit is often meant to be a combination of states of memory elements. Their output levels encode the present state of the circuit.

The choice of the state assignment of the 'abstract' states of a circuit to be designed is influenced by several factors. In the level mode avoiding transient phenomena in state transitions plays a dominant role. Many assignments have to be excluded on these grounds. In the clock mode all possible assignments are allowed. The only restriction is that a unique code must be assigned to each state. It is then possible to keep the cost of the circuit in mind when choosing the state assignment.

In order to get some insight into the state assignment problem in the level mode we shall go into the state assignment of the toggle flip-flop.

Example

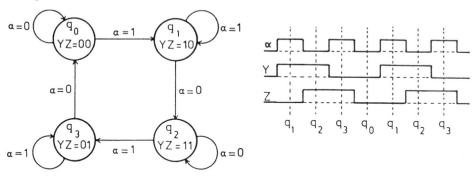

Figure 10.33 State assignment of a toggle flip-flop.

The state diagram in Figure 10.33 describes the realisation of the toggle flip-flop of Figure 10.7. The four states required to describe the logical behaviour are encoded with two D latches Y and Z. The state diagram follows directly from an analysis of the circuit. For each input transition of α the diagram describes the reaction of the circuit. It is a level mode state diagram.

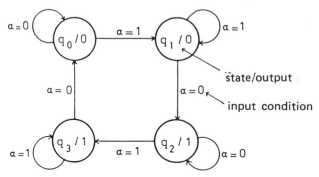

Figure 10.34 State diagram of a toggle flip-flop (external description).

Figure 10.34 describes the external behaviour of the toggle flip-flop. The internal details, e.g. those involving the state assignment, are omitted. The four internal states (it cannot be done with less) are denoted by q_0 through q_3. We will now prove that the assignment of the states q_0 through q_3 cannot be chosen at random. We will do this by using the assignment in Figure 10.35.

We assume that the circuit is initially in the state q_0, encoded with YZ = 00. The state is stable as long as $\alpha = 0$. If α changes from 0 to 1 (an input transition), then the circuit has to change into state q_1 according to the given specification. This state is encoded in Figure

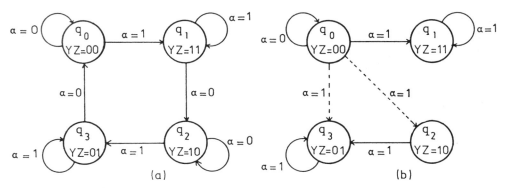

Figure 10.35 State assignment with critical race conditions.

10.35.a with YZ = 11. Both latches, Y and Z, have to change at this transition. When Y and Z change simultaneously, the new state is indeed q_1. The circuit remains in state q_1 until α becomes 0 again. This corresponds to the desired function according to the logic specification.

If, however, latch Y is considerably slower than Z, the combination YZ = 01 occurs as a transient state. This is the state code of q_3. However, according to the specification, q_3 is stable under $\alpha = 1$. There is a risk that the circuit will not arrive in the desired state q_1 at all, but will remain in q_3. This depends on how much the speeds of Y and Z differ. Similar problems arise if the transient state YZ = 10 occurs during the state transition from q_0 to q_1.

From the above it becomes clear that, depending on the operation speed of Y and Z, q_1 or q_3 occurs as the next stable state. See Figure 10.35.b. Which of the two is the actual stable state after the transition depends on the physical realisation of the circuit. In fact, it depends, for one thing, on the spread in the parameters of the components used. □

Without further analysis of the state assignment selected we can already conclude that the state assignment of the states in a level mode circuit cannot be chosen at random. Some assignments, such as that in Figure 10.33, do not create any problems. Others, such as the assignment in Figure 10.35 do.

These problems arise because several memory elements change simultaneously during the state transition between q_0 and q_1 in the state diagram of Figure 10.35. Then transient states may occur. The fact that these may occur is called a *race condition*. When a race condition may lead to wrong stable states after a state transition, it is said to be *critical*. When the intended final state is always reached, the race condition is said to be *non-critical*.

The actual cause of the above problems is that during a state transition the circuit has already lost its information about the old state before the new state has been reached.

This problem can be solved in several ways:

– By choosing a state assignment in which no transient states can occur (a progressive state assignment). For many state diagrams such an assignment is not possible.

- By choosing a state assignment such that any transient state occurring during a state transition always leads to the intended final state. Assignments like this can always be found (Tracey). They sometimes require extra memory elements.
- By using memory elements based on the master-slave principle.

The first two approaches are based on the level mode. They require a detailed analysis of the timing of the components to verify whether the behaviour described is actually being realised. All consequences of transient phenomena have to be checked. For some problems, for example in the synchronisation of (sub)systems, this is absolutely mandatory.

The state assignment in the clock mode

When the circuit is based on flip-flops operating on the master-slave principle, critical race conditions cannot occur. After all, in the preparation of the next state (Y sections are reading in) information about the old state is available (Z sections are disabled).
Consequently it can always be determined, together with the present input combination, what the corresponding new state is. This means that one is, in principle, completely free in the choice of the state assignment of clock mode circuits. One may, for example,

- choose a state assignment that leads to a minimal circuit;
- combine the state assignment, completely or partially, with the encoding of the output signals;
- choose the state assignment such that the states can easily be interpreted, for example in counters by counting immediately in the desired output code.

All of this will be explained in the following chapters. For a more detailed treatment of the state assignment problem in the clock mode the reader is referred to the References.

10.11 Sequential circuits in programmable logic

Until about 1970-1975 it was common practice to realise digital circuits on the basis of 7400 type SSI/MSI logic. Application specific ICs were expensive and only profitable in very large series. Since 1978-1980 programmable ICs (below and Chapter 8) which are sufficiently flexible to be able to realise larger designs have been made available. This has many advantages, including a higher density of gate equivalents per printed circuit board, fewer ICs and less external wiring, which results in a more reliable and a reasonably flexible realisation mode.

As an introduction we compare two realisations of the logic design of Figure 10.32, in 7400 SSI/MSI and in a PAL with an internal register (*registered PAL, sequential PAL*).

Directly mapped on 7400 SSI/MSI ICs this design for a 2-bit binary counter requires four ICs:

        one 7476 Dual J-K flip-flop (requires 1/1 IC);
        one 7404 Hex Inverter (requires 1/6 IC);
        one Quadruple 2-input AND (requires 3/4 IC);
        one Quadruple 2-input OR (requires 1/4 IC).

This type of realisation takes up a *lot of space* and uses *many different ICs*. For many years digital circuits have been realised in this way. Nevertheless, this realisation mode has been outdated because, to cite an example, the many external connections between ICs form a weak link in the reliability of the circuit.

Figure 10.36 shows a realisation of the same design, but now in one sequential PAL of the 16R4 type (Monolithic Memories Inc., MMI). The output buffers invert the signal, so

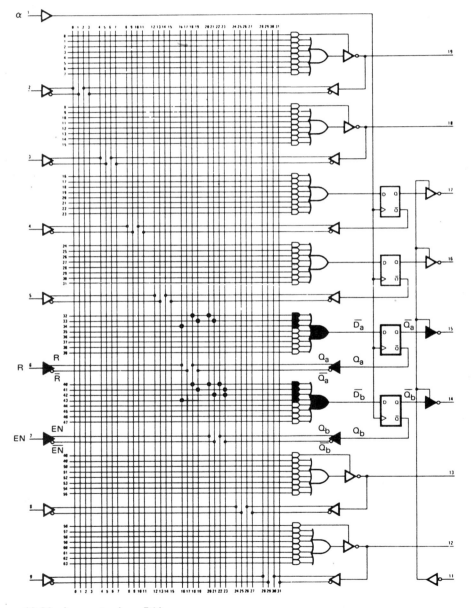

Figure 10.36 A counter in a PAL.

that the realisation must be based on the formulas for $\overline{Q}^n$ instead of on those for Q^n:

$$\overline{Q}^{n+1}_a = [E \cdot Q_a + \overline{E} \cdot \overline{Q}_a + R]^n, \tag{10.6}$$

$$\overline{Q}^{n+1}_b = [E \cdot Q_a \cdot Q_b + \overline{Q}_a \cdot \overline{Q}_b + \overline{E} \cdot \overline{Q}_b + R]^n. \tag{10.7}$$

Such minor adaptions are often necessary in this type of component.

In the '*AND-OR plane*' of the PAL in Figure 10.36 eight logic functions may be formed, each of which comprises a maximum of seven/eight product terms. In principle, functions in more product terms are possible, by cascading in two levels, although this greatly increases the propagation delay time of the signals through the IC. To facilitate cascading the four (direct) outputs are internally fed back to the AND-OR plane. Behind four of the eight functions there is a positive edge-triggered D flip-flop. So at most 16 internal states can be encoded. The encoding of the current state is directly fed back to the AND-OR plane. In this way sequential circuits can be realised efficiently.

PALs are available in many types, with or without an output register. A programmable *universal output block* around the flip-flops is becoming more and more common. Figure 10.37 illustrates this. The output selector enables the user to select the output signal of the AND-OR plane *or* that of the flip-flop, either inverted or not.

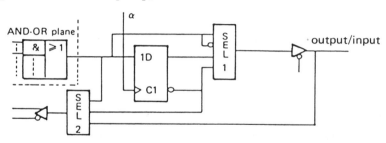

Figure 10.37 Universal output block.

For the feedback one can choose from the output of the relevant AND-OR plane output, the flip-flop output or the external input. With such output modules extremely flexible components have become available (e.g. by Cypress Semiconductor, Advanced Micro Devices, Altera).

As the size increases one eventually encounters a number of 'stringent' restrictions, the most important of which is the maximum number of product terms per external input. In addition, a large part of the IC area remains unused, as a rule, due to the fixed internal structure and also because not all functions have the maximum number of terms.

Because of the stringent restrictions designers are inclined in practice to focus on mapping and fitting problems with these types of programmable logic and to subordinate the other problems, including testability, to them. The 'device' already makes such stringent demands that the designer has little room in which to manoeuvre.

In addition to programmable logic based on AND-OR planes (e.g. PALs, PLAs) so-called *registered PROMs*, a PROM with an output register, are available. The logic functions can then be read in on the basis of their minterm forms (see Sections 8.1 and 8.2).

Registered PROMs come to about 12 inputs. These components are very suitable as well for realising sequential circuits. The only restriction is the total number of input variables.

The market for programmable logic is on the increase at the moment. Other techniques, e.g. *gate arrays* and *logic cell arrays*, are still more flexible because the wiring between the cells can also be chosen by the designer.

References

1. J. Aguilo and E. Valderrama, *A five-input flip-flop*, IEE Proc. on Computers and Digital Techniques, Vol. 127, part E, no. 6, Nov. 1980, pp. 259–262.
2. T.A. Dolotta and E.J. McCluskey, The Coding of Internal States of Sequential Circuits, IEEE Tr. on Electronic Computers, Vol. EC-13, Oct. 1964, pp. 549–562.
3. A.D. Friedman and P.R. Menon, *Synthesis of Asynchronous Sequential Circuits with Multiple-Input Changes*, IEEE Tr. on Computers, Vol. C-17, June 1968, pp. 559–566.
4. F.J. Hill and G.R. Peterson, *Introduction to Switching Theory and Logical Design*, Wiley, New York, 1974.
5. D.A. Huffman, *The Synthesis of Sequential Switching Circuits*, J. Franklin Inst., Vol. 257, nos. 3 and 4, March and April 1954, p. 161–190 and 275–303.
6. G.A. Maley and J. Earle, *The Logic of Transistor Digital Computers*, Prentice-Hall, Englewood Cliffs, N.J., 1963.
7. G.H. Mealy, *A method for Synthesising Sequential Circuits*, Bell System Techn. J., Vol. 34, Sept. 1955, pp. 1045–1079.
8. E.F. Moore, *Gedanken-experiments on Sequential Machines*, Annals of Mathematical Studies, no. 34, Princeton Univ. Press; Princeton, N.J., 1956, pp. 129–153.
9. P.S. Noe and V.T. Rhyne, *Optimum State Assignment for the D flip-flop*, IEEE Tr. on Computers, Vol. C-25, March 1976, pp. 306–311.
10. J.R. Smith and C.H. Roth, *Analysis and Synthesis of Asynchronous Sequential Networks Using Edge-Sensitive Flip-Flops*, IEEE Tr. on Computers, Vol. C-20, Aug. 1971, pp. 847–855.
11. J.R. Story, H.J. Harrison and E.J. Reinhard, *Optimum State Assignment for Synchronous Sequential Circuits*, IEEE Tr. on Computers, Vol. C-21, Dec. 1972, pp. 1365–1373.
12. H. Taub and D. Schilling, *Digital Integrated Electronics*, McGraw-Hill, New York, 1977, pp. 278–321.
13. A.P. Thijssen, *The Assignment of State Variables in Feedback Loops of Sequential Switching Circuits*, Delft Progress Report 3, 1977, pp. 39–53.
14. J.H. Tracey, *Internal State Assignments for Asynchronous Sequential Machines*, IEEE Tr. on Electronic Computers, Vol. EC-15, no. 4, Aug. 1966, pp. 551–560.
15. *The TTL Data Book for Design Engineers*, Texas Instruments, 1977.
16. S.H. Unger, *Hazards and Delays in Asynchronous Sequential Switching Circuits*, IRE Tr. on Circuit Theory, Vol. CT-6, March 1959, pp. 12–25.
17. S.H. Unger, *A Row Assignment for Delay-Free Realizations of Flow Tables without Essential hazards*, IEEE Tr. on Computers, Vol. C-17, Febr. 1968, pp. 146–151.
18. S.H. Unger, *Asynchronous Sequential Switching Circuits with Unrestricted Input Changes*, IEEE Tr. on Computers, Vol. C-20, Dec. 1971, pp. 1437–1444.

11
Timing

11.1 Timing of gated latches

In Chapter 10 the *logic behaviour* of flip-flops and clock mode sequential circuits has been discussed. The *next-state formula* of a flip-flop describes how the state Q^{n+1} of a flip-flop during the (n+1)-th clock period is determined by the state Q^n and the value of the input signals during the n-th clock period. The input signals are thought to be present as long as the clock pulse lasts. In this view a clock pulse is the moment at which the n-th clock period passes into the (n+1)-th clock period.

This *logic model* does not specify in detail when the flip-flop outputs react at the clock pulse. Moreover, input signals may change earlier in practice than the width of the clock pulses suggests.

In this chapter we will study the *timing behaviour* of flip-flops and of clock mode sequential circuits in more detail. It will turn out that in well-designed circuits the timing can be studied separately from the logic behaviour. This makes the problem easier to manage. In *clock mode circuits*, which operate fully synchronously on one clock, the timing can easily be analysed, and it is not difficult to determine how the parameters, such as the maximum clock frequency f_{max}, are derived.

In studying the timing properties of components and systems the relation between logical values, such as 0 and 1, on the one hand and physical values, such as L and H and time, on the other, is defined and we see how the desired function can be physically implemented in a reliable way. In doing so, logical and physical concepts have necessarily been mixed together in the text. In this chapter the *positive logic convention* is used, unless stated otherwise. When, to describe a link between logical and physical properties, we speak of $0 \rightarrow 1$ and $1 \rightarrow 0$ transitions of signals, then $0 \rightarrow 1$, for example, corresponds to $L \rightarrow H$. Properties and conditions are usually formulated 'logically' in the text. With the positive logic convention the link has been defined unambiguously.

It further holds that if the 'unit of time' has not been mentioned explicitly in the following, the unit is ns (10^{-9} s).

The gate model

The study of the timing of flip-flops will be based on the previously introduced gate model (Section 6.12). According to this model a gate (or a combinational logic circuit) consists of a *black box*, with logical and timing properties.

The *logic function* specifies the link between the logic value of the output signals and the logic value of the input signals. This relation only holds in the *static state*.

After an input transition it takes a while before the outputs react. This time is called the *propagation delay time* t_{PD} of the gate or combinational circuit. The exact value of t_{PD} depends on several physical parameters.

Consequently, the reaction time will be uncertain. In many applications one therefore uses *minimum* and *maximum propagation delay times*, $t_{P(min)}$ and $t_{P(max)}$ respectively. According to this model the output value between $t_{P(min)}$ and $t_{P(max)}$ is assumed to be *uncertain*. Signal transitions *might* occur anywhere in between. Figure 11.1 gives a summary.

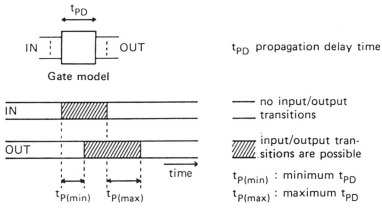

Figure 11.1 The gate model.

Timing of the gated D latch.

The timing of *gated latches* will also be based on this gate model. The gated D latch described in Figure 11.2 is composed of AND gates and an S-R latch. The function of this D latch is described by the equation

$$Z_{new} = \overline{EN} \cdot Z_{old} + EN \cdot D.$$

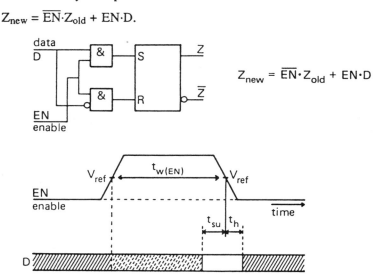

Figure 11.2 Input timing of a gated D latch.

This D latch is disabled when the enable signal EN has the value EN = 0. During this time D can be set to a new value. This does not influence the internal logic state of the latch, nor the output. During EN = 1 the latch is enabled. The new state corresponds to the present value of D. The output adapts to the new present state. If EN subsequently becomes 0 the latch has entered its new state. The memory function is then active.

This description of the function is only correct if the latch has sufficient time to react. The *enable pulse* must have a certain width. The *minimum pulse width* $t_{w(min)}$ is determined by the propagation delay time of the gates of the latch. In the following it is assumed that the width $t_{w(EN)}$ of the enable signal is greater than $t_{w(min)}$. Information about this value can be found in the data books. For a gated D latch with the structure of Figure 11.2 $t_{w(min)}$ roughly equals

$$t_{w(min)} = 4 \cdot t_{P(max)},$$

in which $t_{P(max)}$ is the maximum propagation delay time of one gate.

When EN = 1 the D latch of Figure 11.2 follows the signal at the D input. In many applications output transitions are permitted during EN = 1. Only the desired (internal) logic state after the enable pulse is then specified. In that case it is allowed, in principle, for the value of the D signal to be present only just before the $1 \to 0$ transition of the enable pulse (see the dashed interval in Figure 11.2).

However, D must be able to set the new state of the latch in a reliable way. To do so the D signal must be present in advance a minimum time t_{su}. In a latch this time t_{su}, called the *setup time*, roughly corresponds to the maximum time the latch needs to react to an input transition.

Normally speaking, the D signal must also be present for a short time after the $1 \to 0$ transition of EN. This time is called the *hold time* t_h.

In general the threshold may vary strongly with temperature, for example. The moment a $0 \to 1$ or $1 \to 0$ transition is detected is then uncertain. This also depends on the rise time of the controlling enable signal. This uncertainty leads us to conclude that the best advice is to keep an additional hold time for the input and control signals.

Conclusion

In most applications of gated latches only the internal logic state *after* the enable pulse is of importance. To get a D latch in a predefined state it is *necessary* and *sufficient* for the D signal to be present around the $1 \to 0$ transition in the enable signal EN. The width of this interval is specified with two timing parameters, the setup time t_{su} and the hold time t_h. *The internal construction of the latch determines the value of these parameters.* The times t_{su} and t_h are denoted with regard to the so-called *reference level* V_{ref} of the enable pulse.

Note

When the output signal Z has to be as stable as possible during EN = 1 as well, the dashed interval in Figure 11.2 is not used. D must then be present during the entire enabling period. □

Example

The gated D latch in Figure 11.3 has an internal structure that differs from that in Figure

11.2. In this implementation the setup time t_{su} and the hold time t_h are specified around the $0 \rightarrow 1$ transition of EN. □

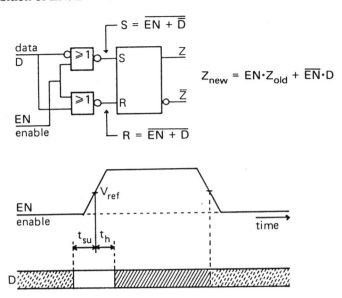

Figure 11.3 Input timing of a gated D latch.

The timing of the gated S-R latch

Figure 11.4 depicts the internal structure of a gated S-R latch. The latch is disabled when EN = 0. With EN = 1 S and R have the following influence on the internal state Z of the latch:

$\quad$ SR = 00: inhibit $\Rightarrow Z_{new} = Z_{old}$,

$\quad$ SR = 01: reset $\Rightarrow Z_{new} = 0$,

$\quad$ SR = 10: set $\Rightarrow Z_{new} = 1$,

$\quad$ SR = 11: excluded as input combination.

The demands to be imposed on the signals S and R are specified in Figure 11.4.b. The set command SR = 10 and the reset command SR = 01 are *dominant*, i.e. they determine the state of the latch regardless of the previous state. For these input combinations it is sufficient when they are present around the $1 \rightarrow 0$ transition of EN.

More stringent requirements apply to the inhibit command SR = 00. If the input combination SR = 00 is alternated with SR = 01 or SR = 10 during the period that EN = 1, even for a very short period, the latch interprets this as a reset or set command. The information on the previous state is then immediately lost. The inhibit combination SR = 00 must therefore be required to be present during EN = 1, *without interruptions*. The setup time t_{su} of the input combination SR = 00 is then defined just before the $0 \rightarrow 1$ transition of the enable signal and the hold time t_h directly after the $1 \rightarrow 0$ transition. In between t_{su} and t_h, when EN = 1, the input combination SR = 00 must be permanently present.

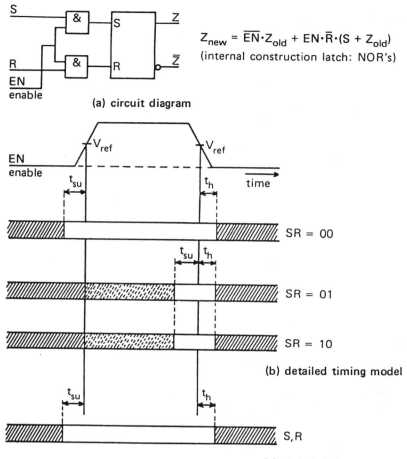

Figure 11.4 Input timing of a gated S-R latch.

In practice, it cannot be said in advance what the next input combination will be. The designer must therefore see to it that the timing of the signals S and R meets the requirements indicated in Figure 11.4.c. In all cases the worst case timing is then used.

Conclusion

Other, stricter requirements are imposed on the timing of the input signals of the gated S-R latch, compared to those of the gated D latch. The reason is that for the D latch both values of the input signal D are *dominant*, i.e. they completely determine the internal state of the latch. With the S-R latch this only holds for SR = 01 and SR = 10, but not for the input combination SR = 00.

For the input signals S and R one therefore usually uses the so-called *global timing model* of the S-R latch, i.e. all the requirements for all input combinations are combined into a worst case timing specification. One of the consequences of the timing properties of gated S-R latches is that they are more *susceptible to noise* than gated D latches.

Summary

- The internal structure of gated latches determines when the input signals are required to be present. The application may also impose requirements.
- The interval in which an input signal must be present is specified by the setup time t_{su} and the hold time t_h.
- In addition, requirements are also put on the (minimum) width t_w of the enable pulse.
- The global timing model, i.e. a worst case timing model, is generally used.
- The gated D latch imposes fewer strict requirements than the gated S-R latch. The latter is more susceptible to noise.

Although these conclusions have been drawn on the basis of the gate model of Figure 11.1, they are valid in general.

Note
In the IEC symbols the global timing model is usually referred to. However, it is possible to detail the timing properties. This increases the complexity of the symbols. □

Propagation delay times

Just like combinational logic circuits latches and gated latches have a certain propagation delay. After an input transition the *minimum/maximum propagation delay times* $t_{P(min)}$ and $t_{P(max)}$ of a gated latch indicate the beginning and end of the interval in which the latch outputs react. These delay times depend on the internal structure of the gated latch, on the technology used to realise the circuit and on operating conditions such as temperature and fanout.

Figure 11.5 indicates when the D latch in Figure 11.5.a reacts to a transition of the D input. If D changes while EN = 1 we once more find the *uncertainty interval* in D at the Z output, but a short time later and somewhat longer. This situation is described in Figure 11.5.b. However, if D changes while EN = 0 the gated latch will only react after the next $0 \rightarrow 1$ transition of EN. In that case, the uncertainty interval in the output signal Z only depends on the parameters of the gated latch; it is found from $t_{P(min)}$ to $t_{P(max)}$ after the $0 \rightarrow 1$ transition of EN.

Note
In the above the discussion was mainly about 'the propagation delay time' of a gated D latch. More accurate observation shows that, depending on the internal structure, the propagation delay time t_{PD} of the D input to the Z output may have another value than t_{PD} of the enable input to Z. There is also a difference between the $1 \rightarrow 0$ and $0 \rightarrow 1$ transitions. Such details can be incorporated in the more detailed timing model, although this makes the timing verification data-dependent. □

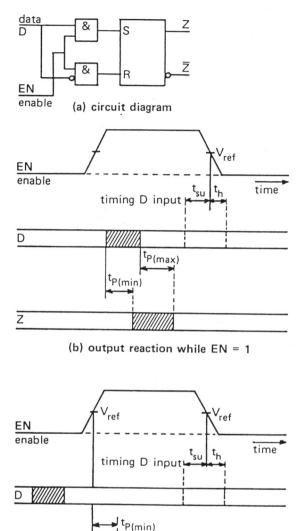

Figure 11.5 Signal propagation in gated latches.

11.2 Edge-triggered timing of flip-flops

Flip-flops are based on two gated latches, even though this structure cannot always be recognised in an implementation diagram. The timing properties of flip-flops therefore have much in common with those of gated latches. As an introduction to the discussion of timing properties of flip-flops we will investigate the timing of the D flip-flop in Figure 11.6. In a D flip-flop with the structure described in Figure 11.6.a the input signal is not allowed to change (Section 11.1) within the interval determined by t_{su} and t_h. This requirement is met, for example, when the signal at the D input changes during the shaded area in Figure 11.6.b.

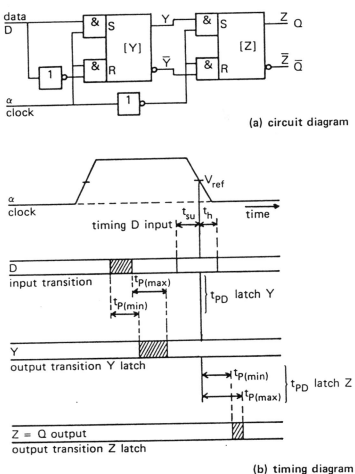

(a) circuit diagram

(b) timing diagram

Figure 11.6 D flip-flop with negative edge-triggered timing.

After a transition of D within the area indicated the Y latch is set. Its output Y reacts between $t_{P(min)}$ and $t_{P(max)}$. Within these restrictions Y is said to react 'directly'. After all, the clock signal α then has the value of 1 and enables the Y latch.

If there is subsequently a $1 \rightarrow 0$ transition of α the Z latch is enabled and takes on the state of Y. Z is set during an interval determined by its $t_{P(min)}$ and $t_{P(max)}$. Between $t_{P(min)}$ and $t_{P(max)}$ the state of the Z latch is uncertain.

The *external Q output* of the flip-flop has the same timing properties as the signal Z.

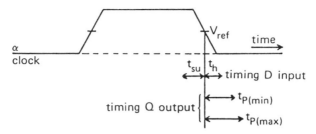

Figure 11.7 Negative edge-triggered timing.

Figure 11.7 summarises the *external timing properties* of the D flip-flop of Figure 11.6. The signal on the D input has to be stable between the *setup time* t_{su} and the *hold time* t_h and also has to have the specified logic value. Immediately after the $1 \rightarrow 0$ transition of the clock signal α the Q output gets the value of D. Q reacts in the *interval* between $t_{P(min)}$ and $t_{P(max)}$.

A flip-flop, of which the timing parameters t_{su}, t_h, $t_{P(min)}$ and $t_{P(max)}$ are specified around the same clock edge, is called an *edge-triggered flip-flop*. When they are specified around the $1 \rightarrow 0$ transition such as in Figure 11.7 the timing is said to be *negative edge-triggered*. With another internal structure a so-called *positive edge-triggered timing* is also possible. Figure 11.8 describes the properties. Positive edge-triggered timing has become the standard in modern ICs.

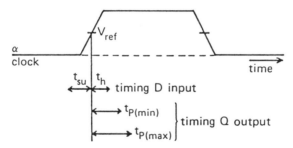

Figure 11.8 Positive edge-triggered timing.

The specification of D^n, Q^n and Q^{n+1}

The *logic operation* of a D flip-flop, as we have seen in Chapter 10, can be described by the formula

$$Q^{n+1} = D^n.$$

Here D^n is the value the D input has reached during the n-th clock period. Q^{n+1} is the resulting value the Q output assumes during the (n+1)-th clock period. Now we can specify these descriptions more precisely.

Definition

D^n *is the value of the signal at the D input at a time (or interval) that determines the effect of D on the state* Q^{n+1} *of the flip-flop during the* (n+1)-*th clock period.*

Based on this definition D^n is the value of the signal at the D input within the interval determined by t_{su} and t_h in Figures 11.7 and 11.8.

Definition

The values Q^n and Q^{n+1} of the flip-flop output Q are understood to be the values that Q has obtained after $t_{P(max)}$ *around respectively the* n-th *and* (n+1)-th *clock pulse and holds until just before* $t_{P(min)}$ *around respectively the* (n+1)-th *and* (n+2)-th *clock pulse.*

Between $t_{P(min)}$ and $t_{P(max)}$ the Q output is *uncertain* in principle, also when $Q^n = Q^{n+1}$. (Normally, the Q output remains stable when $Q^n = Q^{n+1}$.) Figure 11.9 explains the given definitions in a timing diagram for a positive edge-triggered flip-flop.

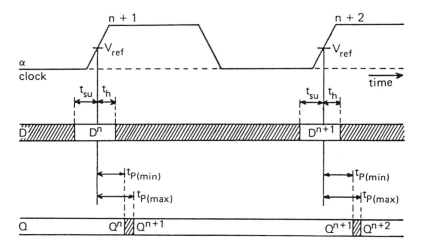

Figure 11.9 Definition D^n, Q^n and Q^{n+1}.

Another diagram like that in Figure 11.9 can also be specified for a negative edge-triggered timing. In that case all reference values are around the $1 \rightarrow 0$ transition of the clock pulse.

Note

When studying the timing of flip-flops one usually talks in terms of the *active* or *passive edge of the clock pulse*. The active edge of the clock pulse is the edge to which $t_{P(min)}$ and $t_{P(max)}$ are referred. It is the edge at which the *output reacts*. □

Note

IC manufacturers as a rule specify $t_{P(typical)}$ rather than $t_{P(min)}$. The *typical value* of the propagation delay time is a kind of mean value of the propagation delay time. For the design process this time has no practical meaning. We will discuss the consequences of this policy at a later stage. □

Note

All timing parameters t_{su}, t_h, $t_{P(min)}$ and $t_{P(max)}$ are usually referred to a fixed level V_{ref}. The thresholds $L \rightarrow H$ and $H \rightarrow L$ of the inputs are thought to be on this *reference level*. In general, the threshold varies because of, say, temperature and it is also technology dependent. The uncertainty introduced in this way must be taken into account while specifying the values of t_{su} through $t_{P(max)}$. See Section 6.1, Figure 6.3. □

11.3 Pulse-triggered timing of flip-flops

In Section 11.1 it has been shown that there are significant differences between the timing of a gated D latch and a gated S-R latch. These differences recur in the timing of flip-flops. The input section of the D flip-flop in Figure 11.6 is based on the gated D latch. The input section of the J-K flip-flop in Figure 11.10 is based on the gated S-R latch.

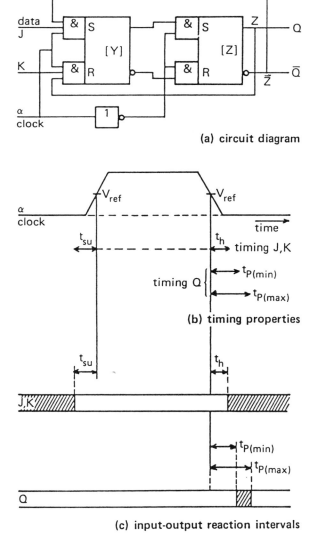

Figure 11.10 J-K flip-flop with pulse-triggered timing.

The *internal feedback* of the J-K flip-flop prevents the input combination SR = 11 from being given at the S-R inputs of the Y latch (Section 9.4). Consequently, in this feedback none of the input combinations is dominant, i.e. has the ability to set the new state Q^{n+1} of the flip-flop independently of the state Q^n. For that reason the timing of Figure 11.10.b applies to the input signals J and K. *J and K must be present during the entire duration of the clock pulse and must have their intended logic value at that time.*

Explanatory note
When the J-K flip-flop of Figure 11.10 is in the state $Q^n = 1$ and the desired new state Q^{n+1} is 1 as well, the input combination necessary for this is JK = -0. See Section 10.5. If, during $\alpha = 1$, the input combination becomes JK = -1 for just a moment, this results in a reset command for the Y latch. During that clock pulse the effect of this reset command cannot be undone. The applied feedback prevents this ($\overline{Q} = 0$ at the set gate).
The timing properties of the gated S-R latch recur in the timing properties of this J-K flip-flop. See also Figure 11.4. Figure 11.10.c specifies in detail when J and K are allowed to change (in the shaded interval) and when they have to be stable. The interval of the corresponding output reaction of Q is also indicated. □

Pulse-triggered timing

Flip-flops with a timing such as in Figure 11.10 are said to be *pulse-triggered*. For a pulse-triggered flip-flop one or more of the input combinations must be present during the entire duration of the clock pulse. Input transitions and noise during the clock pulse can influence the state Q^{n+1} after the clock pulse, and can be disastrous.

Pulse-triggered flip-flops can be classified as 'positive' and 'negative' pulse-triggered. The flip-flop in Figure 11.10, for example, is a *positive pulse-triggered flip-flop*. The inputs then have to be present during a positive going pulse, the $0 \rightarrow 1 \rightarrow 0$ pulse. During this pulse the input signals must be present the entire time.

For a *negative pulse-triggered flip-flop* the input setting must be ready around the $1 \rightarrow 0 \rightarrow 1$ pulse. The words 'positive' or 'negative pulse-triggered' refer to that part of every clock cycle within which the *input signals* must be present.

Edge-triggered or pulse-triggered timing

The property that a flip-flop has an edge-triggered or a pulse-triggered timing is not related to the logic type, D or J-K for example. The timing is only based on the internal structure, in particular on whether the input latch is of the D or S-R type or whether the flip-flop has an internal feedback that controls the input gates of the Y latch directly. When all input combinations and/or input signals are dominant, i.e. have the ability to set the state of the Y latch directly and independently of previous transitions in that same clock period, the flip-flop is said to be edge-triggered.

Figure 11.11 shows how the implementation of a J-K flip-flop with the formula

$$Q^{n+1} = [J\overline{Q} + \overline{K}Q]^n$$

can be based on a positive edge-triggered D flip-flop. The positive edge-triggered timing holds for this J-K flip-flop as well.

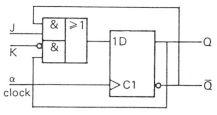

Figure 11.11 Positive edge-triggered J-K flip-flop.

Of course the gates, which derive the control signal for D from the input signals J, K and Q, influence the timing of the external input signals. Figure 11.12 shows this influence qualitatively. Section 11.8 will give more details about the relation between t_{su} and t_h of the J and K inputs externally and t_{su} and t_h of the D input internally.

The property of being edge or pulse-triggered is a *physical property* of flip-flops. When the terms 'positive' or 'negative' are used in the specifications, the following correspond

$$0 \rightarrow 1 \quad \Leftrightarrow \quad L \rightarrow H \quad \Leftrightarrow \quad \text{'positive'}$$
$$1 \rightarrow 0 \quad \Leftrightarrow \quad H \rightarrow L \quad \Leftrightarrow \quad \text{'negative'},$$

in accordance with the previously made agreement to follow the positive logic convention. With pulse-triggered flip-flops the correspondence is

$$0 \rightarrow 1 \rightarrow 0 \quad \Leftrightarrow \quad L \rightarrow H \rightarrow L \quad \Leftrightarrow \quad \text{'positive'}$$
$$1 \rightarrow 0 \rightarrow 1 \quad \Leftrightarrow \quad H \rightarrow L \rightarrow H \quad \Leftrightarrow \quad \text{'negative'},$$

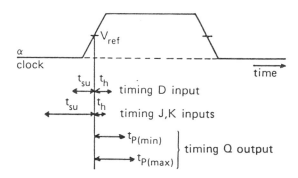

Figure 11.12 Timing properties of edge-triggered J-K flip-flop.

11.4 Properties of edge-triggered or pulse-triggered timing

A comparison of edge-triggered and pulse-triggered timing results in the following:

- With edge-triggered timing the *input signal* must be *present* within the interval determined by t_{su} and t_h. For this flip-flop the time t_c, during which the input signals are not allowed to change, is equal to

$$t_c = t_{su} + t_h. \tag{11.1}$$

- With pulse-triggered timing

$$t_c = t_{su} + t_w + t_h \qquad (11.2)$$

holds for the interval t_c, in which t_w is the *width of the clock pulse*. The width t_w is measured between the reference levels V_{ref} of the thresholds. (With positive pulse-triggered flip-flops the 'positive' going pulse is measured, and with negative pulse-triggered flip-flops the 'negative' going pulse.) Pulse-triggered flip-flops demand higher requirements with respect to t_c of the input signal.

- With edge-triggered flip-flops the *external signals have more time to set* than with pulse-triggered flip-flops. The period these signals must be present is

$$t_i = T - t_{su} - t_h \qquad (11.3)$$

with edge-triggered flip-flops and

$$t_i = T - t_{su} - t_w - t_h \qquad (11.4)$$

with pulse-triggered flip-flops. T is the period of the clock.

- In Sections 11.8 - 11.10 it is shown that with the same pulse/pause ratio of the clock edge-triggered flip-flops can have *a higher clock frequency* (provided the technologies are similar).

- Glitches and other signal transitions may influence the state of the flip-flop when they occur within t_c, the period in which the input signals must be present. The *noise susceptibility* of pulse-triggered flip-flops is therefore much greater than that of edge-triggered flip-flops.

These differences between edge-triggered and pulse-triggered flip-flops become less significant as one uses a more *asymmetrical clock*. However, this has some other disadvantages. *With newer designs of ICs one therefore sees a decided preference for a (positive) edge-triggered timing.*

Indication of timing properties in symbols

Edge-triggered timing is denoted by a triangle at the input. This triangle is called the *dynamic input indicator*. The dynamic input indicator indicates that input signals affected by this command input have an edge-triggered timing with respect to this command or clock input. See Figures 11.13.a and 11.13.b. The *negation indicator* in logic diagrams (and the *polarity indicator* in detailed logic diagrams) indicates that the active clock edge is the trailing edge. See Figure 11.13.b.

Pulse-triggered timing is not denoted by any special input indicator. The *postponed output indicator* is then used at the outputs. It indicates that the output does not react before the clock edge following that edge on which the inputs must be stable *for the first time*. See Figure 11.13.c. A negative pulse-triggered timing is denoted by drawing the negation indicator or, where appropriate, the polarity indicator in front of the clock input. This notation of a pulse-triggered timing can only be applied when the input signals do *not* change in the forbidden intervals.

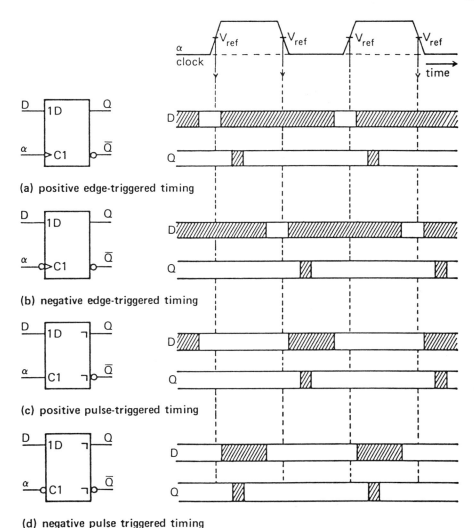

Figure 11.13 Indication of timing properties in symbols.

Note
The combination of edge-triggered inputs and postponed outputs is used as well, with data lockout flip-flops. See Section 11.10. □

11.5 Preparatory and direct acting inputs

In many flip-flops, counters, shift registers and other components not every data and control input is affected by the clock. Figure 11.14 shows the logic diagram of a D flip-flop which, in addition to the D input, also has a *direct set input* and a *direct reset input*. These inputs are often called *asynchronous preset* and *asynchronous clear*. The set and reset inputs in Figure 11.14 determine the logic state of the flip-flop directly,

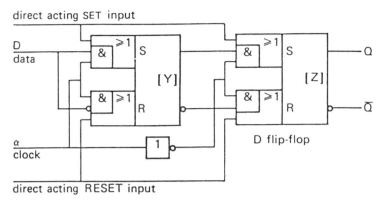

Figure 11.14 D flip-flop with a direct acting set and reset input.

independently of the clock. The flip-flop reacts *directly*. Inputs like these are therefore called *direct acting inputs*. Other inputs, operating under control of the clock, are called *preparatory inputs*.

Noise susceptibility

Direct acting inputs are always *susceptible to noise*. On printed circuit boards and within ICs they influence the reliability negatively. In newer designs of ICs direct acting inputs are interchanged more and more with preparatory inputs, with the same behaviour logically. If, however, a selected component has direct acting inputs *it is advisable to inactivate these inputs*, e.g. by placing a fixed 1 or 0 on them.

Notation of timing

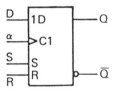

Figure 11.15 Symbol with a direct set and reset.

Figure 11.15 gives an example of how direct acting and preparatory inputs are denoted in the IEC symbols. Preparatory inputs have an *identification number*, which originates in the corresponding clock or command input. The direct acting inputs always perform their function independently, according to the way they have been implemented. Their labels are not preceded by an identification number.

Adaptation of direct set and reset inputs

Many of the present generation of ICs have a direct or asynchronous reset. Their noise susceptibility is great. Besides, as we will see, direct acting inputs do not fit in well within structured design methods, based on the *design model with data path and control*.

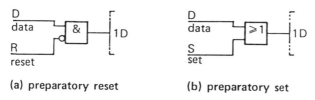

(a) preparatory reset   (b) preparatory set

Figure 11.16 Set and reset implementation.

Figure 11.16 shows how data inputs can be adapted, if necessary, so that a preparatory set or reset is possible. A preparatory reset can be implemented with ANDs, and a preparatory set with ORs. See also Figure 12.8.

11.6 Direct data transfer between flip-flops

In the above we have investigated which requirements the timing of data and control inputs of flip-flops must meet. When the output reaction occurs has also been discussed. These requirements are explained on the basis of the gate model, but the conclusions are appropriate to most of the technologies (static NMOS, CMOS, or TTL, etc.). It is one of the tasks of the logic designer to see to it that all specified timing requirements are met. We will now investigate the consequences for the timing of circuits in which every flip-flop is controlled by the same clock. We will distinguish between data transfer from flip-flop to flip-flop, so-called direct data transfer, and data transfer with combinational logic between the outputs and the affected inputs.

Direct data transfer

In Figure 11.17.a the Q_1 output of flip-flop FF1 directly drives the data input D_2 of flip-flop FF2. Both flip-flops are of the same type and are controlled by the same edge of the clock α. For a correct data transfer it is necessary that the Q_1 output of FF1 remains stable within the interval of the D_2 input of FF2 determined by t_{su} and t_h.

The timing of the flip-flops is specified in Figure 11.17.b. Figure 11.17.b indicates when Q_1 changes and where the interval of D_2, determined by t_{su} and t_h, is situated. We see that the transition of Q_1 is outside this interval. The intended data transfer progresses correctly.

There is even a slight margin between the interval in which Q_1 is allowed to change and the hold interval of D_2. This time is called the *skew time* t_{skew}. This parameter defines the margin by which the clock of FF2 may be postponed, referring to the clock of flip-flop FF1. Within t_{skew} correct data transfer is possible. (See Section 11.10.)

Figure 11.17.c shows how the skew time t_{skew} can be derived directly from the timing parameters of the flip-flops. For t_{skew} it holds that

$$\boxed{t_{skew} = t_{P(min)} - t_h.} \qquad (11.5)$$

For a correct data transfer it is necessary that an output does not change until the affected inputs have read their data:

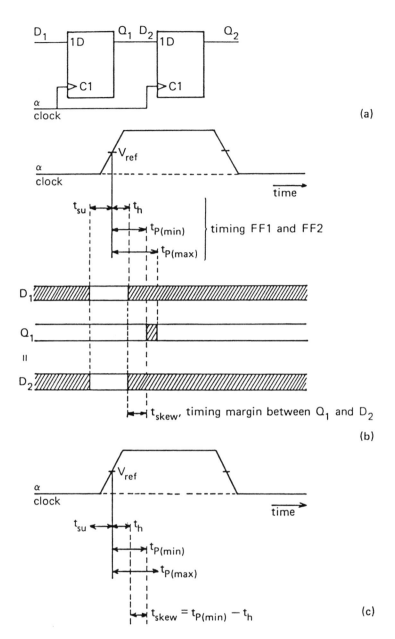

Figure 11.17 Direct data transfer between identical flip-flops.

$$\boxed{\text{correct data transfer} \;\Rightarrow\; t_{skew} > 0.} \qquad (11.6)$$

Figure 11.18 specifies the timing of a fast type and a low-speed type of flip-flop. We see that with fast flip-flops t_{skew} is significantly smaller than with the low-speed types. In Section 11.8 we will prove that this can be disadvantageous.

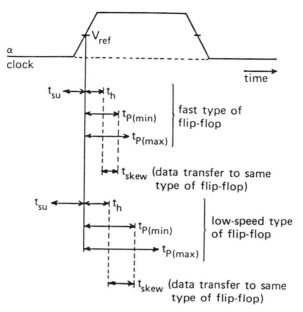

Figure 11.18 Skew time with fast and low-speed flip-flops.

Data transfer between different types of flip-flops

Figure 11.19 specifies the timing parameters of a fast type of flip-flop FF1 and of a low-speed type of flip-flop FF2. When data are transferred from FF1 to FF2 it holds that:

$$t_{skew} = t_{P(min)}(FF1) - t_h(FF2). \qquad (11.7)$$

In Figure 11.19 we see that the margin t_{skew} has become smaller, even when compared to data transfer between flip-flops of type 1. When the difference in speed between FF1 and FF2 becomes too high, it is even possible for t_{skew} to become negative. Reliable data transfer is no longer possible in that case.

With data transfer from a low-speed type to a faster type of flip-flop, from FF2 to FF1 in Figure 11.19, the margin t_{skew} becomes *larger* in data transfer:

$$t_{skew} = t_{P(min)}(FF2) - t_h(FF1). \qquad (11.8)$$

Correct data transfer between flip-flops with different timing parameters is possible, but (11.6) must hold *for all interconnections between the flip-flops*.

Conclusions

- For data transfer between flip-flops of the same type the margin for the timing is specified by the parameter t_{skew}.
- The parameter t_{skew} is usually smaller in faster types of flip-flops than in low-speed types. The value also depends on the internal structure of the flip-flops.

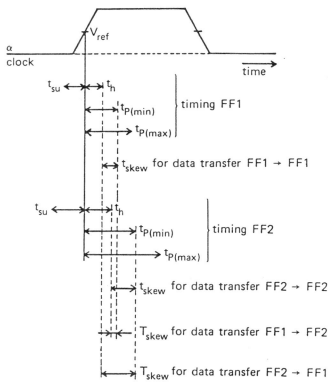

Figure 11.19 Determining t_{skew} with data transfer between different types of flip-flops.

- For data transfer between flip-flops of different types with regard to the operating speed the timing margin t_{skew} can be larger or smaller, depending on the direction of the data transfer.

- With a difference in speed between flip-flops that is too high t_{skew} can become negative. Reliable data transfer is no longer possible in that case.

If (11.5) and (11.6) have been satisfied for all interconnections, direct data transfer between flip-flops of the same type is possible. For flip-flops of different types (11.7) and (11.8) must be used. The parameter $t_{P(min)}$ of the flip-flops is found in these formulas. In most of the data books this parameter is *not* listed. The parameter $t_{P(typical)}$ which is listed does not have the same meaning. For practically all flip-flops it holds that

$t_{P(min)} > t_h$

when the flip-flops in one IC are concerned. The conditions as specified have then been met. With different ICs and without further investigation one does not know whether the data transfer can be carried out correctly! In Section 11.10 it will be proved that this problem can be solved, for example, by placing extra logic (propagation delay time) between the outputs and inputs. In equipment that has to be extremely reliable and/or interchangeable this is common practice.

11.7 Indirect data transfer between flip-flops

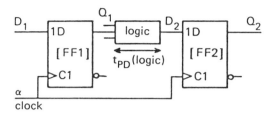

Figure 11.20 Data transfer via combinational logic.

In Figure 11.20 flip-flop FF1 controls the D input of flip-flop FF2 via combinational logic. The output reaction of FF1 is delayed a time t_{PD}(logic) before controlling the input of FF2. This propagation delay time must not exceed a certain limit. The input D_2 of FF2 must be present before the setup interval of FF2 begins.

Figure 11.21 explains this in a timing diagram. After the active clock edge of FF1 its Q output adapts to the new internal state within the interval determined by $t_{P(min)}$ and $t_{P(max)}$. After $t_{P(max)}$ Q_1 has been set to its new value. Subsequently, the combinational logic is set definitely. To do this it needs a time $t_{P(max)}$(logic) at the most. After this the input D_2 of flip-flop FF2 is set. To do this it needs a time t_{su}(FF2) maximally.

All inputs must be ready before the next active clock edge. Depending on the period T of the clock a certain time is left (see Figure 11.21). Reliable data transfer via logic is only possible when

$$T > t_{P(max)}(FF1) + t_{P(max)}(logic) + t_{su}(FF2). \tag{11.9}$$

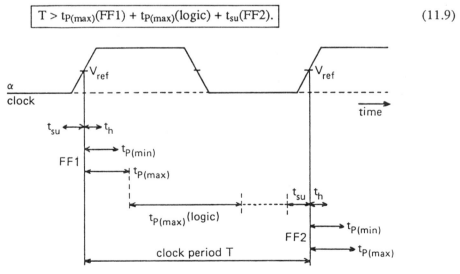

Figure 11.21 Data transfer FF1 → FF2.

In general, the combinational logic which controls a flip-flop input is set by outputs of several flip-flops. Equation (11.9) must then be verified for *all signal paths*. Further, the combinational logic also influences t_{skew}. This influence is analysed in more detail in Section 11.10, where we will investigate the effects of clock skew.

Note

Besides the propagation delay time in the logic the signals are also delayed by the interconnect wiring. Signals have a finite propagation delay time in the wiring. Roughly speaking, one assumes 1 ns of delay time per 15 cm of wire or track. Normally, as the propagation delay time in the wiring is smaller than the gate delays, it may be neglected. However, when the circuit to be designed must be able to run at a high clock frequency, the propagation delay time in the wiring must be incorporated in (11.9). One must take the propagation delay time of signals into account internally in ICs. The timing model is then dependent on the layout of the IC and its technology (MOS). □

The maximum clock frequency

The *repetition frequency* f of a *periodical clock signal* is a parameter indicating how many pulses are given per second. The frequency f is inversely proportional to the period T of the clock:

$$f = \frac{1}{T} \text{ Hertz.} \tag{11.10}$$

From (11.9) and (11.10) it follows that:

$$f \leq \frac{10^3}{t_{P(max)}(FF1) + t_{P(max)}(\text{logic}) + t_{su}(FF2)} \text{ MHz.} \tag{11.11}$$

The maximum clock frequency of a device can be found by checking (11.11) for all signal paths. The minimum value of (11.11) determines the maximum clock frequency.

Example

With edge-triggered flip-flops with

$$t_{su} = 15 \text{ ns}, \; t_{P(max)} = 40 \text{ ns}$$

and combinational logic with a maximum propagation delay time of

$$t_{P(max)} (\text{logic}) = 100 \text{ ns}$$

the maximum clock frequency of the device is

$$f_{max} = \frac{1}{(40 + 100 + 15) \cdot 10^{-9}} = 6.5 \text{ Mhz.}$$

This maximum can be *guaranteed* with the components used. The maximum frequency measured in a prototype may be significantly higher, because of a spreading of parameters. □

Example

When the clock frequency is given, the maximum propagation delay time allowed in logic can be determined. From (11.9) it follows for a clock frequency of 1 MHz with the timing parameters of the flip-flops as specified that:

$$t_{P(max)}(\text{logic}) = T - t_{P(max)}(FF1) - t_{su}(FF2)$$
$$= 1000 - 40 - 15$$
$$= 945 \text{ ns.}$$

The propagation delay time of the combinational logic may not exceed this value. □

Note
In data transfer between pulse-triggered flip-flops, that part of the clock period in which $t_{P(max)}$ and t_{su} are defined must be taken instead of T in (11.9). One must also take into account the minimum width of the clock pulse and/or clock pause. Pulse-triggered flip-flops are generally slower. □

11.8 Internal and external timing

There usually is combinational logic between the outputs of flip-flops and the inputs of other flip-flops driven by them. The propagation delay time of the logic influences the data transfer. Figure 11.22 explains this in more detail.

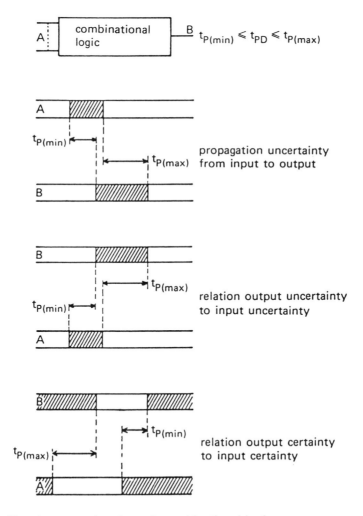

Figure 11.22 Signal propagation through combinational logic.

Figure 11.22.b shows how the *uncertainty interval* in the input signals of combinational logic is transferred to its output. The uncertainty interval shifts in time and also lasts a period of time

$$t_{P(max)} - t_{P(min)}$$

longer.

The reverse is described in Figure 11.22.c. It shows how an *allowable uncertainty interval* in the output of a combinational circuit must be transferred into requirements to be met by the input signals.

Figure 11.22.d shows how a *required certainty interval* in the output signals of combinational logic must be translated into requirements of the input signals. The required certainty interval of the input signals is

$$t_{P(max)} - t_{P(min)}$$

longer and has shifted in time.

This case applies, for example, when the t_{su} - t_h interval of a flip-flop is known and when one is interested in the timing requirements for the inputs of the control logic.

Internal and external setup and hold times

An application of the above is described in Figure 11.23. A positive edge-triggered D flip-flop is driven by combinational logic with a propagation delay time

$$t_{P(min)} \leq t_{PD} \leq t_{P(max)}.$$

For the D input t_{su} and t_h are specified. Figure 11.23.b shows the timing requirements for the input signals of the logic. When we refer these requirements to the clock α it holds that

$$\boxed{t_{su}(\text{logic}) = t_{su}(\text{FF}) + t_{P(max)}(\text{logic}),} \qquad (11.12.\text{a})$$

$$\boxed{t_h(\text{logic}) = t_h(\text{FF}) - t_{P(min)}(\text{logic}).} \qquad (11.12.\text{b})$$

Sometimes $t_h(\text{logic})$ is specified as 0 ns when (11.12.b) gives a negative result. A negative hold time is often found in counters and shift registers. The logic is then built up in several levels, in which the minimum propagation delay time is greater than t_h of the flip-flops. For the calculation of t_{skew} the exact value of t_h should be used, even when it is negative. See Section 11.10.

Clock buffer

With several flip-flops in one IC an *internal clock buffer* is often used to restrict the fanin of the clock input. Usually a buffer/inverter is applied. See Figure 11.24.

When t_{su} and t_h of the D signals are specified in relation to the internal clock $\alpha_{internal}$ they must be redefined with regard to $\alpha_{external}$. With a propagation delay time t_{PD} of the buffer this means

$$t_{su}(D/\alpha_{ext}) = t_{su}(\text{FF}) - t_{PD}(\text{inv}),$$

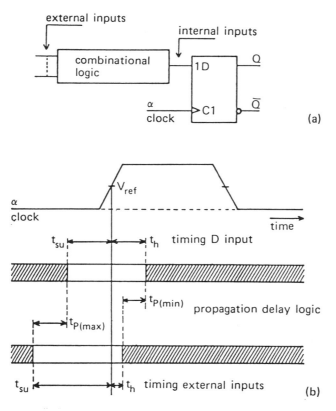

Figure 11.23 Specification of external setup and hold times.

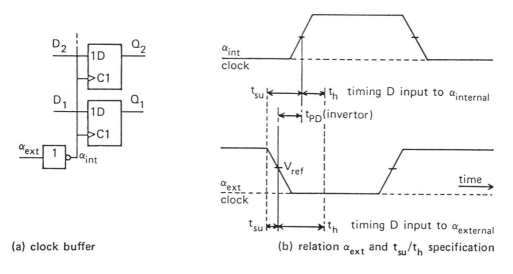

(a) clock buffer

(b) relation α_{ext} and t_{su}/t_h specification

Figure 11.24 Specification of external setup and hold times.

$$t_h(D/\alpha_{ext}) = t_h(FF) + t_{PD}(inv).$$

When $t_{P(min)}$ and $t_{P(max)}$ of the buffer are known it holds that

$$t_{su}(D/\alpha_{ext}) = t_{su}(FF) - t_{P(min)}(inv), \qquad (11.13.a)$$

$$t_h(D/\alpha_{ext}) = t_h(FF) + t_{P(max)}(inv), \qquad (11.13.b)$$

This is explained in Figure 11.25.

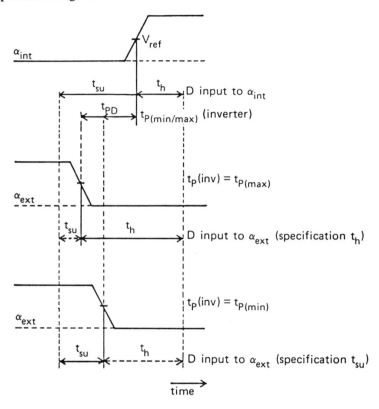

Figure 11.25 Specification of external setup and hold times.

When both a clock buffer and logic between data outputs and data inputs are used Equations (11.12) must be combined with Equations (11.13). This results in

$$t_{su}(data/\alpha_{ext}) = t_{su}(FF) + t_{P(max)}(logic) - t_{P(min)}(inv), \qquad (11.14.a)$$

$$t_h(data/\alpha_{ext}) = t_h(FF) - t_{P(min)}(logic) + t_{P(max)}(inv). \qquad (11.14.b)$$

Example

The logic diagram in Figure 11.26 belongs to a circuit that detects whether a signal a has the value of 1 during two or more successive clock periods (i.e. a = 1 within the $t_{su} - t_h$ interval of the flip-flops, following the clock mode interpretation). The timing parameters of the input signal a and the output signal U are derived in Figure 11.26.b.

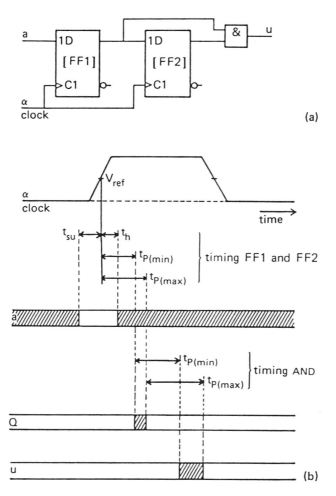

Figure 11.26 Timing calculation.

The logic behaviour of the circuit in Figure 11.27 is identical. This circuit, too, performs the desired test on signal a. The signal timing in this circuit differs significantly from that of Figure 11.26. The input signal a must now meet two requirements, that of the $t_{su} - t_h$ interval of FF1 and that set by the input of the AND gate. This is explained in Figure 11.27.b.

In the realisation of Figure 11.27 the output signal U changes earlier than in that of Figure 11.26. The uncertainty interval of U is also shorter. The preference for a particular circuit realisation depends on which timing requirements are the most important, those of a or those of U. □

With the above example it has been shown that timing requirements often allow several approaches. One of these is to take faster circuits. This has many disadvantages, for example with respect to the circuit layout and dissipation. The other approach is an adaptation of the internal structure.

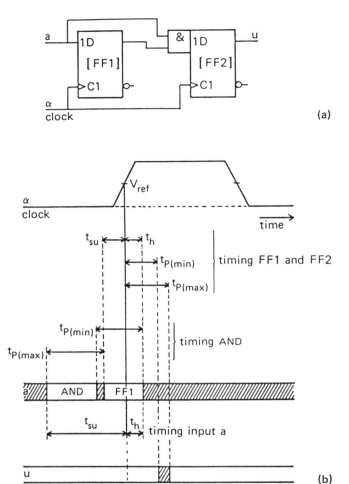

Figure 11.27 Timing calculation.

11.9 Data transfer at different clock phases

When all flip-flops in a circuit are of the same type and react to the same clock edge the verification of the timing is not difficult. In such cases the timing verification is reduced to the *numeric summation* of (maximum) propagation delay times and taking t_{su} and t_h into account as well. In practice, one sometimes deviates from this system, usually because a signal is not present in time, related to a certain clock edge.

In systems with edge-triggered flip-flops it is possible, in principle, for two flip-flops FF1 and FF2 to transfer data at *different edges* of the same clock signal. See Figure 11.28.

In Figure 11.28 flip-flop FF2 takes over the data of flip-flop FF1. For the logic in between a maximum propagation delay time of

$$\boxed{t_{P(max)}(\text{logic}) < t_w(\text{clock}) - t_{P(max)}(\text{FF1}) - t_{su}(\text{FF2})} \qquad (11.15)$$

is permitted. This is less than with data transfer on the same clock edge. See (11.9). On the other hand, FF2 has set its output half a clock period earlier. Flip-flop FF2 can then start addressing a memory sooner, for example. Such decisions influence the *timing verification* negatively. However, when one has to choose between a lower clock frequency for an entire system or for a local adaptation, the choice is obvious.

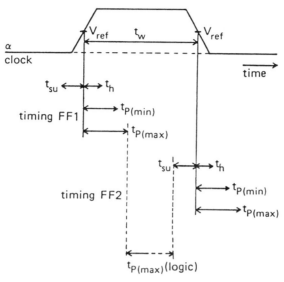

Figure 11.28 Data transfer between edge-triggered flip-flops on different clock edges.

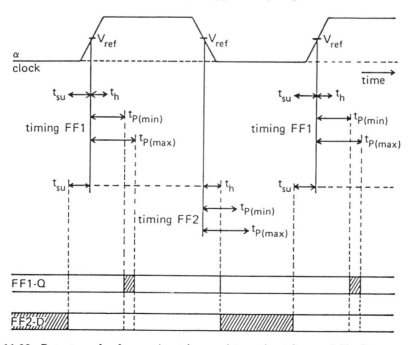

Figure 11.29 Data transfer from edge-triggered to pulse-triggered flip-flops.

In data transfer from edge-triggered to pulse-triggered flip-flops serious problems may arise. Figure 11.29 demonstrates that when the pulse-triggered flip-flop FF2 accepts data from the edge-triggered flip-flop FF1, the Q output of FF1 changes in the $t_{su} - t_h$ interval of flip-flop FF2. Reliable data transfer is *not possible*. Data transfer from edge-triggered to pulse-triggered flip-flops must therefore take place on the *same clock edge*. Reliable data transfer is then possible, in principle, provided that t_h and $t_{P(min)}$ meet the requirements specified in (11.6) through (11.8).

Data transfer from pulse-triggered flip-flops to edge-triggered flip-flops is always possible, even on different clock edges. Edge-triggered flip-flops have lower requirements.

Data transfer between pulse-triggered flip-flops is never possible on different clock edges. This can easily be proved by a figure similar to that of Figure 11.29.

Table 11.1 summarises the possibilities of data transfer between different types of flip-flops (on the condition that $t_{skew} > 0$ in all cases).

		to	A	B	C	D
		from				
positive edge-triggered:	type A	A	+	+	−	+
negative edge-triggered:	type B	B	+	+	+	−
positive pulse-triggered:	type C	C	+	+	+	−
negative pulse-triggered:	type D	D	+	+	−	+

Table 11.1 Possibilities of data transfer.

Note

The above statements apply to direct data transfer. Combinational logic between outputs and inputs may have a negative or a positive influence. □

11.10 Clock skew

The above timing discussion was based on the condition that all flip-flops in a circuit be clocked simultaneously. All flip-flop outputs then change at or just after the same edge of the clock pulse. In practice, the condition of simultaneous clocking is only roughly approximated. Some reasons for this are:

− The real *threshold* between H and L or between L and H is not at the same level in all ICs. The operating temperature of a circuit has a distinct influence on the value of the threshold. Also, the integration technology applied has influence. Compare Figures 6.1 and 6.11.

− The *rise and fall times* of the clock pulse have an influence on the moment the threshold is passed. See Figure 6.3.

− *Long clock lines* delay the clock pulses and distort them. Through a capacitive load of a clock line the edges become less steep. This also causes a shift in the moment at which the flip-flop reacts.

- *Gates in clock lines* influence the time of arrival of the clock edges. Sometimes gates in the clock lines are necessary, e.g. to adapt flip-flops having an output reaction on different clock edges to each other. (There are special ICs which generate both α and $\overline{\alpha}$, with the delay time below a specified minimum.)

Because of the above-mentioned effects, the flip-flops in a circuit do not recognise the clock edges simultaneously. This phenomenon is called clock skew.

For the designer the above discussion implies that in a design all phenomena resulting in clock skew have to be investigated and that their effects must be summed up, with a certain weight factor. This results in *requirements for the timing* of the device. Subsequently, the designer has to verify whether all the requirements are met or whether a device can be made to meet them.

Direct data transfer and clock skew

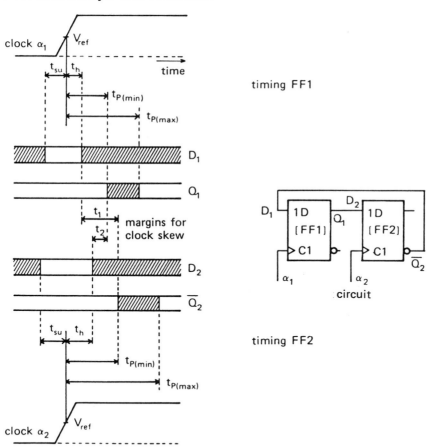

Figure 11.30 Margins available for clock skew with direct data transfer.

In Figure 11.30 the clock pulse α_2 of flip-flop FF2 may at most be 'recognised' a time t_1 earlier than α_1 of flip-flop FF1. After all, should this time be more than t_1 ns, the uncertainty interval of $\overline{Q}_2$ has an overlap with the setup and hold interval of the D_1 input. The same reasoning leads to the conclusion that α_2 may at most be recognised t_2 later. The *available margin for the skew* of α_2 with respect to α_1 is

$$-t_1 < t_{skew} < +t_2, \qquad (11.16.a)$$

or

$$\boxed{-[t_{P(min)}(FF2) - t_h(FF1)] < t_{skew} < [t_{P(min)}(FF1) - t_h(FF2)].} \qquad (11.16.b)$$

When the parameters of the flip-flops are known, the available margins for *clock skew* follow from (11.16.b). If the estimate of the maximum clock skew in the device is lower than found in (11.16), data transfer as indicated in Figure 11.30 is possible.

With fast types of flip-flops the margin resulting from (11.16) is often very small. As a consequence it is difficult for the circuits to meet it. The circuit then does not operate reliably, or not *under all operating conditions*, for example with a rising temperature. The layout of clock lines must be done carefully and extra delays in clock lines should be avoided as much as possible. This is essential to *any design*, even for circuits operating at low clock frequencies.

Indirect data transfer and clock skew

The difference between $t_{P(min)}$ of the driving flip-flop and t_h of the driven flip-flop determines the margin for clock skew. This margin grows with additional logic in between the flip-flops. With respect to the driven flip-flop, $t_{P(min)}$ of the signal received will be proportionally greater. See Figure 11.31.

Conclusions

From the above it follows that:

− Clock skew is caused by
  • a shift or distortion of clock edges;
  • the non-mutual recognition of a clock edge by different modules in a device;
  • a combination of both phenomena.

− Clock skew cannot be avoided completely.

− With direct data transfer the margins are very small.

− With indirect data transfer clock skew is usually not a problem.

− As logic in clock lines may take up the whole margin for clock skew, one is seriously advised not to use it.

− The effect of switching logic in clock lines may be compensated for by gates or other logic in the data lines.

Note

The above conclusions hold for edge-triggered flip-flops. For pulse-triggered flip-flops similar conclusions can be drawn. □

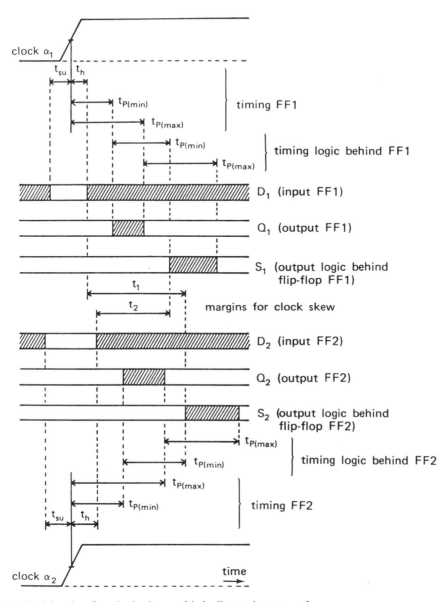

Figure 11.31 Margins for clock skew with indirect data transfer.

In general, fast series of integrated circuits are more susceptible to clock skew than low-speed series. The clock distribution must then be carried out with many precautions. Low-speed circuits have an advantage here. It is therefore advisable not to select the components for a circuit of a faster type than is strictly necessary.

So far clock skew has been discussed as a phenomenon of importance for data around one clock pulse. With the determination of the maximum clock frequency (11.11) some margin for clock skew must be added as well. Generally, its influence on the maximum of the clock frequency will be small.

Edge-triggered flip-flops with data lockout

In data transfer between flip-flops it is not always possible to guarantee that the clock will remain within the available margin under all circumstances. In that case *edge-triggered flip-flops with data lockout* are used. These flip-flops read the data signals on one clock edge, for which the timing of an edge-triggered input is applicable. Subsequently, these flip-flops store the information just read internally until the next clock edge. The output does not adapt before this next clock edge. Any input transition after the sampling of the input is ignored and has no effect. See Figure 11.32.

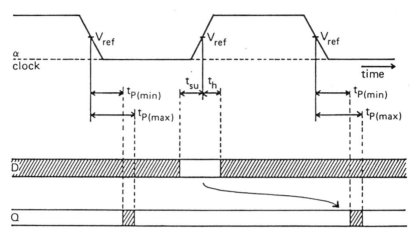

Figure 11.32 Data lockout timing.

With this type of flip-flop the margin for clock skew is specified in (11.17) and (11.18):

$$-[t_w(H) - t_w(FF1) + t_{P(min)}(FF2)] < t_{skew} < [t_w(L) - t_{P(max)}(FF2) - t_{su}(FF1)] \qquad (11.17)$$

and

$$-[t_w(L) - t_{P(max)}(FF1) - t_{su}(FF2)] < t_{skew} < [t_w(H) - t_h(FF2) + t_{P(min)}(FF1)]. \qquad (11.18)$$

Both equations must be met. The parameters $t_w(L)$ and $t_w(H)$ indicate the width of the pause (L) and that of the pulse (H) of the clock.

With data lockout flip-flops clock skew can be suppressed effectively. As far as their internal structure is concerned, data lockout flip-flops may be thought of as being an edge-triggered input flip-flop, followed by a gated latch that is enabled at the next clock edge.

In principle the three types of flip-flops mentioned, with edge-triggered, pulse-triggered or with data lockout timing, can be applied intermixed. However, their outputs normally have to change at the same clock edge, and it must furthermore be verified in each application that $t_{P(min)}$ of one flip-flop is not shorter than t_h of all the flip-flops driven by it ($t_{skew} > 0$!). With regard to *noise susceptibility and cross-talk on data lines* edge-triggered flip-flops with or without data lockout are to be preferred over pulse-triggered flip-flops.

Only the transitions during the $t_{su} - t_h$ interval are then troublesome, and this interval is only a short part of the clock period (at least with lower frequencies).

Clock distribution

A circuit usually has more circuits than can be controlled by one clock driver circuit. It is then necessary to apply a buffer, usually an amplifying inverter. Figure 11.33 shows how this problem is usually solved. A network structure such as in Figure 11.33.b should be avoided. In such a structure the clock pulses are distributed with unequal delays. This results in unnecessary *clock skew*.

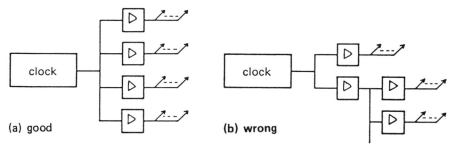

Figure 11.33 Clock distribution.

In structures such as the one in Figure 11.33.a one must try to balance the load of all clock lines. In fast circuits this requires a certain adjustment.

In a digital circuit there is often some cross-talk between the wiring. By cross-talk we mean the phenomenon that some noise is measured on signal lines not supposed to have any transitions. This is the result of transitions on neighbouring lines. These are transferred by an *electromagnetic* and/or a *capacitive coupling* between the wiring. As a rule cross-talk is difficult to prevent. One of the principles on which a *well-designed digital circuit* is based is therefore that the next clock pulse should only be given when all transitional phenomena, including cross-talk, have faded away, i.e. when *the entire circuit is in a stable state*. The easiest and most certain way to accomplish this is by having the circuit controlled by one clock and having all transitions started at or just after the same clock edge. (As has been said before there may be compelling reasons to deviate from this principle.)

Note

When a component (flip-flop) changes its state this always results in a short *change in the current flow*. When all flip-flops are being controlled by the same clock, a certain voltage drop on the power supply occurs at the switching clock edge. The effects of this can be compensated for by using *decoupling capacitors*. □

11.11 Synchronisation of external signals

The model for the timing of flip-flops has been introduced on the basis of master-slave flip-flops, composed of two gated latches (Figure 11.6). In principle, the model and the conclusions hold for all types of flip-flops whose functioning is based on *one external clock pulse.* However, the internal structure does have an influence on the *numerical value* of the parameters. If a designer conforms to the limits given for the parameters, a correctly operating circuit is guaranteed (as far as its timing is concerned).

The model is a *worst-case* model. If one incorporates layout and data-dependence as well, stricter limits for the parameters are possible. This will then make the verification and simulation more time-consuming, which increases the cost of a design and is therefore only worthwhile for larger series.

The model is sufficiently detailed for the design of *printed circuit boards*. For IC designs the model should be further refined and adapted, depending on the technology used. When ICs of different families (such as MOS, TTL) are applied on PCBs the model has to be extended as well. The term 'compatible' does not mean much in this context (cf. Sections 6.1 and 6.12).

In practice, systems with two- and sometimes even four-phase clocks are found. A development of a more recent date is that of *self-timed logic*. Here synchronisation information is incorporated in the signal encoding. Within a system with self-timed logic each module has its own clock and a module reads in data when its environment is 'stable'. See the References. The problems with respect to the *testability* of this kind of design are very great.

Synchronisation

Signals do not always enter clock-synchronously. In principle, such signals may change within the t_{su}-t_h interval of the driven flip-flops. They must therefore be 'synchronised'. One usually applies the setup of Figure 11.34.a. By sampling with an edge-triggered D flip-flop one can ensure that its Q output will not change arbitrarily, but just after the active clock edge. The transition interval is then fixed. The output signal of the D flip-flop can then be used to set memory elements and/or combinational logic.

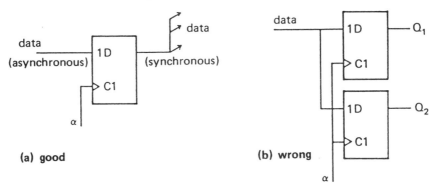

Figure 11.34 Synchronisation of external signals.

If the input signal changes in the diagram of Figure 11.34.a within the t_{su}-t_h interval, the result of the sampling will, of course, be uncertain. At the output the old signal value is still available or the new value is already present. In the former case the new signal value is recognised one clock pulse later.

To get more certainty about the signal value repeated sampling is sometimes applied. A signal transition is only accepted if the new value has been present for an (adjustable) number of clock periods. Sampling at clock frequencies higher than the internal frequency also occurs or, with an edge-triggered system, sampling at the other clock edge.

Synchronisation as shown in Figure 11.34.b should be avoided. In this setup it is possible for FF1 to read in another value than FF2. An external signal is therefore only allowed to be distributed *after synchronisation*.

Data lines are normally synchronised with a *data valid signal*. This signal then arrives somewhat later than the data itself, so that if 'data valid' is recognised the data signals are certainly stable and can be read in. This requires only one signal to be synchronised and/or sampled.

Synchronisation problems

With synchronisation the timing model, especially the parameters $t_{P(min)}$ and $t_{P(max)}$, has to be verified. When signals are entering asynchronously one cannot guarantee that D will not change within the t_{su}-t_h interval. In the literature this problem is given ample attention. Measurements have shown that, as external signal transitions occur closer to the moment at which the clock passes the threshold between High and Low, the uncertainty interval in the output value is longer. It is even possible for the output to remain between the specified H/L regions for some time. See Figure 11.35. The probability of this phenomenon, however, decreases as more time passes after the sampling moment.

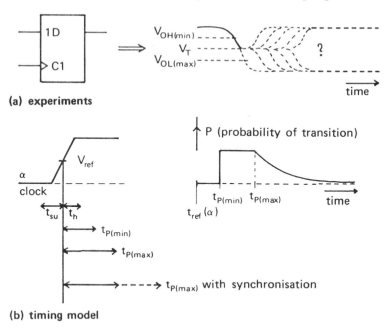

Figure 11.35 The parameter $t_{P(max)}$ with synchronisation.

As far as our timing model is concerned this effect means that a much longer $t_{P(max)}$ should be used for synchronisation flip-flops. With high-frequency circuits one therefore sometimes reserves more than one clock period to have a synchronised signal stabilised. *Synchronisation then requires more time than normal data transfer.*

If sufficient time is available for synchronisation, one often interconnects two D flip-flops in series. The output Q_2 of the second flip-flop then meets the $t_{P(min)}$ and $t_{P(max)}$ as specified for that type. When setting combinational logic the timing can easily be verified again.

11.12 Conclusions

Practice has shown that it is necessary to separate the logic design of a circuit and its timing verification at as early a stage as possible. The *required controllability of the design process* (e.g. simulations) is one of the major reasons for this. It is not possible to simply introduce a separation between logic design and timing verification. If one conforms to a fully (positive) edge-triggered timing system, based on one clock, there will be few problems. The noise susceptibility of the entire system will then be low and, in any case, it will be better than with a (partial) pulse-triggered timing. With the parameters t_{su} and t_h as introduced for the inputs and $t_{P(min)}$ and $t_{P(max)}$ for the outputs the process of timing verification is reduced to formulating and checking simple numerical relations.

With this approach the timing problem is partly shifted to a lower design level. For, when all memory elements have to switch simultaneously, this places high demands on the power distribution in ICs. This is a reason why, with large ICs, one often deviates more from the model used here. On printed circuit boards decoupling capacitors over the power supply of ICs intercept transitions in the power supply sufficiently. In that case there is no reason to deviate from fully edge-triggered timing.

The problem of the very small margins for *clock skew* has yet to be solved. In the above we have ascertained that clock skew *greatly depends on the components used* and to a degree also on, for example, the layout of a circuit. The (maximum) operating frequency of a circuit, in principle, does not have an influence on the margins for clock skew. With data transfer between flip-flops the susceptibility to clock skew is determined by the difference between $t_{P(min)}$ of the driving flip-flop and t_h of the driven flip-flop (with direct data transfer). In order to keep clock skew manageable we must therefore first of all look at the *internal structure* of the flip-flops.

Example

The flip-flop model in Figure 11.36.a is characterised by a relatively short hold time t_h at the input and relatively high values of $t_{P(min)}$ and $t_{P(max)}$ respectively at the output. This is advantageous with regard to clock skew.

The flip-flop model in Figure 11.36.b can, when controlled by α (in exact counter phase with $\overline{\alpha}$), be interchanged with the model of Figure 11.36.a. Both flip-flops react with their outputs on the falling edge of α. However, when we look to respectively t_h and $t_{P(min)}$ of the latter model, t_h turns out be longer and $t_{P(min)}$ shorter. This is *very disadvantageous* with regard to clock skew! □

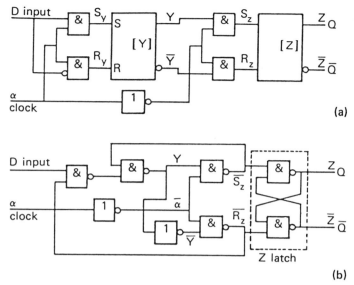

Figure 11.36 Flip-flop designs.

Example

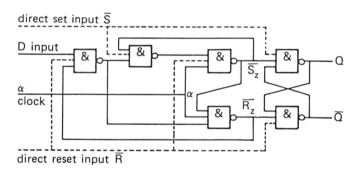

Figure 11.37 Detailed logic diagram of a D flip-flop.

Figure 11.37 shows the detailed logic diagram of an edge-triggered D flip-flop that is used in, for example, TTL. With some manipulation of formulas this diagram can be deduced from the diagrams of Figure 11.36. The clock pulse α immediately controls the output section and, through a feedback (delay!), the inputs. This design is *disadvantageous* with regard to clock skew. □

Two-phase clocking

In the design of memory elements one should keep a large number of aspects in mind, including, for example,
− technology dependence,
− gate or transistor model (e.g. TTL versus NMOS/CMOS),
− layout rules,

– desired operational frequency,
– susceptibility to clock skew.

One may wonder whether clock skew should be suppressed with an adapted internal structure of flip-flops. An alternative may be to use a *two-phase clock*.

In VLSI designs two-phase clocks are frequently used. The two phases are then denoted as φ_1 and φ_2. Figure 11.38 explains this.

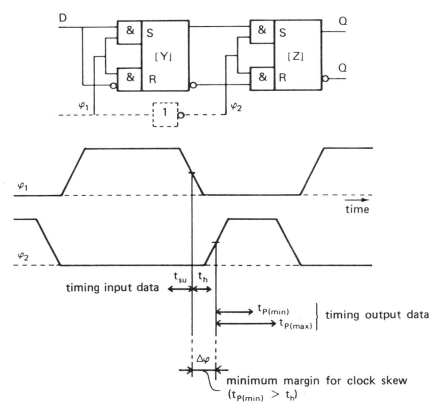

Figure 11.38 Flip-flop with a φ_1/φ_2 clock system.

The clock pulse for the Z latch is then not derived by an inverter, but is generated separately. With data transfer to a similar type the margin for clock skew becomes

$$t_{skew} = t_{P(min)} - t_h + \Delta\varphi, \qquad (11.19)$$

in which $\Delta\varphi$ is the shift between the falling edge of φ_1 and the rising edge of φ_2. Of all parameters that are able to influence t_{skew} $\Delta\varphi$ can be adjusted, independently of the technology applied. In this way the consequences of clock skew can be sufficiently suppressed. Of course, a thorough study is required to determine the minimal value of the time $\Delta\varphi$. One might wish for some information about this in application notes!

Note

In VLSI designs two-phase clocks are usually not introduced *because of* the clock skew phenomenon. Compare, for example, 'dynamic logic' in Chapter 6. Clock skew then only makes demands on the φ_1/φ_2 clock. The above shows that a two-phase clock has more advantages to offer. □

The subject of timing as discussed in this chapter is the basis for the timing of data transfer at the flip-flop and register transfer level. At the system level many other aspects are involved, for example, the concept 'bus' and data transfer based on 'hand shaking', to name a few.

References

1. L.C. Bening, T.A. Lane and C.R. Alexander, *Developments in Logic Network Path Delay Analysis*, Proc. ACM IEEE 19th Design Automation Conference, Las Vegas, June 1982, pp. 605–615.
2. T.J. Chaney and F.U. Rosenbergen, *Characterization and Scaling of MOS Flip-flop Performance*, Caltech Conf. on VLSI, 1979, pp. 357–374.
3. T.J. Chaney and C.E. Molnar, *Anomalous Behavior of Synchronizer and Arbiter Circuits*, IEEE Tr. on Computers, Vol. C-22, 1973, pp. 421–422.
4. D.M. Chapiro, *Reliable High-Speed Arbitration and Synchronization*, IEEE Tr. on Computers, Vol. C-36, 1987, pp. 1251–1255.
5. S.T. Flannagan, *Synchronization Reliability in CMOS Technology*, IEEE J. of Solid-State Circuits, Vol. SC-20, 1985, pp. 880–882.
6. R.B. Hitchcock, *Timing Verification and the Timing Analysis Program*, Proc. ACM IEEE 19th Design Automation Conference, Las Vegas, 1982, pp. 594–604.
7. D.J. Kinniment and J.V. Woods, *Synchronization and Arbitration Circuits in Digital Systems*, Proc. IEE, Vol. 123, 1976, pp. 961–966.
8. L. Kleeman and A. Cantoni, *Metastable Behavior in Digital Systems*, IEEE Design & Test of Computers, Dec. 1987, pp. 4–19.
9. B. Magnhagen, *Practical Experiences from Signal Probability Simulation of Digital Systems*, Proc. ACM IEEE 14th Design Automation Conference, Las Vegas, 1977.
10. L.R. Marino, *General Theory of Metastable Operation*, IEEE Tr. on Computers, Vol. C-30, 1981, pp. 107–115.
11. C. Mead and L. Conway, *Introduction to VLSI Systems*, Addison-Wesley, Reading, Mass., 1982, Ch. 5.
12. M. Muraoka e.a., *ACTAS, An Accurate Timing Analysis System for VLSI*, Proc. ACM IEEE 22nd Design Automation Conference, Las Vegas, 1985, pp. 152–158.
13. M. Pechoucek, *Anomalous Response Times of Input Synchronizers*, IEEE Tr. on Computers, Vol. C-25, 1976, pp. 133–139.
14. J.H. Shelly and D.R. Tryon, *Statistical Techniques of Timing Verification*, Proc. ACM IEEE 20th Design Automation Conference, Las Vegas, 1983, pp. 396–402.
15. P.A. Stoll, *How to Avoid Synchronization Problems*, VLSI Design, 1982, pp. 56–59.
16. D.M. Tavana, *Metastability, A Study of the Anomalous Behavior of Synchronizer Circuits*, Programmable Logic Handbook, MMI, Santa Clara, 1985.
17. S.H. Unger and C.-J. Tan, *Clocking Schemes for High-Speed Digital Systems*, IEEE Tr. on Computers, Vol. C-35, 1986, pp. 880–895.
18. H.J.M. Veendrick, *The Behavior of Flip-flops Used as Synchronizers and Prediction of their Failure Rate*, IEEE J. of Solid-State Circuits, Vol. SC-15, 1980, pp. 169–176.

12
Registers

12.1 The internal structure of registers

Registers are frequently used in digital circuits. A *register* is a *group of flip-flops* with additional combinational circuits, which have a *common control*. Registers are used for the temporary storage of data. Special registers can also perform some operations on the stored data, such as shifting the bits one position to the left or to the right. This is often applied in mathematical operations. Another example is the series-parallel or parallel-series conversion of data in systems with serial data transport.

An IC data book describes many types of registers. Not every type is relevant to new designs. Many types date from a period when insight into subjects such as timing was still in its infancy. The number of 'gate equivalents' of a design had a significant influence on the yield of the production process. Reducing the number of gates was an important goal in everyday design practice, often at the expense of, for example, noise immunity.

The number of gate equivalents has now lost its dominant role. Reliability, noise immunity and suitability for a top-down design process have become more important. In this chapter the design of registers is discussed and attention is paid to the selection of a proper type of register for a specific application. Examples are given.

Registers

Figure 12.1.a describes the *basic logic diagram* of a 4-bit register. At the clock pulse (in this case at the 0 → 1 transition) this register loads new data. This data is stored until the next clock pulse and is available during that time at the Q outputs. The IEC symbol for this register is shown in Figure 12.1.b. The clock input is drawn on the *common control block*.

Figure 12.2.a describes, as an example, the *detailed logic diagram* of the 4-bit register SN74175. This register differs in two ways from the logic diagram as specified in Figure 12.1.a. It has:

– a direct reset input on all flip-flops;
– a buffer/inverter in the clock line and in the reset line.

In both cases the input adapter has been added to adapt the fanin. When the logic level on the direct reset input is LOW the register is reset. This is indicated by R(L). When the logic level on the reset input is HIGH this input has *no effect*. The register SN74175 is an old design. In modern ICs a *preparatory reset* is preferred, because of the *poor noise immunity* of *direct acting inputs*.

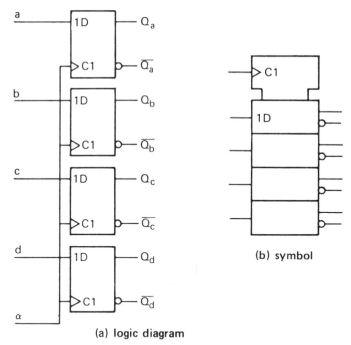

Figure 12.1 Logic diagram of a 4-bit D type register.

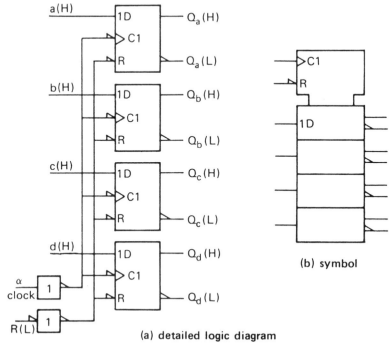

Figure 12.2 4-Bit D type register SN74175.

In the realisation phase the designer must map a logic design (Figure 12.1) onto available components. For Figure 12.2 this means that a constant level HIGH must be fed to the reset input. Because of the noise susceptibility this signal should be connected as closely as possible to the reset input pin of the IC. The clock input needs no adjustment. The circuit SN74175 may then be considered as a *realisation* of the *logic design* in Figure 12.1.

The internal structure of registers

Many registers can perform various functions, for example:

LOAD (LD):	load the data presented on the data inputs;
INHIBIT (INH):	store the data just read;
SHIFT RIGHT (SHR):	shift data one position to the right/down;
SHIFT LEFT (SHL):	shift data one position to the left/up;
RESET (R):	reset, or LD 0...0.

These registers have an *internal organisation* comparable to that depicted in Figure 12.3. Each flip-flop obtains its data input signal from the selector/multiplexer in front of it. All

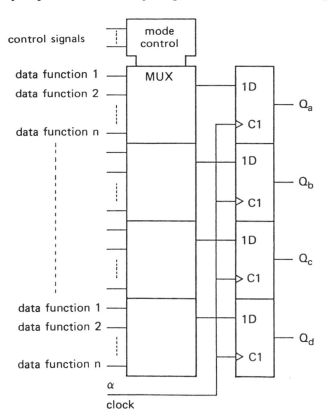

Figure 12.3 Internal structure of registers.

selectors have a *common control*. After the control signals have selected a specific mode, the corresponding data is selected and loaded at the next clock pulse.

With this internal structure flexible registers can be designed. If necessary, additional logic can be placed in front of the selectors, for preprocessing the input data. This enables operations such as

$$Q_dQ_cQ_bQ_{aBIN} \Rightarrow Q_dQ_cQ_bQ_{aBIN} + 1,$$

which is one of the operations of 4-bit counters.

12.2 How to use programmable registers

The register in Figure 12.4 depicts a design of a *programmable* 4-bit shift register. Two mode control signals S_1 and S_0 select the desired mode. The possible functions are:

$S_1S_0 = 0\ 0\ \Rightarrow$ function INHIBIT;

$S_1S_0 = 1\ 1\ \Rightarrow$ function LOAD;

$S_1S_0 = 0\ 1\ \Rightarrow$ function SHIFT RIGHT;

$S_1S_0 = 1\ 0\ \Rightarrow$ function SHIFT LEFT.

How these functions are selected is illustrated in Figures 12.4.a through 12.4.d. That part of the circuit given in bold is necessary to perform the selected function. The remaining components and connections are drawn with thinner lines, as they are not used at that moment. The mode control signals S_1 and S_0 *do the programming*. The clock pulse determines the moment at which the selected function is executed by the register.

The function table

The relation between the values of the mode control signals and the corresponding function is usually specified in a *function table*. For the register in Figure 12.4 this is Table 12.1. Once such a table has been drawn up one no longer has to study the internal structure of the register every time it is used.

S_1	S_0	function
0	0	INHIBIT
0	1	SHIFT RIGHT
1	0	SHIFT LEFT
1	1	LOAD

Table 12.1 Function table of a 4-bit shift register.

Longer registers can be composed by using the 4-bit register in Figure 12.4, with n circuits up to 4n-bit registers. To that end the control inputs S_1 and S_0 of every section have to be connected to the *external control inputs* S_1 and S_0, and the serial inputs with respectively the Q_d and Q_a outputs of the adjoining sections. Table 12.1 then turns out to be applicable to the 4n-bit register as well. In counters, however, the function table of a 4n-bit counter sometimes differs from that of the 4-bit section. In general, extensions necessitate the specification of a *new function table*.

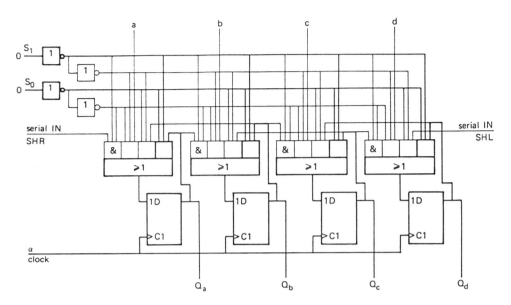

(a) $S_1S_0 = 00 \Rightarrow$ function INHIBIT

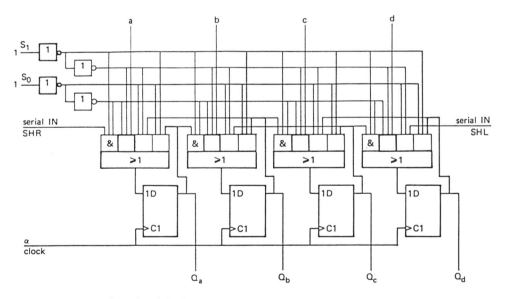

(b) $S_1S_0 = 11 \Rightarrow$ function LOAD

Figure 12.4 Modes of a shift register (*continued on next page*).

Figure 12.4 (continued).

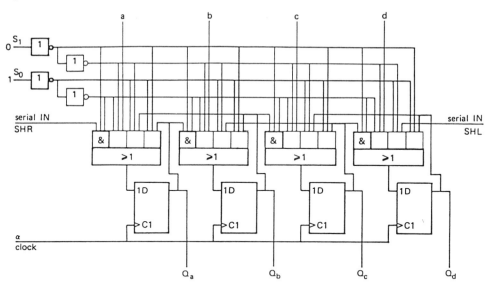

(c) $S_1S_0 = 01 \Rightarrow$ function SHIFT RIGHT

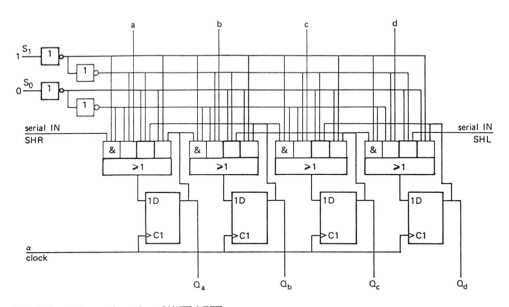

(d) $S_1S_0 = 10 \Rightarrow$ function SHIFT LEFT

Figure 12.4 Modes of a shift register.

In many applications use is not made of every function of the above register. A simpler register can then be chosen. The alternative is to restrict the options of a given programmable register, by combining input signals and/or using a constant level H or L at some of the inputs. The register is then easier to program by the control unit.

Example
In order to obtain a register with only the functions LOAD and INHIBIT we see from Table 12.1 that:

$$S_1 S_0 = 0\,0 \;\Rightarrow\; \text{function INHIBIT};$$
$$S_1 S_0 = 1\,1 \;\Rightarrow\; \text{function LOAD}.$$

It follows that only one control signal is needed; this signal can program S_1 and S_0 simultaneously. (Note: the combined control signal then has a fanin of two inputs!) □

Example
We need a register with the functions LOAD and SHIFT RIGHT. The relevant input combinations are

$$S_1 S_0 = 1\,1 \;\Rightarrow\; \text{function LOAD};$$
$$S_1 S_0 = 0\,1 \;\Rightarrow\; \text{function SHIFT RIGHT}.$$

Only one external control signal S_1 is necessary to select the appropriate function. S_0 is set to 1. □

Note
The function tables in data books are usually specified in terms of H and L. These describe existing circuits. In the design process we often use function tables in 0 and 1 at the logic design level. When the logic functions are specified, they must be mapped onto the H/L levels; this sometimes requires a level conversion. □

12.3 Timing properties of registers

An important criterion in the selection of the proper register type is the way in which the functions are implemented. Sometimes two registers with the same function table have totally different timing properties.

Example
Figure 12.4 specified the *internal structure* of a 4-bit shift register with the facilities:

> LOAD;
>
> SHIFT LEFT/RIGHT;
>
> INHIBIT.

These functions are realised by a 1-out-of-4 selector per flip-flop. For every combination of two mode control signals S_1 and S_0 one of the input gates of each selector is enabled to allow the corresponding data signal to pass. The other gates, which have a 0 on at least one of the inputs, are then disabled.

Figure 12.4.a shows which part of the circuit is used to select the function INHIBIT. This function is realised by feeding back the output of each flip-flop to its input. The input combination is then $S_1 S_0 = 00$.

All of the flip-flops in Figure 12.4 have positive edge-triggered timing. The data signals a through d, the series IN data signals and the mode control signals S_1 and S_0 exercise their influence only on the D input of the flip-flops. Figure 12.5 specifies the timing of all inputs and outputs. No hold time t_h is specified for the input signals. It is 0 ns or even negative, because of the propagation delay time of the selector (see Section 11.8).

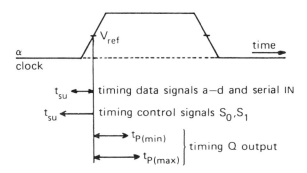

Figure 12.5 Timing specification.

Figure 12.6 shows an older design of this shift register. The logic behaviour of both shift registers, expressed in the formula of Q^{n+1}, is identical. In Figure 12.6 the function INHIBIT is realised by interrupting the clock. With the input combination $S_1 S_0 = 00$ the internal clock line remains at a constant value of 1. The external clock does not come through.

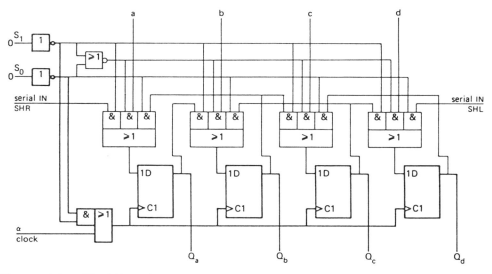

Figure 12.6 4-Bit shift register.

In the design of Figure 12.6 the input signals S_1 and S_0 are only allowed to change when the clock pulse α has the value of 1. If they were able to change while $\alpha = 0$, the transient phenomena in S_1 and S_0 might lead to unintended clock transitions.

In Figure 12.6 the data signals a through d and the series IN signals do not influence the internal clock circuit. They merely control the data inputs. The relevant timing diagram for Figure 12.6 is specified in Figure 12.7. The flip-flops are edge-triggered, but externally the 4-bit shift register has a *pulse-triggered timing*, as far as the control signals S_1 and S_0 are concerned. The data signals have an edge-triggered timing.

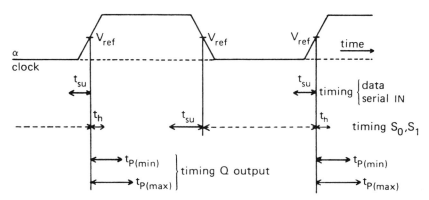

Figure 12.7 Timing specification.

The above-described difference in implementation of the function INHIBIT is, for example, found in the shift registers SN74LS194A and SN74194 (Texas Instruments). The A type is a newer version, with completely edge-triggered timing. □

Note

The edge-triggered or pulse-triggered properties are always specified in relation to the external clock input. When, as above, certain inputs have an edge-triggered timing and others a pulse-triggered timing, the entire circuit is usually called pulse-triggered. The timing analysis can be accomplished more quickly because fewer cases have to be distinguished. Because of this 'simplification' the clock frequency may be lower than strictly necessary. □

Such differences in the timing properties of the components are important in selecting, for example, the type of registers and counters. The data books give little explicit information on this point. In most cases one has to formulate the criteria for the selection of components oneself.

A direct acting reset input is encountered in many register designs. How a reset on a preparatory base can be implemented is described in Section 11.5. Sometimes a *counter* is used instead of a register. Counters often have a preparatory reset input and can also load and store data.

Example

Figure 12.8 shows an example of an implementation of a preparatory set and reset in a 4-

bit shift register (one section shown). The reset does not have to control all of the input gates of the selector. It is sufficient to repress the mode control signals S_1 and S_0. (Note: as shown the reset has a fanin of three inputs.) □

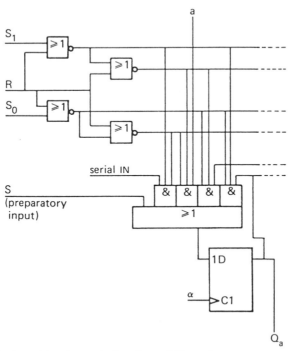

Figure 12.8 Set/reset implementation in a shift register.

12.4 Applications of shift registers

Multiplication and division

The shift operation is part of many processes, e.g. of *multiplication*. When the binary number N of k bits is given,

$$N = a_{k-1}2^{k-1} + a_{k-2}2^{k-2} + ... + a_2 2^2 + a_1 2^1 + a_0 2^0, \tag{12.1}$$

multiplying by 2 results in 2N:

$$2N = a_{k-1}2^k + a_{k-2}2^{k-1} + ... + a_2 2^3 + a_1 2^2 + a_0 2^1 (+ 0.2^0). \tag{12.2}$$

When the bits $a_{k-1}, a_{k-2}, ..., a_0$ are placed in this order in a register, multiplying by 2 means shifting the contents one place to the left. The right-hand side is filled up with a 0. The multiplication of a binary encoded number by 4 is a shift two places to the left, etc. When multiplying by a factor 6, for example, the contents of the register shifted one place must be added to the contents shifted two places. (Note that the result is k+3 bits wide.)

Shifting is an important operation in *division* as well. Division by 2 can be done by shifting the contents of the register in which the binary encoded number is stored one place to the right. The right-hand bit, the least significant bit, is then lost.

From these two examples it follows that the shift operation is important in executing mathematical operations.

Counting to n with shift registers

Simple cyclic counters can be designed by making use of shift registers. The best known is the *ring counter*, consisting of a shift register with overall feedback. The principle of this counter is shown in Figure 12.9.a, while Figure 12.9.b specifies the timing diagram if a single 1 is rotated in the register.

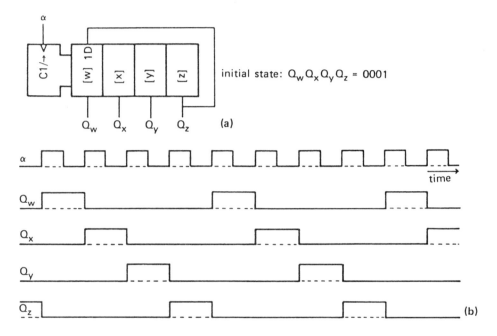

Figure 12.9 Principle and timing diagram of a ring counter.

With n sections one can count cyclically to n in this way. With an initial state other than 10...00 *degeneration of the cycle length* is possible. Thus, a register of six sections with an initial state of 100100 is back in its initial state after three clock pulses.

Application ring counter: divide-by-n counter

In many applications one pulse out of every series of n pulses must be selected. Ring counters are very suitable for this purpose. The original series of pulses is then put on the clock input of the shift register. One of the flip-flop outputs is the output signal. The width of the output pulse varies from minimally one period to maximally $n - 1$ periods of the input pulses, the width being determined by the initial state of the shift register. With the contents 10...00, for example, the width of the output pulse is minimal.

Counting to 2n with shift registers

Counting to 2n is also possible with a feedback shift register of n sections. In this case the $\overline{Q}$ output of the last section rather than the Q output must be fed back (see Figure 12.10). This cyclic counter is known as the *Johnson counter* or *Mobius counter*, or as a *twisted ring counter*. In the initial state WXYZ = 1000 this counter has a cycle length of 2n = 8 steps, as shown in Figure 12.10, for example. Here, too, degeneration of the cycle length is possible.

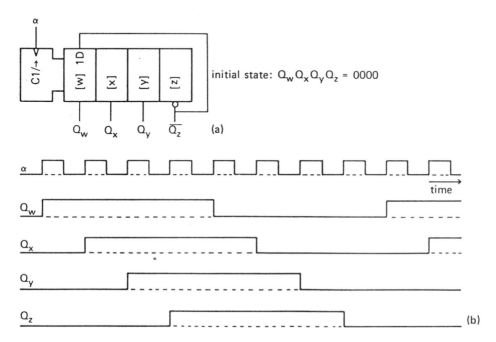

Figure 12.10 Principle and timing diagram of a twisted ring counter.

An important application of the *twisted ring counter* is its use as a *multi-phase clock pulse generator*. Multi-phase clocks are used in larger digital circuits. Each of the clock phases must then remain in accurate relationship to the central clock, and the various clock phases have to be in a fixed mutual relation. With circuits such as those in Figures 12.9 and 12.10 this is done automatically, even when the frequency of the incoming clock varies. These clock pulse generators sometimes have a detection circuit which indicates that the register has reached a wrong internal state due to a fault. Certainly for a clock generator, the heart of every circuit, such a detection facility is advisable.

Modulo-2 feedback shift registers

With a properly selected feedback network shift registers can go through a cycle longer than 2n. Well-known examples are the *modulo-2 feedback shift registers*. In these registers the input signal D_{in} on the input of the first section is

$$D_{in} = \sum_{k=1}^{n} c_k Q_k \text{ (mod-2)}, \tag{12.3}$$

in which $c_k = 1$ if the output Q_k is part of the sum. Otherwise $c_k = 0$. The numerical value of 0 or 1 is assigned to every flip-flop output, depending on its logic state (either the logic 0 or the logic 1 state). The *summation modulo-2* then results in 0 if the number of ones in the summed output signals is even, and 1 if it is odd.

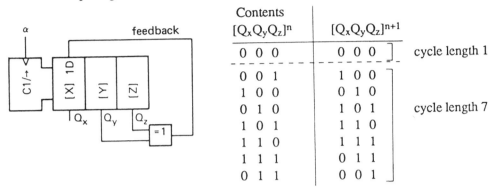

Figure 12.11 Modulo-2 feedback shift register.

The register in Figure 12.11 runs through one of two cycles, one of length 1 (the *zero cycle*) and one of length $2^3 - 1 = 7$ (see the table). The literature gives, for each length n of the register, feedback networks for which the register has the maximum cycle length of $2^n - 1$. Modulo-2 feedback shift registers are frequently used for error correction and for error detection in data transmission and data storage.

12.5 Parallel-series and series-parallel converters

Figure 12.12 sketches the principle of a *parallel-series converter* for an 8-bit data word. After the data has been loaded at clock pulse 0 the first bit a_0 is at the series output. Then the register is switched to the shift mode. At each subsequent clock pulse the next bit appears at the series output. When bit a_7 has appeared at the output the register is reloaded, and a_0 of the next data word appears, etc.

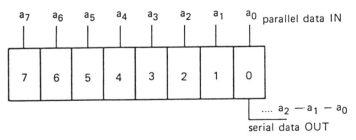

Figure 12.12 Parallel-series converter.

The need to convert a 2n-bit word into two n-bit words often arises in digital systems, for example when a channel is smaller (is fewer bits wide) than a storage medium. Figure

12.13 shows how this problem can be solved by an alternating LOAD and SHIFT RIGHT command.

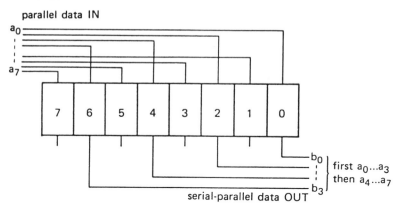

Figure 12.13 Parallel-parallel converter of 8 to 4 bits.

The opposite of parallel-series conversion, *series-parallel conversion*, is also made possible by using registers. Figure 12.14 shows the concept of a circuit which converts two 4-bit words into one 8-bit word, while preserving the order of the bits.

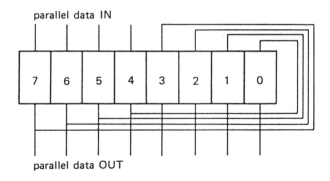

Figure 12.14 Parallel-parallel converter of 4 to 8 bits.

Each of the converters in Figures 12.12 through 12.14 must be made complete by adding a *control unit*, which determines which function (LD, SH, etc.) must be performed. The specification and the realisation of control units will be treated in a second volume.

Counting in parallel-series converters

During a parallel-series conversion of data the contents of the register shifts out and the register becomes empty. This space can be used to count the number of conversion steps. Figure 12.15 illustrates the principle involved.

To count the number of conversion steps one extra section is added to the shift register. This section is initially set to 1 at the same time as the 8-bit data word is loaded. At each SHIFT RIGHT command a 0 shifts into the register on the left-hand side. After loading,

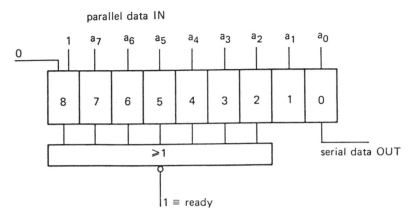

Figure 12.15 Counting shift register.

the output signal of the NOR gate is 0. Seven clock pulses later its output becomes 1. Data bit a_7 is then at the output. Now the next 8-bit data word can be loaded.

The principle of the *counting shift register* can be applied in series-parallel conversion as well.

References

1. A.C. Davies, *Properties of Waveforms Obtained by Nonrecursive Digital Filtering of Pseudo-Random Binary Sequences*, IEEE Tr. on Computers, Vol. C-20, 1971, pp. 270–281.
2. J.W. Golomb, *Shift Register Sequences*, Aegean Park Press, Laguna Hills Ca., 1982.
3. J.H. Lindholm, *An Analysis of the Pseudo-Randomness Properties of Subsequences of Long m-Sequences*, IEEE Tr. on IT, Vol. IT-14, 1968, pp. 569–576.
4. R.M.M. Oberman, *Disciplines in Combinational and Sequential Circuit Design*, McGraw-Hill, New York, 1970, Ch. 16.
5. J.B. Peatman, *Digital Hardware Design*, McGraw-Hill, New York, 1980.
6. W.W. Peterson and E.J. Weldon, *Error-Correcting Codes*, Wiley, New York, 1972.
7. H.H. Roth, *Linear Binary Shift Register Circuits Utilizing a Minimum Number of Mod-2 Adders*, IEEE Tr. on IT, Vol. IT-11, 1965, pp. 215–220.
8. G.H. Tomlinson and P. Galvin, *Generation of Gaussian Signals from Summed m-Sequences*, Electronics Letters, Vol. 21, pp. 521–522.

13
Counters

13.1 Introduction

Some form of counting occurs in many processes. In some of these, events have to be counted, either with or without a signal, when a certain number of events have passed. In other processes the passing of a period of time has to be detected. This form of counting is usually based on a periodic signal with a fixed repetition frequency. Counters are also applied to keep track of how far a process has progressed. A well-known example is the program counter in a computer. Consequently, counters come in a great variety, depending on the application.

In designing counters, concepts such as architecture, implementation and realisation are used. The *architecture specification* lays down which requirements the counter has to meet, i.e. how the counter should manifest itself to the user. The architecture describes the *external behaviour* of the counter. The architecture specification not only includes items such as the desired counting frequency and the capacity of the counter, but also indicates how the counter should interact with its environment.

During *implementation* a structure that can realise the architecture is determined. If necessary, the internal code is specified as well.

Finally, in the *realisation phase* the components are selected and their characteristics are examined in relation to the chosen structure. It is then investigated whether the specification can be met.

The sequence of activities described here is followed in the design of most digital circuits. In what follows, we start by introducing components for counters and their structures. How counters interact with their environment is also given a great deal of attention.

Counters can be described by the *finite state model*. The counter is then specified in a state diagram. This diagram describes how and under which conditions the counter goes through its successive states. The realisation procedure consists of:

– selection of the state assignment;

– forming the next-state formulas;

– mapping these formulas onto hardware.

This procedure, based on the finite state model, has been explained in Chapter 10. For larger counters this procedure is impractical. In that case one opts for modular design.

The modular structure of counters

Larger counters are usually composed of *modules*, based on the base of the number system applied. In digital technique all digits are represented in a binary code. Compare this to the BCD code. With n bits per module numbers d, which lie between

$$0 \leq d \leq 2^n - 1,$$

can be represented.

The following describes how a cyclic counter of three decades runs through the cycle of 000_{DEC} through 999_{DEC}.

Start:

$000 \to 001 \to 002 \to 003 \to 004 \to 005 \to 006 \to 007 \to 008 \to 009 \to$
$010 \to 011 \to 012 \to 013 \to 014 \to$ ....
.... $096 \to 097 \to 098 \to 099 \to$
$100 \to 101 \to 102 \to 103 \to 104$ ....
....
.... $996 \to 997 \to 998 \to 999 \to$
$000 \to$ repetition of the cycle.

This counter can be composed of modules, one for each decimal digit. For m modules of base g which complete a full cycle it holds that:

– Each module cyclically goes through the same subcycle $0 \to g-1$.

– The least significant module, module no. 1, runs through the subcycle g^{m-1} times.

– Module no. k runs through the subcycle g^{m-k} times.

– A module counts one step if all preceding modules have the content g-1.

If a counter goes downwards through its code, module k counts if all preceding modules have the content 0.

Conclusion

A counter can be composed of modules which all run through the *same subcycle*. Besides, information is *transferred* between the modules. Depending on the contents of the preceding modules this indicates whether or not a module has to step.

Figure 13.1 illustrates the structure of a decimal counter. The events to be counted are signalled to the least significant decade as count pulses. There is a carry between the decade sections. This modular structure is found in most counters. The differences have to do with the way in which the *carry mechanism* has been implemented, which can be done in several ways. The carry mechanism largely determines the properties of the counter, as will be shown.

In the BCD code the size of a module is more or less fixed. A *4-bit module* is an obvious choice, with each module representing one decimal digit. For the BIN code the basic module consists of one bit. Binary counters on printed circuit boards are usually composed of 4-bit or 8-bit modules. These have the advantage of a higher packing density.

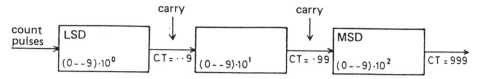

Figure 13.1 Global structure of a 3-decade counter.

Counters and their environment

Counters interact strongly with their environment. To begin with, the events to be counted originate in that environment. Besides, the environment reacts to the counters, for example to their full/empty reports. When discussing the properties of counters, the way they interact with their environment should also be considered. To that end the following concepts are introduced.

Stand alone conditions

The counter is regarded as being separate from its environment. The maximum counting frequency is determined entirely by the rate at which the counter is able to process the count pulses. The fact that the environment also needs some time to communicate with and to adjust to the counter is left out of consideration.

IC manufacturers follow this view in their data books. The parameter f_{max} (maximum counting frequency) specified in this way does not give a realistic impression of the possibilities in a particular application.

Clock mode environment

The moments at which the events to be counted appear are related to the system clock of the environment. The following cases are distinguished:

– events represented by the conditional appearance of the clock pulse;

– the clock pulses (edges) indicate the moments at which events are possible, while a *count enable* signal indicates whether or not the event occurred.

This outline implies that a counter must have processed a command before the next clock pulse is given *and* that the environment is able to react to a message from the counter before that clock pulse, for example an overflow indication of the counter.

Pseudo clock mode environment

Interaction as with the clock mode. Some settings require more time than one clock period. More clock cycles are reserved for these when the control is specified. The rate at which commands are given to the counter should be adjusted to take account of this. We will evaluate the pros and cons of this form of interaction in Sections 13.9 and 13.12.

The rest of this chapter is divided as follows:
– Simple asynchronous and synchronous counters and their interaction with their environment.
– Modular structures of 4n-bit counters and their interaction with the environment.
– The derivation of formulas and characteristics of 4n-bit counters from those of standard 4-bit counter modules.
– Examples of applications.

13.2 Asynchronous binary counters

By way of introduction to this subject, we shall examine the design of a 4-bit binary counter. The timing diagram in Figure 13.2 specifies its function. In the diagram all shifts resulting from propagation delay times have been omitted. The flip-flops composing the counter react around the $0 \to 1$ transition of the clock or count pulse.

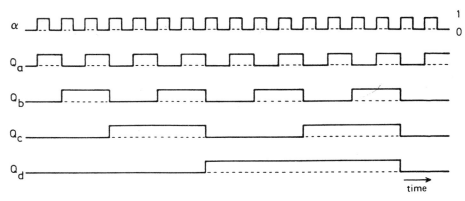

Figure 13.2 Timing diagram of a binary counter.

Section A of the counter with output Q_a is the least significant section. It follows from the timing diagram that:

- Flip-flop A with output Q_a toggles on the clock/count pulse α, on its $0 \to 1$ transition to be exact.

- Flip-flop B with output Q_b toggles on the $1 \to 0$ transition of Q_a or, which boils down to the same thing, on the $0 \to 1$ transition of $\overline{Q}_a$.

- Each subsequent flip-flop toggles on the $0 \to 1$ transition of $\overline{Q}$ of its immediate predecessor.

This principle can be extended to binary counters in n bits. Every next flip-flop toggles on the $0 \to 1$ transition of $\overline{Q}$ of its predecessor. The circuit in Figure 13.3 realises the behaviour described above. The circuit consists of a chain of J-K flip-flops that have been switched so that they toggle on each clock pulse. For $J = K = 1$ the equation

$$Q^{n+1} = [J\overline{Q} + \overline{K}Q]^n = \overline{Q}^n$$

holds. For $J = K = 0$ the flip-flop inhibits, $Q^{n+1} = Q^n$.

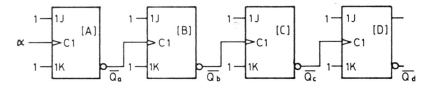

Figure 13.3 Asynchronous binary counter.

The counter in Figure 13.3 has the following characteristics:

- The counter consists of *identical (1-bit) modules* and can be extended immediately.
- The *internal information transport* is based on pulses to the clock input of the next section. Owing to this construction an internal 'ripple' may arise, which passes through a number of consecutive sections.
- The *maximum counting frequency* is determined by the characteristics of flip-flop A. All subsequent flip-flops in the chain count at a lower frequency. Under stand alone conditions these counters are extremely fast.
- For *longer counters* the next count pulse may, in principle, be given before the tail of the counter has completely processed the previous count pulse.

The advantage of this counter structure lies in its simplicity. The counter requires very few components. Nevertheless, this setup has its disadvantages. Counters with this internal structure fit badly into a synchronous (clock mode) environment (see Figure 13.4).

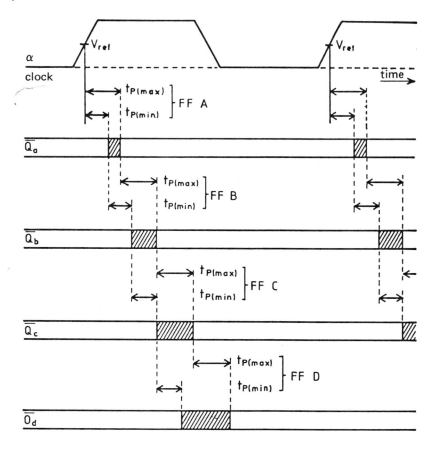

Figure 13.4 Timing of an asynchronous binary counter.

All flip-flops react one after the other. After the count pulse the first to react is the output of flip-flop A, between $t_{P(min)}$ and $t_{P(max)}$. Flip-flop B, in turn, reacts between $t_{P(min)}$ and $t_{P(max)}$ later, with regard to flip-flop A. The largest reaction time is, for example, found at the transition of 01...11 into the content 10...00. In order to be able to detect a certain state the counter has to be fully stabilised at that moment. If the counter has been applied within a clock mode environment, it should at least be stable during the setup time around the next clock pulse. One should also keep in mind the time required for the detection circuit.

Example

An asynchronous binary counter is embedded in a clock mode environment. The function of the counter is to detect how long a specific pattern or condition is present. To that end a count pulse is derived from the clock pulse α, as long as this pattern is detected. This count pulse arrives at most $t_{P(max)}$(clock-to-counter) later. Assume a maximal delay of

$$t_{P(max)}(\text{clock-to-counter}) = 25 \text{ ns.}$$

The specifications of the flip-flops composing the counter are:

$$t_{su} = 10 \text{ ns,} \quad t_{P(min)} = 8 \text{ ns,} \quad f_{max} = 30 \text{ MHz,}$$
$$t_h = 5 \text{ ns,} \quad t_{P(max)} = 20 \text{ ns.}$$

The detection circuit for the 'final state' takes $t_{P(min)} = 10$ ns, $t_{P(max)} = 20$ ns. It is subsequently assumed that the environment must be able to process the final state at the next clock pulse. To that end the detection circuit sets a flip-flop, clocked by clock α. This flip-flop has a setup time $t_{su} = 10$ ns.

Within this clock mode environment Figure 13.5 shows the *maximum clock/count pulse frequency* for $1 \leq n \leq 20$ sections. The maximum frequency f_{max} meets the condition

$$f_{max} < \frac{1}{t_{P(max)}(\text{clock-to-counter}) + t_{P(max)}(\text{counter}) + t_{P(max)}(\text{detection}) + t_{su}(\text{FF})} \text{ MHz}$$
(13.1)

in which $t_{P(max)}$(counter) is:

$$t_{P(max)}(\text{counter}) = n \cdot t_{P(max)}(\text{FF}). \tag{13.2}$$

For comparison the maximum frequency of the counter under stand alone conditions has been included in Figure 13.5. It has been set at 30 MHz.

We note that f_{max} in (13.1) is the maximum frequency the environment imposes on the counter. Under all conditions the counter itself has the same maximum frequency. We will show below schemes in which f_{max} of the counter itself also depends on the number of sections of the counter (see Figure 13.8, for example). □

Note

The control method described above is very common in asynchronous counters. With synchronous counters it is better to use counters with an enable input. The delay time

$$t_{P(max)}(\text{clock-to-counter})$$

may then be omitted. □

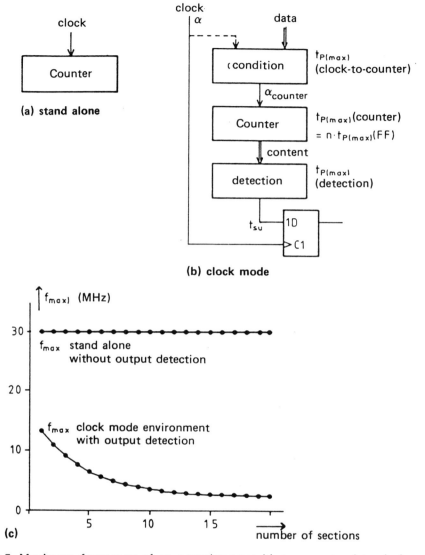

Figure 13.5 Maximum frequency of an asynchronous binary counter (stand alone/clock mode).

Conclusion

The maximum counting frequency of a counter with the internal structure of Figure 13.3 is determined by

– the characteristics of the counter;

– the way it interacts with its environment.

The environment may play a dominant role in this.

13.3 Synchronous binary counters

The counter in Figure 13.6 is a *synchronous binary counter*. The counter has been designed in accordance with the design method for clock mode sequential circuits as described in Chapter 10. It realises the behaviour specified in Figure 13.2. The formulas of the input signals are

$$J_a = K_a = 1, \tag{13.3.a}$$
$$J_b = K_b = Q_a, \tag{13.3.b}$$
$$J_c = K_c = Q_b Q_a, \tag{13.3.c}$$
$$J_d = K_d = Q_c Q_b Q_a, \tag{13.3.d}$$

etc. The formulas for longer counters can easily be obtained by extrapolation.

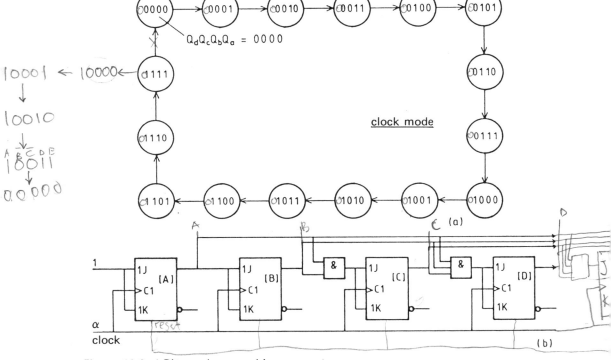

Figure 13.6 4-Bit synchronous binary counter.

The characteristics of this counter are:

- All flip-flops are under *direct control* of the *clock/count pulse*.
- The *output reaction* of all flip-flops lies between $t_{P(min)}$ and $t_{P(max)}$, as specified after each active edge of the clock.
- All information is transferred on a *preparatory basis*.
- With a view to these characteristics the counter fits well into a *clock mode environment*.
- The counter can easily be extended.

When calculating the *maximum clock frequency* one should keep in mind whether the counter is stand alone or used within a clock mode environment. In the first case f_{max} is limited by

$$f_{max} < \frac{1}{t_{P(max)}(FF) + t_{P(max)}(AND) + t_{su}(FF)} \text{ MHz} \quad (13.4)$$

and by the maximum clock frequency of the flip-flops. If it is assumed that there are no restrictions on the number of inputs of an AND gate, f_{max} does not depend on the number of sections.

Within a clock mode environment, with state detection, f_{max} is limited by

$$f_{max} < \frac{1}{t_{P(max)}(FF) + \max[t_{P(max)}(AND), t_{P(max)}(\text{detection})] + t_{su}(FF)} \text{ MHz}. \quad (13.5)$$

This time, too, f_{max} does not depend on the number of sections. When the AND gates do not have an unlimited number of inputs they have to be composed at two or more levels. This causes a reduction of f_{max}. (Nor is $t_{P(max)}$ of an AND gate entirely independent of its number of inputs, in practice.)

When deriving (13.5) it was assumed that the propagation delay time of the counter is the dominating factor in the interaction with the environment.

Example

With flip-flops that satisfy the specification given in Section 13.2 and with

$$t_{P(max)}(AND) = 15 \text{ ns}, \; t_{P(max)}(\text{detection}) = 20 \text{ ns}$$

for f_{max} under stand alone conditions we find

$$f_{max} < \frac{1}{20 + 15 + 10 \text{ (ns)}} = 22.2 \text{ MHz}, \quad (13.6)$$

and in a clock mode environment, with state detection,

$$f_{max} < \frac{1}{20 + 20 + 10 \text{ (ns)}} = 20 \text{ MHz}. \quad (13.7)$$

In principle, these values do not depend on the number of sections of the counter. If we compare these results with those of the previously discussed asynchronous counter, it turns out that asynchronous counters used in the stand alone mode are generally faster. In a clock mode environment, it is just the other way round. See Figure 13.7. (Section 13.9 gives a further explanation of the counter-environment relation.) □

Note

Designing a counter in the BCD code reveals a further difference between the synchronous and the asynchronous approach. An asynchronous counter requires a special design with regard to possible transition phenomena which may be generated by the circuit that propagates a count pulse to the next section. Transition phenomena on the carry signals are disastrous! □

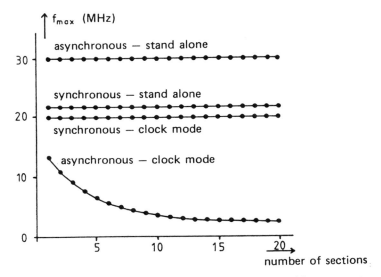

Figure 13.7 Comparison of f_{max} of asynchronous and synchronous binary counters.

13.4 Modular structure of synchronous binary counters

With a structure as in Figure 13.6 the number of inputs on the AND gates increases proportionally to the length of the counter. Figure 13.8 shows an alternative structure, which is modular.

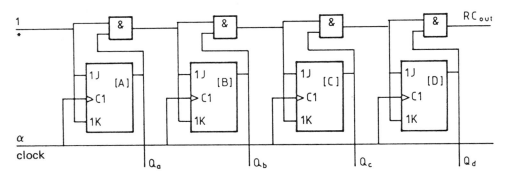

Figure 13.8 Modular synchronous binary counter.

The logic properties are equal to those of the counter in Figure 13.6. For the maximum frequency (stand alone) it holds that:

$$f_{max} < \frac{1}{t_{P(max)}(FF) + (n-1) \cdot t_{P(max)}(AND) + t_{su}(FF)} \text{ MHz.} \qquad (13.8)$$

With an increasing number of sections f_{max} also strongly decreases under stand alone conditions.

Between these extremes, *parallel carry* according to Figure 13.6 and *serial carry* as in Figure 13.8, there are compromises. Usually the carry logic per 4-bit (n-bit) section is implemented in parallel. In that case RC_4, RC_8 etc., are available (see Figure 13.9.a). When the carry between the 4-bit sections is processed in series, Figure 13.9.b, the term $(n-1) \cdot t_{P(max)}$ in the denominator of (13.8) may be roughly divided by 4. There is clear evidence of counter acceleration!

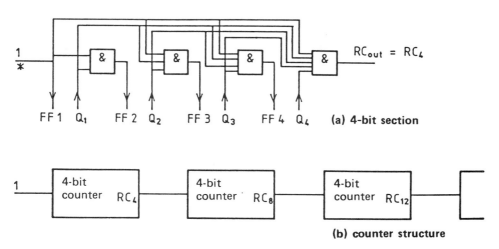

Figure 13.9 Series-parallel structure of the carry.

Note

Instead of a serial transport the network of Figure 13.9.a can be used, again over four 4-bit sections. For a 16-bit counter the entire internal carry is then constructed in two levels. Analogous to a similar construction in adders (Sections 7.3/4) such counters are also called 'counters with *look-ahead carry*'. □

Counters with enable input

Counters in a clock mode environment almost always have an *enable input*. With such an input counting may be started (enabled) or stopped (disabled). In recent designs the enable operates on a *preparatory basis*; with regard to the clock the enable then has a (positive) edge-triggered timing. Older designs often disable with 'clock interruptions'. These designs are very susceptible to noise. (Also see Section 12.3).

In the counters in Figure 13.8 and Figure 13.9 a 'counter enable' input can be implemented by replacing the fixed 1 by the enable signal at the inputs marked *. In Figure 13.8 the result of this is that an enable/disable command has to set the entire carry network. This means that the setup time

$$t_{su}(\text{enable-to-clock})$$

will be rather long. In practice, therefore, the construction of Figure 13.10 is often used.

Each section in Figure 13.10 has two enable inputs. In data books and in the literature these are denoted as 'enable P' and 'enable T'. The enable P input is used to process the EN_{in}-signal in parallel. The enable T input is used to construct the serial part of the carry, beginning with the first section. In practice, the construction of the carry mechanism with

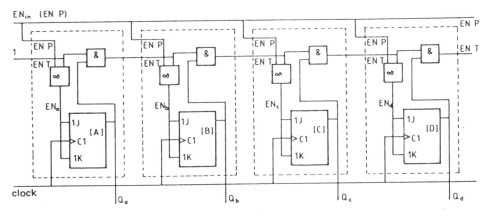

Figure 13.10 Binary counter with parallel enable.

enable P and enable T is especially used in *4-bit counter sections*. Internally, these 4-bit counters then have a parallel carry network. Between these 4-bit sections the carry is formed by means of an enable P/T network. For examples and for the frequency characteristics of the counters see Section 13.5 and further.

Note
The counter in Figure 13.6 can easily be extended with an enable input. This can be done directly, using the input denoted as 1. A small extension here as well makes a network which is *fully constructed in parallel* a possibility. □

Conclusions
Fully synchronous counters (clock mode) can be applied in several ways:

– with the pulses to be counted fed directly to the clock/count pulse input of the counter;

– with counter control by the enable inputs.

In the latter case the system clock (environment) is connected directly at the clock inputs of the counter sections. The events to be counted have to be converted into an enabling signal, viz. per event in an enable signal during one clock period. In addition, the enable signal should be synchronised, i.e. it is not allowed to change within the t_{su}-t_h intervals. In a clock mode environment one almost always applies control by enable inputs.

13.5 4-Bit binary counters

The standard width of counters for PCB designs is four (sometimes eight) bits. We shall discuss some examples of standard 4-bit chips. The following design is the basis of many ICs.

The counter design in Figure 13.11 consists of:

– Four flip-flops, which encode the present state of the counter ($Q_d Q_c Q_b Q_a$ = 0000 through 1111). The flip-flops are under direct control of the clock α.

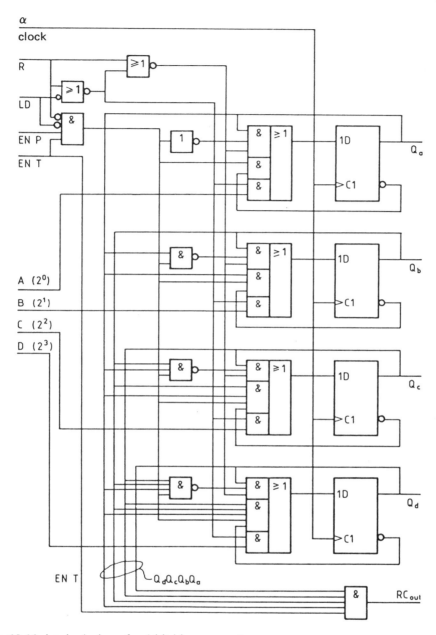

Figure 13.11 Logic design of a 4-bit binary counter.

- Four selectors, which are set with control signals for loading the data corresponding to the selected function.
- Gates, which determine the value of $[Q_d Q_c Q_b Q_{a \text{ BIN}} + 1]^{n+1}$ from the present state $[Q_d Q_c Q_b Q_{a \text{ BIN}}]^n$. Other gates derive the internal control signals for the selectors from the input signals R, LD, EN P and EN T.

All control signals and all data signals influence the state of the flip-flops by means of their D inputs. The counter has a positive *edge-triggered timing*.

Function tables

How this counter is operated can easily be deduced from Figure 13.11. The relation between the desired function and the corresponding value of the control signals is described in a *function table*.

The 4-bit counter in Figure 13.11 has two enable inputs EN P and EN T, with which a carry mechanism as described in Figure 13.10 can be realised. In every 4-bit section the carry is generated in parallel. The EN P/EN T construction can be applied over 4-bit sections as well. This requires EN T to be able to control the RC_{out} output gate directly (for the serial part of the carry mechanism). *The signal EN T therefore directly controls the RC_{out} output*. For the 4-bit counter itself EN T controls on a *preparatory basis* (clock mode).

The function table of the counter is given in Table 13.1. In putting this table together we have assumed that it must always be possible to test the RC_{out} signal, even between the clock pulses. If this is not necessary, for example with a reset command that does not depend on the present state of the counter, Table 13.2 may be used. It contains more don't cares.

Function	R	LD	EN P	EN T	ABCD
Reset/RC_{out}	1	–	–	1	----
Load/RC_{out}	0	1	–	1	data
Count/RC_{out}	0	0	1	1	----
Inhibit/RC_{out}	0	0	0	1	----

Table 13.1 Function table, with the possibility to test RC_{out}.

Function	R	LD	EN P	EN T	ABCD
Reset	1	–	–	–	----
Load	0	1	–	–	data
Count	0	0	1	1	----
Inhibit	0	0	0	–	----
Inhibit	0	0	–	0	----

Table 13.2 Function table, without the possibility to test RC_{out}.

Explanation

When the *reset signal* R has the value R = 1, all input gates of each selector are blocked by at least one 0. At the next clock pulse the counter will then arrive in the state $Q_dQ_cQ_bQ_a$ = 0000. During a *reset command* with R = 1 the other control signals are don't care. The reset signal is *dominant* (Table 13.2). When, during a reset command, another action in the circuit depends on the present state, $Q_dQ_cQ_bQ_a = 1111$ for example, the setting according to Table 13.1 has to be used (EN T = 1). □

The *load signal* LD is, in turn, dominant over EN P and EN T. Then

$$[Q_d Q_c Q_b Q_a]^{n+1} = [ABCD]^n.$$

Should one also want to be able to test on the present state $[Q_d Q_c Q_b Q_a]^n = 1111$, the setting should again be with EN T = 1. Compare Table 13.1 with Table 13.2.

A count command is given with

R, LD, EN P, EN T = 0, 0, 1, 1.

The counter counts *cyclically* from $Q_d Q_c Q_b Q_a = 0000$ through 1111. The input signals of the D flip-flops are:

$$D_a = \overline{Q_a}, \qquad (13.9.\text{a})$$

$$D_b = \overline{Q_a} \cdot Q_b + Q_a \cdot \overline{Q_b}, \qquad (13.9.\text{b})$$

$$D_c = \overline{Q_b Q_a} \cdot Q_c + Q_b Q_a \cdot \overline{Q_c}, \qquad (13.9.\text{c})$$

$$D_d = \overline{Q_c Q_b Q_a} \cdot Q_d + Q_c Q_b Q_a \cdot \overline{Q_d}. \qquad (13.9.\text{d})$$

The formulas (13.9) correspond to a *binary counter*. Each section counts (divides by two) if all its predecessors are in the state Q = 1.

For the function *inhibit* it is sufficient for one of the enable inputs to have the value 0. The other signal is then automatically a don't care, unless the ripple carry output is also being tested.

Note

In practice *double commands* to counters occur frequently. Some commands can be combined, such as 'LD data' and 'test on full'. On the subsequent clock pulse the control unit can then react to 'full', while the counter itself is again initialised. □

The 4-bit binary counter SN74LS163A

The *logic design* described above has been implemented in the circuit SN74LS163A (Texas Instruments), of which Figure 13.13 shows the detailed logic diagram. The counter is realised in TTL technology. For *fanin* and *fanout* reasons some additional inverters have been used. These are marked *. As a result the reset signal R and the load signal LD have *to be inverted*.

Once again we are confronted with the fact that *logic symbols* are used in a detailed logic diagram for components that have been added for *electrical reasons*. This is common practice. It often makes detailed logic diagrams hard to interpret, the more so because it is not stated which components have a logic function and which another (fanin/out, signal delay, etc.).

Table 13.3 describes the function table in H and L. If no tests have to be performed on the RC_{out} signal, Table 13.4 may be applied. This table allows some more don't cares. The H and L added to the signal names in the table heading refer to the signal level at which the corresponding name/proposition is true. In function tables in 0/1 the heading will read, for example, $\overline{R}$, $\overline{LD}$, etc.

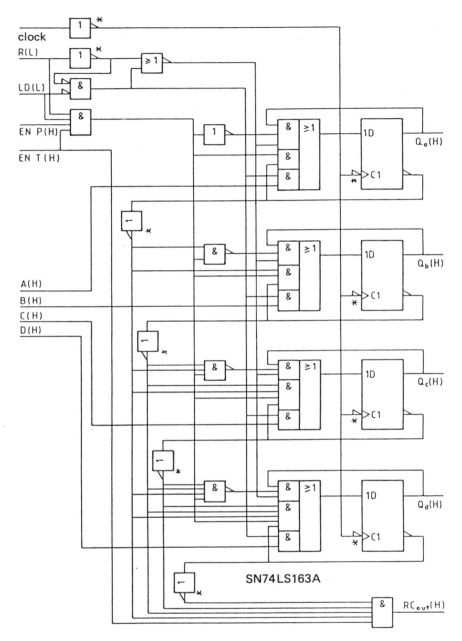

Figure 13.12 Detailed logic diagram of the 4-bit binary counter SN74LS163A.

Timing properties

The counter has two groups of input signals, *data signals* and *control signals*. These exercise their influence through the D input of the flip-flops. The signals of these two groups have *positive edge-triggered* timing with regard to the clock, specified by t_{su} and

Function	Control in L and H				
	R(L)	LD(L)	ENP(H)	ENT(H)	ABCD(H)
Reset/RC$_{out}$	L	–	–	H	----
Load/RC$_{out}$	H	L	–	H	data
Count/RC$_{out}$	H	H	H	H	----
Inhibit/RC$_{out}$	H	H	L	H	----

Table 13.3 Function table SN74LS163A, with the possibility to test RC$_{out}$.

Function	Control in L and H				
	R(L)	LD(L)	EN P(H)	EN T(H)	ABCD(H)
Reset	L	–	–	–	----
Load	H	L	–	–	data
Count	H	H	H	H	----
Inhibit	H	H	L	–	----
Inhibit	H	H	–	L	----

Table 13.4 Function table SN74LS163A, without the possibility to test RC$_{out}$.

t_h. (In Figure 13.13 $t_h = 0$, because of the propagation delay of the logic between the external inputs and internal D inputs.) The Q outputs change between $t_{P(min)}$ and $t_{P(max)}$. See Figure 13.13.

The output signal RC$_{out}$ is controlled by:

– the Q outputs (controlled by the clock);

– the EN T input (directly to RC$_{out}$).

Therefore two sets of timing parameters are specified for the RC$_{out}$ signal:

– $t_{P(min)}$ and $t_{P(max)}$ referring to the clock;

– $t_{P(min)}$ and $t_{P(max)}$ referring to EN T.

All of this is illustrated in Figure 13.13.

Loops

Because of the direct influence of EN T on RC$_{out}$ a *loop* may be formed between a data path in which the counter is used and the control unit that sets the counter. This problem is frequently encountered where control signals have a direct influence on outgoing (test) signals. This influence may be present in the static state, i.e. intended by the designer, and may also result from transition phenomena during setting (dynamic state). See also Section 13.11.

The 4-bit binary up/down counter SN74LS169A

Figure 13.14 describes the detailed logic diagram of the 4-bit up/down counter SN74LS169A (Texas Instruments). This counter is able to count in two directions. The direction is set by an *up/down control signal*.

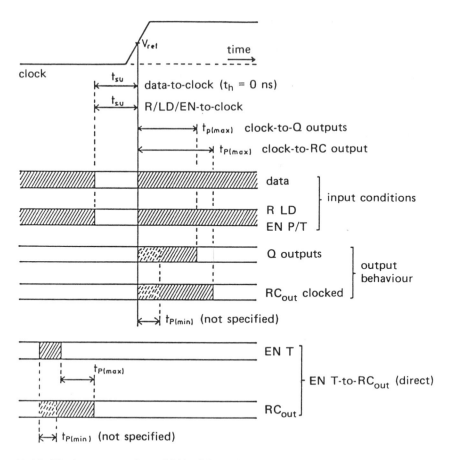

Figure 13.13 Timing properties of SN74LS163A.

In an upward direction the signals on the D inputs conform to formulas (13.9), and downwards to formulas (positive logic) (13.10).

$$D_a = \overline{Q_a}, \qquad (13.10.a)$$

$$D_b = \overline{\overline{Q}_a \cdot Q_b} + \overline{Q}_a \cdot \overline{Q}_b, \qquad (13.10.b)$$

$$D_c = \overline{\overline{Q_b \overline{Q}_a} \cdot Q_c} + \overline{Q}_b \overline{Q}_a \cdot \overline{Q}_c, \qquad (13.10.b)$$

$$D_d = \overline{\overline{Q_c \overline{Q}_b \overline{Q}_a} \cdot Q_d} + \overline{Q}_c \overline{Q}_b \overline{Q}_a \cdot \overline{Q}_d. \qquad (13.10.d)$$

The conversion (compare (13.9) with (13.10)) is done by transferring the signals $\overline{Q}_a$ through $\overline{Q}_d$ instead of Q_a through Q_d. Selectors are used for this conversion. The control circuit of flip-flop A does not change.

In the up mode the RC_{out} signal must have the value 1 if

$$Q_d Q_c Q_b Q_a = 1111,$$

and in the down mode if

$$Q_d Q_c Q_b Q_a = 0000.$$

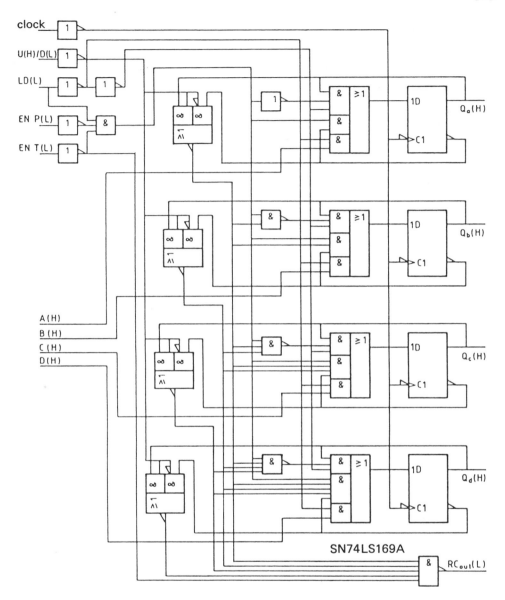

Figure 13.14 Detailed logic diagram of the 4-bit binary up/down counter SN74LS169A.

The formula for the RC_{out} signal is

$$RC_{out} = Q_d Q_c Q_b Q_a \cdot U/D \cdot EN\ T + \overline{Q_d}\,\overline{Q_c}\,\overline{Q_b}\,\overline{Q_a} \cdot \overline{U/D} \cdot EN\ T. \qquad (13.11)$$

In such cases this can imply that changing the counting direction suppresses the outgoing carry! See 'loops' above as well as Section 13.11.

For the rest the circuits SN74LS163A and SN74LS169A are based on the same concept.

Function tables

Table 13.5 describes the function table for control with RC$_{out}$ test. Table 13.6 does the same for those cases in which the control does not test on the outgoing RC$_{out}$.

Function	Control in L and H				
	U(H)/D(L)	LD(L)	EN P(L)	EN T(L)	ABCD(H)
Reset	not present				
Load/up mode	H	L	–	L	data
Load/down mode	L	L	–	L	data
Count up	H	H	L	L	----
Count down	L	H	L	L	----
Inhibit/up mode	H	H	H	L	----
Inhibit/down mode	L	H	H	L	----

Table 13.5 Function table of SN74LS169A, with the possibility to test RC$_{out}$.

Function	Control in L and H				
	U(H)/D(L)	LD(L)	ENP(L)	ENT(L)	ABCD(H)
Reset	not present				
Load	–	L	–	–	data
Count up	H	H	L	L	----
Count down	L	H	L	L	----
Inhibit	–	H	H	–	----
Inhibit	–	H	–	H	----

Table 13.6 Function table SN74LS169A, without the possibility to test RC$_{out}$.

Note
Some counters, such as the SN74LS169A, have inverted inputs for the enable signals EN P and EN T and an inverted output for the ripple carry signal RC$_{out}$. In logic diagrams and in the corresponding function tables the designation (L) is added to the name. Other counters, such as the SN74LS163A, have normal inputs/outputs.
All of this is not without consequences for an external carry propagation network. Some level adjustments have to be made to the *logic diagrams* shown previously. It is recommended not to mix counters of different types (with respect to input/output polarity). □

13.6 4-Bit BCD counters

The detailed logic diagram of a BCD counter does not deviate much from its binary equivalent. In an 'up counter' in the BCD code there is a jump from state 9 (1001) to state 0 (0000). The RC$_{out}$ signal (positive logic) is then

$$RC_{out} = Q_d \overline{Q_c} \overline{Q_b} Q_a, \tag{13.12}$$

usually realised as

$$RC_{out} = Q_d Q_a \tag{13.12.a}$$

because states 10 (1010) through 15 (1111) do not occur. The latter is not advisable because of *error propagation*. If the counter, for any reason, comes into states 10 through 15, a RC_{out} signal may be produced.

Figure 13.15 shows the detailed logic diagram of the up/down BCD counter SN74LS668 (Texas Instruments). The corresponding function tables are the same as Tables 13.5 and 13.6 for its binary equivalent.

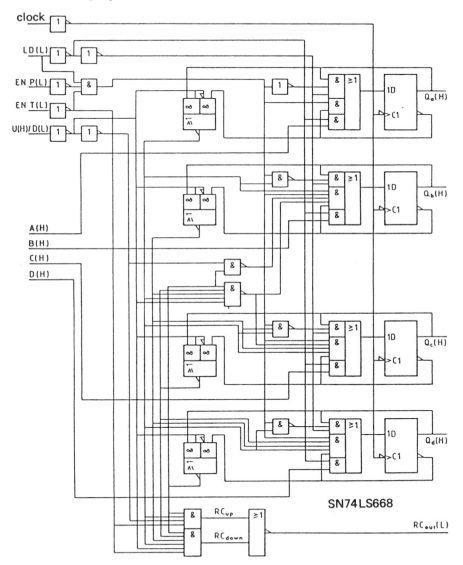

Figure 13.15 Detailed logic diagram of the BCD up/down counter SN74LS668.

The carry network has the *logic structure* of Figure 13.16. With respect to the *timing* of the RC_{out} output it should again be noted that not only EN T but also the control signal U/D has a direct influence on the RC_{out} output.

The use of a series-parallel carry network (EN P-EN T) has consequences for the setup time during commutation. Examples are discussed in the following section.

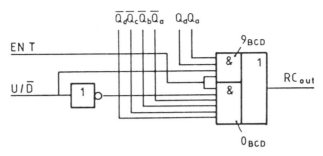

Figure 13.16 Logic structure of RC_{out} network.

13.7 The structure of 4n-bit counters

Before examining 4n-bit counters we shall first introduce a model for a 4-bit counter, in a clock mode environment.

The model of a 4-bit counter

Figure 13.17 shows the *internal logic structure* of a 4-bit counter, as well as its relation with the environment. We can see that:

– All flip-flops are under direct control of the clock.

– The counter exchanges data with other components in the so-called data path.

– Control information comes from and goes to the circuit's control unit.

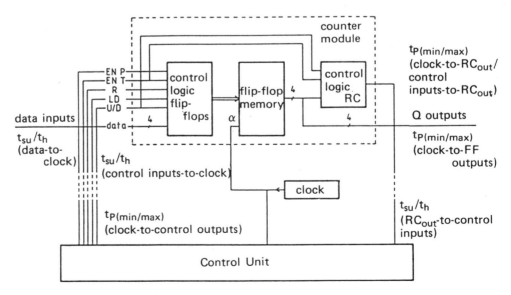

Figure 13.17 4-Bit counter, in clock mode environment.

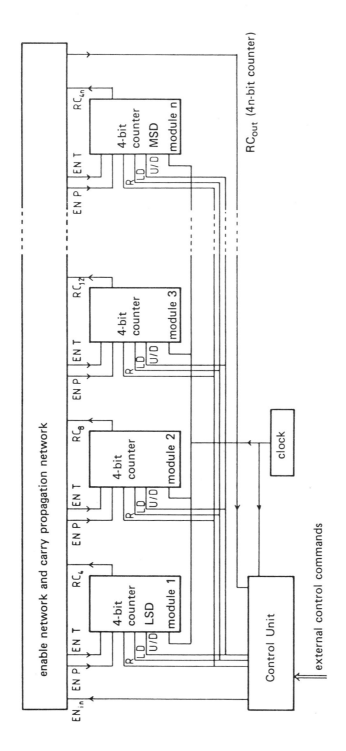

Figure 13.18 Structure of a 4n-bit counter.

The given *timing parameters* define the requirements and indicate the margins for the interaction between the counter/control and the rest of the system. The system has one clock generator, to which all times/settings are referred.

In the 4-bit counter described an EN P/EN T construction is assumed. With only one enable input the EN P input is omitted. Further, the reset input (R) and/or an up/down input (U/D) is not always present.

Note
In this counter there is a direct correspondence between *function* and *control input*. This is also indicated by the names of the control inputs. Commands to counters can be *encoded* more efficiently if the function-control input relation is abandoned. For an example see circuit SN74AS867 by Texas Instruments, in which the functions Reset, Load, Count Up and Count Down are encoded by two control signals S_0 and S_1. □

The model of a 4n-bit counter

The characteristics of a 4n-bit counter, constructed from 4-bit modules, are determined by:

– the characteristics of the 4-bit modules,

– the way the 4n-bit counter is structured.

Figure 13.18 roughly shows how a 4n-bit counter can be used in a clock mode environment. The counter is set by the control unit. The control signals for the functions reset, load and up/down (R, LD, U/D) are usually distributed in parallel to all 4-bit modules.

Above all sections there is a communication network, having the following functions:

– take care of the internal carry propagation in the counter;

– process and distribute the 'counter enable' signal EN_{in};

– generate the RC_{out}(counter) signal (interaction with the control unit).

In many realisations the control unit tests whether the counter is full/empty by means of the RC_{out} signal of the counter. This signal should therefore be sent to the control unit as a test signal. The data inputs and outputs of the 4-bit modules have been omitted. To begin with we study the counter-control relation, because the worst case timing usually occurs there.

Structures for the enable and carry propagation network

The enable and carry propagation network of 4n-bit counters can be composed in many different ways, some of which we shall examine. Figure 13.19 describes the first of these.

The structure of Figure 13.19 is applied in counters in which the 4-bit modules have only one *enable input*. It can also be used in counters with an EN P/EN T construction. This structure does not require extra hardware. Its principal characteristics are the following:

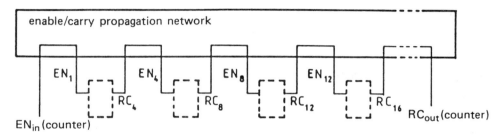

Figure 13.19 Serial enable/carry propagation network for counters with one enable input.

- Setting (by the control unit) of the EN_{in}(counter) signal may cause a ripple carry through the entire counter. A U/D switch can cause this as well.
- During counting ($EN_{in} = 1$) the carry ripples through the entire counter. As the number of sections increases, f_{max} gets lower and lower.
- The commands 'Reset' and 'Load' require a short setup time. The execution of these commands after the clock pulse may, however, be followed by a delay in order to settle the ripple carry network
- With a disabled counter (EN_{in}(counter) = 0) the RC_{out}(counter) signal is *not available*. The control unit is consequently not able to test on it.
- The combinational relation between EN_{in}(counter) and RC_{out}(counter) carries the risk of creating loops between the data path (counter) and control unit. See Section 13.11.

Conclusion: *This counter structure is only suitable for low-speed applications.*

Structure 13.20 is found in counters in which the 4-bit modules have an EN P/EN T enable construction. See the previous sections. The EN_{in}(counter) signal sets all sections in parallel. Further information transport is based on the ripple carry and is processed in series.

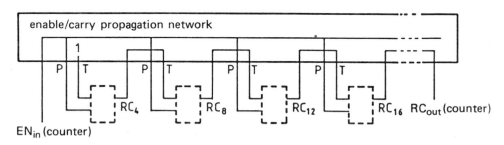

Figure 13.20 Parallel enable and ripple carry in counters with EN P/EN T inputs.

The principal characteristics are the following:

- Setting EN_{in}(counter) by the control unit no longer causes a ripple carry through the circuit. The enable signal has a short setup time. In addition, EN_{in}(counter) no longer has any direct influence on the RC_{out}(counter) output (in the static state). When the counter is disabled the RC_{out} signal is available.

- A U/D conversion still causes a ripple carry through the circuit. The setup time of this command consequently depends on the number of sections.

- For the rest the characteristics of the counter do not deviate very much from those in Figure 13.19. (We shall discuss a special application of this construction for fast counting later.)

The structure of Figure 13.21 requires extra hardware in the enable and carry propagation network. All of the information of EN_{in}(counter) and of the 4-bit sections (RC_4, RC_8, etc.) is processed *in parallel*. This speeds up the counter considerably.

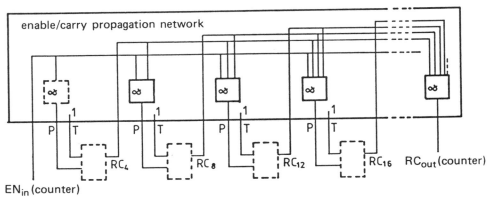

Figure 13.21 Parallel enable and carry network in counters with EN P/EN T construction.

With this structure all functions have practically the same setup time. All operations are equally fast (parallel information transport). It further holds that:

- The maximum (counting) frequency is independent of the number of sections (as long as the propagation time of AND gates with more inputs does not increase).

- The signal RC_{out}(counter) continuously provides information about the end-around carry. The signal EN_{in}(counter) does not directly influence it.

- A U/D switch still has a direct influence (which may create a loop).

Application

In practice one often uses a variant of Figure 13.10 and Figure 13.20. This is shown in Figure 13.22 for a 4-bit counter. In order to be able to see the advantage of this structure we shall first observe the next part of the cycle of this 4-bit binary counter.

Flip-flop D changes if *all* preceding flip-flops are in the state $Q = 1$ ($EN_{in} = 1$). Of the preceding flip-flops flip-flop A arrives in this state last, after which Q_a still has to set the entire network, when it has a serial structure as in Figure 13.20. In Figure 13.22 the information of Q_a is distributed *in parallel*. The further information, about Q_b and other outputs, is available at least one clock period sooner than Q_a and is processed *in series*.

Q_d	Q_c	Q_b	Q_a
.	.	.	.
.	.	.	.
0	1	0	0
0	1	0	1
0	1	1	0
0	1	1	1
1	0	0	0
.	.	.	.
.	.	.	.

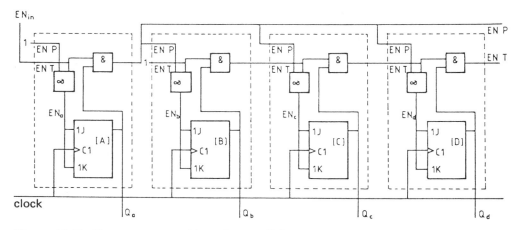

Figure 13.22 Binary counter with series-parallel transport.

This construction is especially used in 4n-bit counters. Here, too, this setup is accelerating with respect to counting. We note that there is hardly any acceleration with regard to 'Load' and 'Up/Down' functions. See also Sections 13.9 and 13.12.

Function tables of 4n-bit counters

Depending on:

– the choice of the 4-bit module and

– the choice of the structure of the 4n-bit counter

the function table of the 4n-bit counter has to be adjusted with respect to that of the 4-bit counter. In general the function table of the 4n-bit counter differs from the function table of the n-bit counter. IC data books do not supply any information on this point. It is up to the user to draw up these tables from the specifications.

13.8 The parameter specification of 4n-bit counters

For the analysis of the properties of 4n-bit counters it is necessary to derive external parameters, comparable to those of Figure 13.17. We shall examine how these parameters depend on the parameters of the 4-bit counter. The following groups of parameters are to be distinguished:

- t_{su} and t_h of external data inputs.
  As a rule these parameters are equal to those of the 4-bit counter. They do not require further consideration.
- $t_{P(min)}$ and $t_{P(max)}$ of the Q outputs of the flip-flops.
  These do not have to be adjusted either.

This describes the relation with the other part of the data path.
Further:

- t_{su} and t_h of the signals R, LD, EN P and/or EN T, insofar as they are used in parallel and do not influence the (ripple) carry network.
  They then do not require adjustments either.
- t_{su} and t_h of the signals EN T and U/D and others, where appropriate. These signals may control the flip-flops in parallel or through the enable/carry propagation network (in parallel or through a ripple carry). All paths along which they are able to exercise any influence have to be checked, which results in a 'worst case' timing.
- $t_{P(min)}$ and $t_{P(max)}$ of the Q outputs to the RC_{out} output of the 4n-bit counter.
  These parameters depend on the structure chosen for the carry propagation network. The relation may be different for each configuration.
- t_{su} and t_h of the signals EN T and U/D to the RC_{out} output of the 4n-bit counter.
  These parameters, too, depend on the structure chosen.

In the following examples we shall confine ourselves to fully synchronous 4-bit/4n-bit counters. Others, such as the SN74190-series and counters constructed with them may, as far as we are concerned, be relegated to a museum after about 20 years of faithful service. An analysis would immediately show their poor noise immunity.

The 'Load' command in 4n-bit counters

In the setup of Figure 13.18 the 'Load' command is distributed in parallel. This command, given through the LD signal, has no direct influence on the Q outputs, nor on the RC_{out} of each section. For the 'Load' command it therefore holds that:

$$t_{su}(\text{LD-to-clock, 4n-bit counter}) = t_{su}(\text{LD-to-clock, 4-bit counter}), \qquad (13.13)$$

$$t_{h}(\text{LD-to-clock, 4n-bit counter}) = t_{h}(\text{LD-to-clock, 4-bit counter}). \qquad (13.14)$$

A small adjustment might possibly be necessary, because of the effect of the higher fanin of the load input.

Upon *execution* of the 'Load' command all flip-flops are loaded with external data. The carry propagation network then begins to react again. Depending on its structure this may take quite some time.

For the structure according to Figure 13.19, for example, the following holds:

$$t_{P(max)}(\text{clock-to-}RC_{out}/4n\text{-bit counter}) \tag{13.15}$$
$$= t_{P(max)}(\text{clock-to }RC_{out}/4\text{-bit counter})$$
$$+ (n-1)\cdot t_{P(max)}(\text{EN T-to-}RC_{out}/4\text{-bit counter}).$$

Note that $t_{P(max)}$ *increases in proportion* to the number of 4-bit sections of the counter.

In a parallel enable/carry propagation network according to Figure 13.21 the following holds:

$$t_{P(max)}(\text{clock-to-}RC_{out}/4\text{-bit counter}) \tag{13.16}$$
$$= t_{P(max)}(\text{clock-to-}RC_{out}/4\text{-bit counter})$$
$$+ t_{P(max)}(\text{AND}).$$

There is no cumulative effect when the number of sections increases.
With slow counters the above will not pose a problem. Fast counters, however, are not possible with an internal structure as in Figure 13.19.

Switching the U/D input in 4n-bit counters

The *switching* of the U/D input of counters has two consequences, one involving the setup time with respect to the counter's clock and one involving the RC_{out} of the counter. Both parameters change in 4n-bit counters. For the structure in Figure 13.19 it holds that:

$$t_{su}(\text{U/D-to-clock}/4n\text{-bit counter}) \tag{13.17}$$
$$= t_{P(max)}(\text{U/D-to-}RC_{out}/4\text{-bit counter})$$
$$+ (n-2)\cdot t_{P(max)}(\text{EN T-to-}RC_{out}/4\text{-bit counter})$$
$$+ t_{su}(\text{EN T-to-clock}/4\text{-bit counter}),$$

and

$$t_{P(max)}(\text{U/D-to-}RC_{out}/4n\text{-bit counter}) \tag{13.18}$$
$$= t_{P(max)}(\text{U/D-to-}RC_{out}/4\text{-bit counter})$$
$$+ (n-1)\cdot t_{P(max)}(\text{EN T-to-}RC_{out}/4\text{-bit counter}).$$

For the structure in Figure 13.21 it holds that:

$$t_{su}(\text{U/D-to-clock}/4n\text{-bit counter}) \tag{13.19}$$
$$= t_{P(max)}(\text{U/D-to-}RC_{out}/4\text{-bit counter})$$
$$+ t_{P(max)}(\text{AND})$$
$$+ t_{su}(\text{EN P-to-clock}/4\text{-bit counter}),$$

and

$$t_{P(max)}(\text{U/D-to-}RC_{out}/4n\text{-bit counter}) \tag{13.20}$$
$$= t_{P(max)}(\text{U/D-to-}RC_{out}/4\text{-bit counter})$$
$$+ t_{P(max)}(\text{AND}).$$

The *processing time* of the U/D switch in a counter depends on the internal structure.

After the next clock pulse a ripple carry arises in Figure 13.19, while Figure 13.21 gives a fast result.

Counting with 4n-bit counters

Once all inputs have been settled and $EN_{in} = 1$, counters count their normal cycle. To calculate the maximum counting frequency all paths having any influence on it have to be checked, viz.:

– the timing of the flip-flops of all 4-bit sections;

– the setup time which applies for the RC_{out}-signal of the 4n-bit counter.

For Figure 13.19 the relevant formulas (period T of the clock) are:

$$T > t_{P(max)}(\text{clock-to-}RC_{out}/\text{4-bit counter}) \qquad (13.21)$$
$$+ (n-2) \cdot t_{P(max)}(\text{EN T-to }RC_{out}/\text{4-bit counter})$$
$$+ t_{su}(\text{EN T-to-clock/4-bit counter}),$$

and

$$t_{P(max)}(\text{clock-to-}RC_{out}/\text{4n-bit counter}) \qquad (13.22)$$
$$= t_{P(max)}(\text{clock-to-}RC_{out}/\text{4-bit counter})$$
$$+ (n-1) \cdot t_{P(max)} (\text{EN T-to-}RC_{out}/\text{4-bit counter}).$$

For Figure 13.20 the lower bound for period T is also formulated by (13.21). The same holds for (13.22), with regard to $t_{P(max)}(\text{clock-to-}RC_{out})$.

For Figure 13.21 it holds that:

$$T > t_{P(max)}(\text{clock-to-}RC_{out}/\text{4-bit counter}) \qquad (13.23)$$
$$+ t_{P(max)}(\text{AND})$$
$$+ t_{su}(\text{EN P-to-clock/4-bit counter}),$$

and

$$t_{P(max)}(\text{clock-to-}RC_{out}/\text{4n-bit counter}) \qquad (13.24)$$
$$= t_{P(max)}(\text{clock-to-}RC_{out}/\text{4-bit counter})$$
$$+ t_{P(max)}(\text{AND}).$$

Conclusions

Depending on
– the 4-bit counter section chosen,
– the internal structure of the 4n-bit counter, and
– the selected function of the counter

a 4n-bit counter will require the same or a longer setup time before the clock pulse or propagation delay time after the clock pulse, with respect to the 4-bit counter section. If all the required functions are known for a certain application, the maximum counting frequency (stand alone) can easily be calculated. In a serial oriented enable/carry propagation network the maximum counting frequency soon becomes very low with an increasing number of sections (see also Sections 13.9 and 13.12).

13.9 Pseudo clock mode use of counters

It appeared in Section 13.8 that some operations require considerably more time than others, especially if there is a cumulative effect in, for example, the ripple carry. In a clock mode environment it is recommended to execute or set all operations within one clock period. The circuit is at rest the moment the clock pulse is given and no *cross talk* is then present. For counters this means that one either has to lower the frequency of the system clock below the longest execution time or that one has to opt for a parallel structure of the counter. The latter costs extra gates and/or space on ICs. See Figure 13.21.

In practice, the clock period is sometimes adapted to the most frequently used operation, which is usually also the fastest. For preparing/executing the other operations two or more clock periods are reserved. This imposes demands on the control unit. *The control signals of the counter have to be stable long enough and must be free of transitions.* A disadvantage of this solution is that (part of) the counter is not at rest when clock transitions occur. *Cross talk* to the rest of the circuit may be disastrous at such a moment. With a carefully designed layout this phenomenon may be suppressed, but cannot be entirely excluded.

This kind of failure is generally data-dependent, and because of this the errors appear to be random. They are extremely difficult to trace. In VLSI designs they occur relatively often. During a simulation, these errors are usually not recognised. It is the responsibility of the logic designer to anticipate them and to take the necessary precautions.

Example
Asynchronous binary counters have a very simple internal structure (Figure 13.3). They take up a relatively small part of the IC's area, a reason why they are applied so frequently. Cross talk makes their application risky. Figure 13.23 shows a solution that functions well with long asynchronous binary counters. Of course, it will take a few clock pulses after an input transition before the counter reaches its final state.

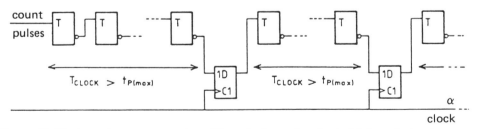

Figure 13.23 Asynchronous counter with a clock mode-adapted internal structure.

At regular intervals the internal clock signal in the asynchronous counter is clocked in an *edge-triggered* D flip-flop. Each interval comprises a maximum number of flip-flops so that the period T of the clock is larger than the maximum propagation delay time of that part of the counter. The parameters $t_{P(max)}$ and t_{su} of the synchronisation flip-flops should also be kept in mind. □

Such a solution, dividing the delay of a ripple over various clock periods, can be used frequently. The fully synchronously oriented timing model with all its advantages can then remain intact outside this counter structure. This is called *pseudo clock mode clocking*.

13.10 Applications

Counters with variable cycle length

In many processes the cycle length is not a multiple of 2^n or 10^n. It is then possible to either design a special counter for this application, or to cut the cycle length of a standard counter. The options are as follows:

- Equip a counter that has a cycle length longer than required with a *detection circuit* that detects the end of the desired cycle. Then reset the counter and let it start again.
- Load a counter that has a cycle length longer than required with the desired cycle length and let it count down to 0 in the 'count down mode'.

The advantage of the second option is that one can use the standard RC_{out} output, eliminating the need for a separate detection circuit.

The disadvantage of both of these is that when the cycle length (k) has to be preset the counter must be loaded with k–1 or the contents k–1 have to be tested, respectively. After all, loading with a cycle length k or a reset costs one clock pulse and the count down procedure to the final state costs k clock pulses: together k+1 clock pulses. For the user this kind of presetting is very awkward and is a possible source of errors.

The following solutions may be used for this problem.

Solution 1
Equip the counter with a subtractor, which transfers k–1 to it instead of k.

Solution 2
Adapt the counter so that a RC_{out} is already given in state 1 (count down mode). For this purpose only the least significant decade has to be adapted.

Both solutions have to be examined to see what happens if the counter is loaded with cycle length '0'. This often has to be detected separately, which means a further adaptation of the counter.

Solution 3
Adapt the process the counter is used for. If possible, include a NOP-step (no operation) in the control unit, in which the data path gets an 'inhibit' command. In many processes it is also possible to load a counter earlier, before the process to be counted starts, after which one step has to be made in the CTDN-mode at the next clock pulse. The counter is then ready in the state k–1 when the count down cycle has to start.

Interrupting a cycle with a direct reset

A tactic which is often seen in practice is that designers interrupt a count cycle via the *direct reset input*, a feature which a number of counters already have. This is used in synchronous as well as in fully or partly asynchronous counters. The final state is detected by the gate circuit. As soon as the output of this detection circuit indicates that the final state has been reached the direct reset input forces the counter into the zero state. Figure 13.24 exemplifies how a 4-bit binary counter section can be reduced in this way to an 11-step cycle. The timing diagram illustrates the transition from position 10 to position 11/0.

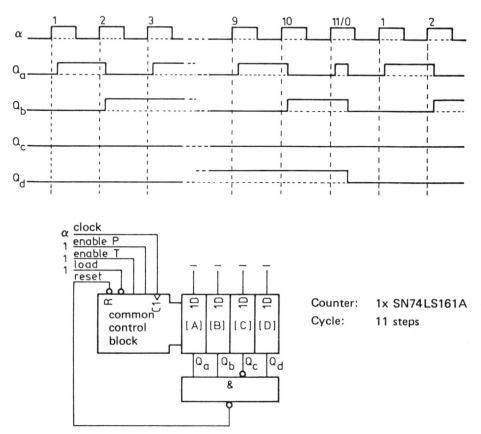

Figure 13.24 Interruption of a count cycle via the direct reset.

For a variety of reasons the solution for interrupting a cycle outlined here has to be rejected.

1. The *effective propagation delay time* $t_{P(max)}$ of the flip-flop outputs strongly increases. As soon as the final state has been reached it is *immediately* changed into the zero state, the desired new state of the flip-flops. Due to the long $t_{P(max)}$ the *maximum counting frequency* of the counter is reduced. This is shown in Figure 13.25. Figure 13.25.a indicates when the outputs are *stable* after a clock pulse if no reset is given. Figure 13.25.b shows the same, but now including a reset. Compare the transition of state 10 to 11/0 in the timing diagram of Figure 13.24.

   In the calculation of the *maximum clock pulse frequency* the time necessary to inactivate the reset should be added. After all, the next clock pulse is not allowed to arrive until the counter is fully prepared internally. This extra time is roughly $t_{P(detection)} + t_{P(reset)}$.

2. The *reset pulse 'kills' itself*. As soon as the counter reacts to a reset command the internal state changes and the reset command consequently disappears immediately. In many cases the reset pulse is then too short; as a result not all flip-flops of the counter are reset (many *intermitting errors*).

3. *Spikes from the detection circuit* reset the counter unintentionally. This occurs especially in asynchronously coupled counter sections.

4. Furthermore, the *other objections* against the use of direct inputs are also valid, e.g. the increased *noise susceptibility*.

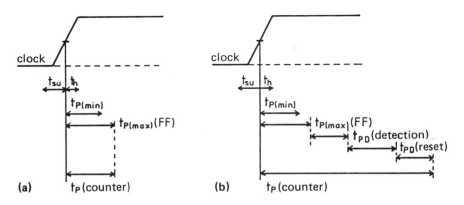

Figure 13.25 Increase of the effective propagation delay time.

Counting in a non-standard code (pattern generators)

In many applications a cyclic pattern of pulses is required. Such pattern generators can be designed with the finite state model. Another possibility is a counter with a code converter *behind* it. This has some drawbacks:

– The propagation delay time after the clock pulse is relatively long.

– The code converters may produce transition phenomena, which may influence the controlled process.

It is, therefore, advantageous to place the code converter/pattern generator *in front of* the state register. Figure 13.26 describes both setups.

Example

Figure 13.27 describes how an 11-step cycle in a register-PROM combination can be programmed. The access time t_{acc} of the PROM is not added to the propagation delay time of the register, as far as the other components in the data path are concerned. □

Miscellaneous

Counters are versatile components. They can be used in many places in circuits, counting (down) being just one of the applications. Another application, for example, is using a

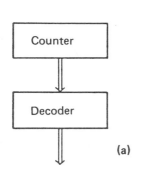
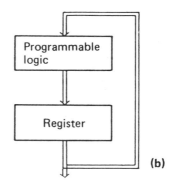

Figure 13.26 Setup of pattern generator.

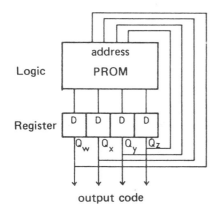

address $Q_w\ Q_x\ Q_y\ Q_z$					contents $D_w\ D_x\ D_y\ D_z$			
0	0	0	0	→	0	0	1	0
0	0	0	1	→	-	-	-	-
0	0	1	0	→	0	1	0	1
0	0	1	1	→	-	-	-	-
0	1	0	0	→	1	1	1	1
0	1	0	1	→	0	1	1	1
0	1	1	0	→	-	-	-	-
0	1	1	1	→	1	0	1	1
1	0	0	0	→	1	1	0	0
1	0	0	1	→	-	-	-	-
1	0	1	0	→	0	0	0	0
1	0	1	1	→	1	1	1	0
1	1	0	0	→	0	1	0	0
1	1	0	1	→	-	-	-	-
1	1	1	0	→	1	0	0	0
1	1	1	1	→	1	0	1	0

Figure 13.27 Pattern generator based on a register-PROM combination.

counter as a *register with preparatory (synchronous) reset*.

Almost all of the registers listed in data books do not have a preparatory reset. By using a counter with preparatory reset and load facilities instead of a register, one has created a register function without any additional gates. Storing data, one of the functions of registers, can be set by the enable input.

Another application is detecting whether a number is equal to 00...00. In division, for example, dividing by zero should be avoided. For the divisor register, too, a U/D counter is then usually applied. The 0...0 state can be detected via the RC_{out} output.

13.11 Counter-control loops

In 4-bit counters with an internal construction as in Figures 13.11 through 13.15 the EN T signal and with U/D counters the U/D signal as well has a direct influence on the RC_{out} output signal. Switching EN T or U/D may suppress the RC_{out}. In counter-control interaction this may create problems.

Example

A counter must cyclically run through the cycle

$$0 - 1 - 2 \ldots 253 - 254 - 255 - 254 - 253 \ldots 2 - 1 - 0 - 1 - 2 \ldots$$

The 8-bit binary U/D counter used for this goes through the first part of it in the 'count up mode'. See Figure 13.28.

state	command	command
253	CTUP	CTUP
254	CTUP	CTUP
255	CTDN*	CTUP/LD 254
254	CTDN	CTDN
253	CTDN	CTDN
.	.	.
.	.	.
.	.	.

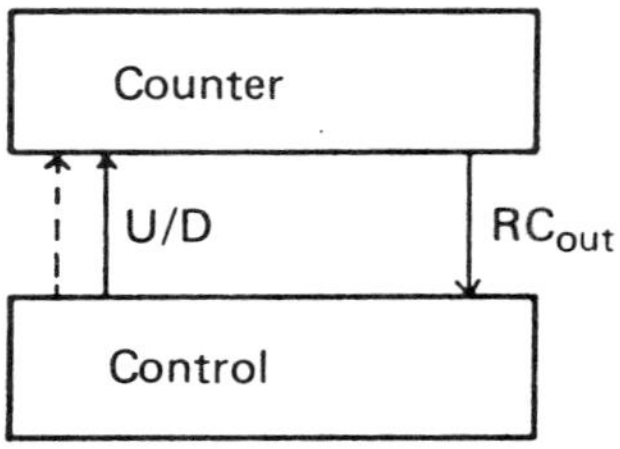

Figure 13.28 A counter-control loop.

When the control unit detects an RC_{out} signal, content = 255, the next state must be 254. If the control unit then switches to the 'count down mode', this will kill the RC_{out} signal in most U/D counters. From this the control unit draws the conclusion that the counter is not yet in state 255, and gives CTUP again. This triggers an RC_{out} signal again, and so on. The control-counter combination oscillates.

This problem can be avoided by giving the command CTUP/LD 254 in state 255. Since LD is dominant over CTUP/CTDN the counter is loaded with 254 at the next clock pulse. In addition, the command 'CTUP' sees to it that the RC_{out} signal remains intact.

The same problem also arises at the transition $\ldots - 1 - 0 - 1 - \ldots$, and can be solved in a similar way. □

The problems observed here occur very frequently in detailing *control specifications*.

Note

The modification of the counter control as shown here does not completely solve the problem as stated! The problem of a *static loop* is replaced by a *dynamic loop*. We will not go into further details here. □

13.12 Summary

In Section 13.1 the concepts of architecture, implementation and realisation have been introduced. These concepts will be elaborated here. All of the information required to do this can be found above.

In the *architecture specification* of a counter, for example, the desired 'external' behaviour of the counter is described, including its range, the maximum frequency of the events to be counted, the counting mode (up/down, etc.), the counting code (if relevant) and the interaction with the environment. In particular, an accurate definition of the environment, the timing, allowable dissipation and other requirements deserve special attention. It makes a great deal of difference whether the design is intended as a standard IC on printed circuit boards, the '7400 series' for example, or whether the design has to be realised as part of a VLSI IC. In the former case a lot of attention should be given to a design which is clear and easy to use, so that the designer quickly knows how the IC has to be applied in a specific application. Every application is 'new'. In older designs one therefore often used to reserve a separate command input for each 'function', with a fixed priority scheme. In addition, standard interfacing with regard to fanin and fanout is an elementary requirement.

Within a VLSI environment the 'logic' design costs hardly affect the price of the product. Instead, the area required by the IC determines the yield of the process and thus the price of the circuit. In principle, the design process is done once, so that it may be useful, for example, to work out technology-bound preferences for certain operators. The 'interfacing' is likewise done once and is allowed to deviate from the standard rules.

Determining how far the freedom of the designer extends lies somewhere between architecture specification and *implementation*. At this stage of the design process structures are sought to fulfil the specifications. Counters are considered circuits in which the structure has a great influence on the external characteristics. This has been explained in Sections 13.7 through 13.9. Particularly the formulas for setup and hold times and propagation delay times as formulated in Section 13.8 make a circuit design *parametric*. Once the specified range has been determined, the external characteristics are easy to deduce for each structure and can subsequently easily be compared with the specifications.

Example

Asynchronous counters have the advantage of a simple internal organisation. Figure 13.5 showed the maximum frequency f_{max} for such a counter under stand alone conditions and in a clock mode environment. Figure 13.29 gives the required maximum frequency for a certain application, within a clock mode environment (dotted line). One can now easily see how many sections the asynchronous counter is marginally allowed to have at most, so that the specification is just met.

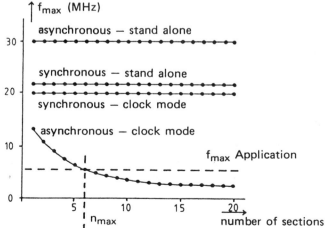

Figure 13.29 Determination of the maximally allowed number of sections.

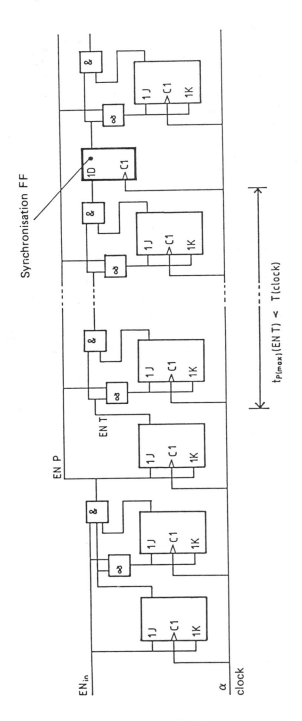

Figure 13.30 Design with synchronisation flip-flops.

If the required range of the binary counter is larger than n_{max}, then a pseudo clock mode structure (Figure 13.23) may offer a solution. A disadvantage of this structure is that it may take several clock periods before the counter has been stabilised. When this time is not available, one has to resort to a fully synchronous design. Although this usually requires an extra number of gates, the performance of the counter is better. □

Lower design levels

It holds in designs of digital circuits as well that prevention is better than the cure. This especially holds for possible transition phenomena at the realisation level, such as *cross talk*. By means of a careful layout and distribution of supply lines the effect of cross talk can be limited. It is better to anticipate the possible occurrence of cross talk and other transition phenomena and to combat the consequences at a *logic design level*. This is the basis of the *clock mode* view in a design. These problems meet each other somewhere between logic implementation and the realisation in hardware.

Example

The application of the EN P/EN T construction in counters in principle allows the EN T signal to be active while the clock pulse is given. The idea behind this application is that the signal EN P of the first section(s) blocks all input gates on which EN T has an influence. In itself this view is correct.

However, transitions in the EN T circuit may induce transitions on other signal lines, especially at the moment that the clock pulse is given and these signals have to be stable. This problem may be circumvented by applying *synchronisation flip-flops* at certain distances in the logic design. See Figure 13.30. □

In this case the logic characteristics of the counter are not affected. During counting the counter always has the correct contents. In the opinion of the authors applying provisions such as those described above is part of the task of the logic designer. It increases the margins during the realisation, which is particularly favourable with respect to the reliability and reproducibility of the designed circuit. Experiences with 'mapping' a logic design onto hardware in this way have been shown to be very favourable.

References

1. T.L. Floyd, *Digital Logic Fundamentals*, C.E. Merrill Publ. Co., Columbus, Ohio, 1977.
2. M.M. Mano, *Digital Logic and Computer Design*, Prentice-Hall, Englewood Cliffs, N.J., 1979.
3. R.L. Morris and J.R. Miller, *Designing with TTL Integrated Circuits*, McGraw-Hill, New York, 1971.
4. R.M.M. Oberman, *A Flexible Rate Multiplier Circuit with Uniform Pulse Distribution Outputs*, IEEE Tr. on Computers, Vol. C-21, 1972, pp. 896–899.
5. R.M.M. Oberman, *Electronic Counters*, Macmillan, London, 1973.
6. J.B. Peatman, *Digital Hardware Design*, McGraw-Hill, New York, 1980.
7. Texas Instruments, *The TTL Data Book for Design Engineers*.

Problems

Introduction

0.1. In determining the value of a variable with a continuous range in a binary code a *rounding error* is made. To keep this error small it is *necessary* to:
  a. choose the number of bits of the binary code as small as possible.
  b. choose the number of bits of the binary code as large as possible.
  c. adapt the size of the interval between two steps to the uncertainty in the continuous signal.
  d. apply an analog-digital converter.

0.2. The statements below concern the advantages of digital systems over systems that operate with analog signals/quantifiers. One of the statements does not apply. Which one?
  a. In digital systems the accuracy of the result need not be less than the result of an operation in a system with quantifiers with a continuous range, despite the fact that a rounding error is made when digitising signals.
  b. Systems operating with discrete quantifiers are automatically more reliable than systems operating with analog quantifiers.
  c. Digital systems are easier to reproduce than analog systems.
  d. Digital systems are less susceptible to parameter variations of components and power supplies.

0.3. Progressive codes are sometimes used in the design of code shafts. Which of the following cycles of code combinations is a progressive code?

a.	b.	c.	d.
000	000	000	000
001	010	001	010
010	011	010	011
011	010	100	111
100	110	101	101
101	111	110	100

0.4. With code shafts the reflected binary code (Gray code) is used:
  a. to improve the mechanical construction of the shaft.
  b. because the reflected binary code is able to encode more positions than a corresponding binary code.
  c. because the reflected binary code has a standard parity bit.
  d. to avoid transient phenomena.

Chapter 1

1.1. In the normal binary code the number 1011011010_{BIN} is:
  a. a *twofold*, but not a *fivefold*.
  b. not a twofold, but a fivefold.
  c. neither a twofold nor a fivefold.
  d. both a twofold and a fivefold.

1.2. The BCD representation of 11100011_{BIN} is:
  a. 0010 0010 0111.
  b. 1110 0011.
  c. 343.
  d. E3.

1.3. The number N is encoded in the BCD code with three decimal digits:

$$N = a_{11}a_{10}a_9a_8 \; a_7a_6a_5a_4 \; a_3a_2a_1a_0 \; BCD.$$

The bit a_0 is the least significant bit. The *sum* of the weights of the bits a_{11} through a_0 in the BCD code is:
  a. 1665.
  b. 2048.
  c. 4095.
  d. 4096.

1.4. The number N is encoded in the BCD code with three decimal digits:

$$N = a_{11}a_{10}a_9a_8 \; a_7a_6a_5a_4 \; a_3a_2a_1a_0 \; BCD.$$

Some of the eight combinations 000 through 111 of the bits $a_{11}a_{10}a_9$ are not used in the BCD code. How many of these combinations do not occur?
  a. One combination does not occur.
  b. Two combinations do not occur.
  c. Three combinations do not occur.
  d. Six combinations do not occur.

1.5. The *binary number* 101011110100011 is notated in the *hexadecimal code* as follows:
  a. 2 11 13 3.
  b. 2 B D 3.
  c. A F 4 C.
  d. 2 5 7 2 3.

1.6. A number G is encoded in the hexadecimal code as G = 237. The *BCD code* corresponding with this number G is:
  a. 0010 0011 0111.
  b. 001 000 110 111.
  c. 0101 0110 0111.
  d. None of the above.

1.7. The number G is given in the hexadecimal code as:

$$G = 472_{HEX}.$$

When this number is represented in the decimal code, the sum of the digits is:
a. ≤ 9.
b. 10.
c. 11.
d. ≥ 12.

1.8. A number G is given in the BCD code as:

G = 1001 0110 0101.

The hexadecimal code of G is:
a. 5C3.
b. 3C5.
c. 4545.
d. 0011 1011 0011.

1.9. Which of the following codes has error detecting properties?
a. The three-out-of-seven code (35 combinations).
b. The 5-bit telegraph code (32 combinations).
c. The BCD code (binary coded decimal code).
d. The hexadecimal code.

1.10. A number of objects have to be numbered. This may, for example, be done in the binary code (BIN), in the octal code (OCT), in the decimal code (DEC) or in the hexadecimal code (HEX). If there are *three* digit positions available, the *maximum number* of objects to be encoded in each code is:
a. $N_{BIN} / N_{OCT} / N_{DEC} / N_{HEX} = 3 / 3 / 3 / 3$.
b. $N_{BIN} / N_{OCT} / N_{DEC} / N_{HEX} = 2^3 / 2^9 / 2^{10} / 2^{12}$.
c. $N_{BIN} / N_{OCT} / N_{DEC} / N_{HEX} = 2^3 / 8^3 / 10^3 / 16^3$.
d. $N_{BIN} / N_{OCT} / N_{DEC} / N_{HEX} = 2^3-1 / 8^3-1 / 10^3-1 / 16^3-1$.

1.11. Two numbers A and B are encoded in the hexadecimal code as:

A = 3789,

B = 8917.

The sum S of A and B, S = A + B, in the hexadecimal code is:
a. C1A1.
b. C0A0.
c. 12706.
d. none of the above.

1.12. One the statements below is false. Which one?
a. The ASCII code is a 7-bit alphanumeric code, which can be extended by an eighth parity bit.
b. A parity bit gives the ASCII code error correcting properties.
c. A letter/figure shift is applied to be able to encode more characters.
d. Progressive codes are applied to suppress transient phenomena in, for example, code shafts.

1.13. A letter/figure and figure/letter shift in a code are used to:
  a. indicate that either numbers or letters are being transmitted.
  b. extend the number of different characters that can be encoded from 2^n to maximally $2^{n+1} - 4$.
  c. give the code error detecting or error correcting properties as desired.
  d. adapt the code for use in serial data transfer.

1.14. Three statements are made about the reflected binary code (Gray code):
  − A reflected binary code for encoding 2^n consecutive elements generally requires more bits than the binary code. This is because only one bit at a time may be changed in the reflected binary code.
  − In the reflected binary code the weights of the bits (from right to left) are proportional to $1 : 2 : 4 : 8$.
  − The reflected binary code is typically meant to be used in arithmetical operations.
  Of these three statements:
  a. none are correct.
  b. one is correct.
  c. two are correct.
  d. three are correct.

1.15. Part of a code shaft in the reflected binary code (Gray code) is given below:

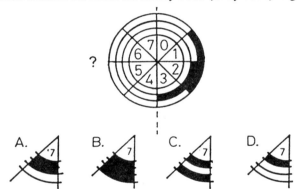

Which segment should be in position seven?
  a. A
  b. B
  c. C
  d. D

Chapter 2

2.1. The following three statements are made about the logic functions F_1, F_2 and T:
1. From $T + F_1 = T + F_2$ it follows that: $F_1 = F_2$ for all values of the variables.
2. From $F_1 = F_2$ it follows that: $T + F_1 = T + F_2$ for all values of the variables.
3. From $T \cdot F_1 = T \cdot F_2$ it follows that: $F_1 = F_2$ for all values of the variables.

Of these statements:
a. none are correct.
b. one is correct.
c. two are correct.
d. three are correct.

2.2. Which term(s) is/are *superfluous* in the function
$$S = b\bar{c} + a\bar{c} + a\bar{b} + \bar{b}c.$$
a. $\bar{b}c$.
b. $a\bar{b}$ *and* $a\bar{c}$.
c. $a\bar{b}$ *or* $a\bar{c}$.
d. None of these terms are superfluous.

2.3. In the product term
$$u\bar{\bar{v}}\bar{w}xyz$$
$\bar{w}$ may be replaced by:
a. $\bar{w} \to \bar{w} + x$.
b. $\bar{w} \to \bar{w} + \bar{x}$.
c. $\bar{w} \to \bar{w} + x + y + z$.
d. $\bar{w} \to u + \bar{v} + x + y + z$.

2.4. One of the following four equations is false. Which one?
a. $x + y \cdot z = (x + y) \cdot (x + z)$.
b. $(x + y) \cdot (x + z) \cdot (y + z) = x \cdot y + x \cdot z + y \cdot z$.
c. $\bar{x} + \bar{y} + \bar{z} = \overline{x + y + z}$.
d. $x \cdot y + \bar{x} \cdot z + y \cdot z = x \cdot y + \bar{x} \cdot z$.

2.5. Of the following three equations
$$\overline{x + y + z} = \bar{x} \cdot \bar{y} \cdot \bar{z}$$
$$(x + y)(\bar{x} + z)(\bar{y} + \bar{z}) = (x + y)(\bar{x} + z)$$
$$(x + y)(x + z)(y + z) = xy + xz + yz$$
a. none are correct.
b. one is correct.
c. two are correct.
d. three are correct.

2.6. Of the following four equations

$$(x + y)(\bar{x} + z)(y + z) = (x + y)(\bar{x} + z)$$

$$w + x + y + \overline{w}\bar{x}\bar{y}z = w + x + y + z$$

$$x + \bar{y}z = x + \bar{y}$$

$$\overline{xy + \bar{x}z + yz} = \bar{x}\bar{z} + x\bar{y} + \bar{y}z$$

    a. none are false.
    b. one is false.
    c. two are false.
    d. three or four are false.

2.7. Of the following three equations

$$(x + y) \oplus (y + z) = \bar{y} \cdot (x \oplus z)$$

$$\bar{x}y + xz + \bar{y}z = \bar{x}y + xz$$

$$\overline{\bar{x}y + \bar{y}z + \bar{z}x} = xyz + \bar{x}\bar{y}\bar{z}$$

    a. none are false.
    b. one is false.
    c. two are false.
    d. three are false.

2.8. The function

$$S = wx + \bar{w}y + \bar{x}y + \bar{w}z + \bar{x}z$$

can be written as:
    a. $S = wx + y + z$.
    b. $S = (w + x + y) \cdot (w + x + z)$.
    c. $S = w + x + y + z$.
    d. $S = wx + \bar{w}y + \bar{x}z$.

2.9. The function S

$$S = wx + \bar{w}xz + \bar{w}y + \bar{y}z + w\bar{x}y$$

can be written as:
    a. $S = w + x + y + z$.
    b. $S = wx + \bar{w}y + \bar{y}z$.
    c. $S = (w + y + z)(x + y + z)$.
    d. $S = z + y\bar{z} + x\bar{z}$.

2.10. The function S is given as:

$$S = (x + y)(x + \bar{y} + \bar{z})(\bar{y} + z) + yz.$$

If S is entered in a truth table as a function of x, y and z, the function value column contains exactly:
    a. two zeros.
    b. three zeros.
    c. four zeros.
    d. five zeros.

2.11. Two functions S and T are given as:
$$S = ab + \bar{a}c + \bar{a}b,$$
$$T = ac + \bar{a}\bar{b} + b\bar{c}.$$
The function V is the AND function of S and T. The simplest formula for V is:
a. $V = ab\bar{c} + \bar{a}\bar{c} + \bar{a}bc$.
b. $V = ab + b\bar{c} + \bar{a}bc$.
c. $V = \bar{a}b\bar{c} + ac + \bar{a}\bar{b}c$.
d. $V = ab + b\bar{c} + \bar{a}c$.

2.12. The minterm form of a logic function is a sum-of-products form for which it holds that:
a. Its number of terms is minimal.
b. In each product term the number of variables is minimal.
c. All variables or their negates occur once in each product.
d. Each term has a minimum number of variables, from which the name *min*imal *term* is derived.

2.13. A function in three variables x, y and z is specified as follows:
$$f_0 = f_4 = f_6 = f_7 = 0;$$
$$f_1 = f_2 = f_3 = f_5 = 1.$$
(The indices i of the function values f_i are based on the following assignment of weights: $x \leftrightarrow 4; y \leftrightarrow 2; z \leftrightarrow 1$.)
The *minterm form* of this function is:
a. $\bar{x}y + \bar{y}z$.
b. $\bar{x}\bar{y}z + x\bar{y}\bar{z} + xy\bar{z} + xyz$.
c. $\bar{y}z + xy$.
d. $\bar{x}\bar{y}z + \bar{x}y\bar{z} + \bar{x}yz + x\bar{y}z$.

2.14. A club has two groups of members, a group of three persons {A, B, C} and the remaining members. The board must, according to its by-laws, consist of *exactly* four members, out of which the group {A, B, C} has to supply *at least half* of the board members.
In the following formulas the variables A/B/C stand for the proposition

'member A/B/C has been appointed to the board'

and O_i for

'i other members have been appointed to the board.'

Which of the following formulas specifies, as correctly and as simply as possible, the conditions under which the board has been made up (S = 1) in accordance with the regulations?
a. $S = AB + AC + BC + ABC$.
b. $S = AB\bar{C} + A\bar{B}C + \bar{A}BC + ABC$.
c. $S = ABO_2 + ACO_2 + BCO_2 + ABCO_1$.
d. $S = AB\bar{C}O_2 + A\bar{B}CO_2 + \bar{A}BCO_2 + ABCO_1$.

2.15. Four persons A, B, C and D are considered for appointment to the board of a club. According to the by-laws the section of which A and B are part must have the absolute majority in the board. C and D are not part of this section.

The following propositions are given:

S: 'The board is made up in accordance with the by-laws',

A: 'Person A has been appointed to the board', etc.

The conditions under which the board can be formed in accordance with the by-laws are put in equation form as:

a. $S = (A + B) \cdot (\overline{C \cdot D}) + A \cdot B \cdot (C + D)$.
b. $S = (A + B) \cdot (\overline{C} \cdot \overline{D}) + A \cdot B \cdot (\overline{C \cdot D})$.
c. $S = (A \oplus B) \cdot (\overline{C + D}) + A \cdot B \cdot (C + D)$.
d. $S = (A \oplus B) \cdot (\overline{C + D}) + A \cdot B \cdot (\overline{C \cdot D})$.

Chapter 3

3.1. Two cells in a Karnaugh map are adjacent cells if and only if the corresponding minterms differ in one variable only. In the Karnaugh map given below one cell contains a cross (×).

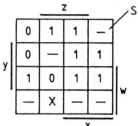

The number of adjacent cells of the cell containing the cross, for which the function value $f_i = 1$ has been specified in the Karnaugh map, is:
a. zero.
b. one.
c. two.
d. three.

3.2. When the function S

$$S = \overline{w}x\overline{y} + \overline{w}xz + w\overline{x}y + w\overline{x}z + wyz + xyz,$$

is placed in a four-variable Karnaugh map, the total number of ones in this map is:
a. six.
b. seven.
c. eight.
d. more than eight.

3.3. The minimal sum-of-products form of the function S specified in the Karnaugh map has:

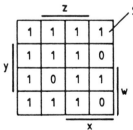

a. four terms with a total of nine letters (= variables).
b. five terms with a total of eleven letters.
c. six terms with a total of thirteen letters.
d. six terms with a total of fourteen letters.

3.4. The minimal sum-of-products form of the function S specified in the Karnaugh map below has:
   a. three terms with a total of six letters (= variables).
   b. four terms with a total of nine letters.
   c. four terms with a total of ten letters.
   d. four terms with a total of eleven letters.

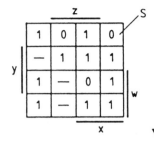

3.5. The minimal sum-of-products form of the function S specified in the above Karnaugh map has:
   a. four terms with a total of nine letters (= variables).
   b. four terms with a total of ten letters.
   c. five terms with a total of eleven letters.
   d. five terms with a total of twelve or more letters.

3.6. A Karnaugh map is given below.
   Of the minimal sum-of-products form of the given function S:
   a. there is one form.
   b. there are two forms with an equal number of terms/letters.
   c. there are three forms with an equal number of terms/letters.
   d. there are four forms with an equal number of terms/letters.

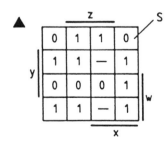 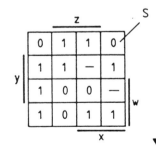

3.7. A Karnaugh map is given above.
   Of the minimal sum-of-products form of the given function S:
   a. there is one form.
   b. there are two forms with an equal number of terms/letters.
   c. there are three forms with an equal number of terms/letters.
   d. there are four forms with an equal number of terms/letters.

3.8. The function S is given as:
$$S = wyz + w\bar{x}\bar{y}\bar{z} + \bar{w}\bar{x}yz.$$
The minimal form of the function $\bar{S}$, written as a sum-of-products, contains:
a. five terms with a total of ten variables.
b. five terms with a total of twelve variables.
c. four terms with a total of eleven variables.
d. three terms with a total of nine variables.

3.9. The negation $\bar{S}$ of the function S,
$$S = \bar{w}y + \bar{x}y + \bar{x}z,$$
in the *minimal sum-of-products* form is:
a. $\bar{S} = \bar{w}y + \bar{x}z + \bar{x}y$.
b. $\bar{S} = x\bar{y} + \bar{y}\bar{z} + wx$.
c. $\bar{S} = \bar{x}\bar{y}\bar{z} + wx + x\bar{y}$.
d. $\bar{S} = \bar{x}\bar{y}\bar{z} + \bar{w}x\bar{y} + wx$.

3.10. The minimal form as *product-of-sums* of S in the Karnaugh map below consists of:
a. three factors with a total of eight letters.
b. four factors with a total of nine letters.
c. five factors with a total of 20 letters.
d. seven factors with a total of 28 letters.

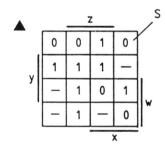
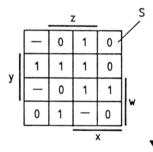

3.11. The above Karnaugh map specifies the function S. The *minimal forms* of S are determined as sum-of-products and also as product-of-sums. For these forms of S it holds that:
a. The sum form contains four terms, while the product form contains five factors.
b. The sum form contains five terms, while the product form contains four factors.
c. The sum form contains four terms, while the product form contains four factors.
d. The sum form contains five terms, while the product form contains five factors.

3.12. A prime implicant of a function S is a product term that belongs to S which cannot be combined with other product terms to form a product term in fewer variables. In the process of combining implicants don't cares may be used. The set of prime implicants forms the basis from which the minimal sum-of-products form can be determined.

The Karnaugh map below specifies a function S, with don't cares. The total number of *different* prime implicants for this specification is:
a. four.
b. five.
c. six.
d. seven.

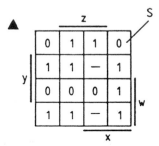
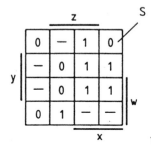

3.13. Four statements are made about the function S specified in the Karnaugh map above. One of them is *not* correct. Which one?
a. For the *minimal forms* S_1 as a sum-of-products and S_2 as a product-of-sums of the function S it holds that $S_2 = S_1$.
b. The minimal product-of-sums form contains a total of six letters/variables (not counting 'S =').
c. The minimal product-of-sums form and the minimal sum-of-products form have the same number of letters.
d. All don't cares can be freely chosen as 0 or as 1. A different choice for different don't cares is permitted.

3.14. The function of a device depends on the present value of four signals w through z, which have been specified as:

$\overline{w}\overline{x} = 1$ ⇒ set action 1;

$yz = 1$ ⇒ set action 2;

$wx\overline{z} = 1$ ⇒ set action 3;

$w\overline{y} = 1$ ⇒ set action 4;

$\overline{w}x\overline{y} + \overline{w}x\overline{z} = 1$ ⇒ set action 5.

Examine this specification with a Karnaugh map. It then follows that:
a. The function has been specified for all combinations of w through z.
   The specification does not contain any contradictions.
b. The function has not yet been specified for one combination of w through z.
   The specification is contradictory in two cases.
c. The function has not yet been specified for one combination of w through z.
   The specification is contradictory in three cases.
d. The function has not yet been specified for two combinations of w through z.
   The specification is contradictory in three cases.

3.15. The binary variables x, y and z can pass through all combinations of values, from xyz = 000 through xyz = 111. In the three Karnaugh maps below it is indicated how functions S, T and U depend on x, y and z.

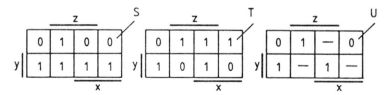

For U one may take:
a. U = S·T.
b. U = S + T.
c. U = S ⊕ T.
d. U is not a simple function of S and T.

Chapter 4

4.1.

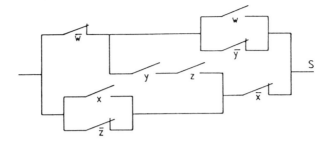

The network represented above consists of make and break contacts. When this network is conducting, its logic function S has the value of 1. S is placed in a four-variable Karnaugh map. This map then contains:
a. seven ones.
b. eight ones.
c. nine ones.
d. ten or more ones.

4.2. Put a 1 in the Karnaugh map if the contact network conducts for the corresponding input combination (w = 0 ↔ w-contact is in the position as drawn; etc.).

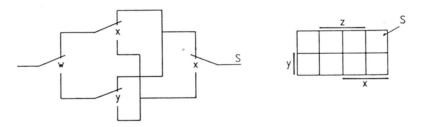

The Karnaugh map then contains:
a. four ones.
b. five ones.
c. six ones.
d. another number of ones.

4.3. A circuit is given below.

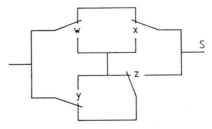

The minimal formula for S (interconnection ↔ 1) is:
a. $S = x + \overline{w} + \overline{y}z$.
b. $S = wx + \overline{wx} + \overline{y}z$.
c. $S = wx + \overline{wx} + xy + \overline{y}z$.
d. $S = wx + \overline{wx} + x\overline{y} + \overline{y}z$.

4.4. A circuit is given below.

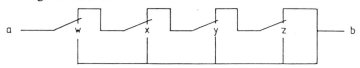

The minimal formula for the function S (between a and b, interconnection ↔ 1) is:
a. $S = 1$.
b. $S = w + x + y + z$.
c. $S = w + \overline{w}x + \overline{wx}y + \overline{wxy}z$.
d. $S = w + \overline{w}x + \overline{wx}y + \overline{wxy}z + \overline{wx}\overline{y}\overline{z}$.

4.5. There is a circuit between two voltage levels H and L, consisting of contacts of two relays Y and Z. The table applies to the output S.

| relay activated | | output voltage |
relay Y	relay Z	on S
no	no	not connected
no	yes	L
yes	no	H
yes	yes	not connected

In this case the minimal circuit is:

a. 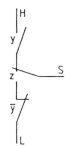 b. 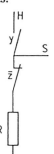 c.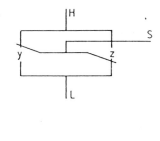

d. None of the above.

4.6. Of the following statements
 – With contact circuits realised in the AND-OR form (parallel interconnections (OR) of series interconnections (AND)) hidden loops never occur.
 – In the non-activated state of a relay the contacts are always open.
 – In a realisation through decomposition only transfer contacts are used.
a. none are false.
b. one is false.

c. two are false.
d. three are false.

4.7. The underlying circuit contains a hidden loop.

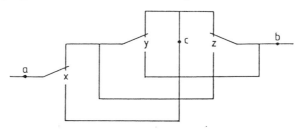

Which of the following statements about this figure is correct?
a. The hidden loop can be eliminated by placing a diode at point a or at point b.
b. The hidden loop can be eliminated by placing a diode at point c.
c. The hidden loop need not be eliminated because, by virtue of another path, the corresponding term already belongs to the function.
d. None of the above.

4.8. Design, as simply as possible, a circuit with contacts which realises the formula
$$S = w + \overline{w}x + \overline{w}\overline{x}y + \overline{w}\overline{x}\overline{y}z.$$
$S = 1$ if the circuit forms an interconnection. The total circuit requires (at least):
a. four make contacts.
b. four transfer contacts.
c. three transfer contacts and one make contact.
d. more than the number of contacts mentioned in a through c.

4.9. Design, as simply as possible, a circuit with contacts that realises the formula
$$S = w\overline{x} + \overline{w}z + x\overline{y} + yz.$$
$S = 1$ if the circuit forms an interconnection. The total circuit requires (at least):
a. three transfer contacts.
b. four transfer contacts.
c. five transfer contacts.
d. more than five transfer contacts, if necessary complemented with make and/or break contacts.

4.10. A circuit built with contacts and a relay is given below.

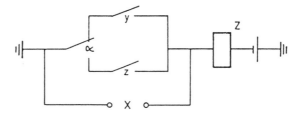

The relay Z of this circuit is activated under the condition $\overline{\alpha}y = 1$. During the transfer of α a so-called hold chain is formed via the contact z, which is then closed. If the transfer of α does not happen quickly enough, relay Z may be deactivated. This is not meant to happen. To bridge the switching of α an extra hold chain is added between the points X. This chain should be:

a. ── /z ──

b. ── /y ──

c. ── /y ── /z ──

d.

Chapter 5

5.1. *Positive logic* has been used in a diagram. Point x in the diagram has a voltage level 'Low' if a certain call signal is present. In the drawing point x is designated as:
 a. $\overline{\text{CALL}}$.
 b. CALL.
 c. CALL(H).
 d. CALL(L).

5.2. A symbol is given below in accordance with the IEC system.

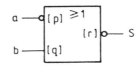

The internal signals have a certain logic value, which depends on the external logic values and the function symbol. The same *external* logic operation is described by the symbol:

 a.   a —[&]— S, b —
 b.   a —[≥1]— S, b —o
 c.   a —[&]— S, b —o
 d.   a —[&]o— S, b —o

5.3. A component description is given by means of the following H/L table.

a	b	c	S
L	L	L	L
L	L	H	H
L	H	L	H
L	H	H	L
H	L	L	H
H	L	H	L
H	H	L	L
H	H	H	H

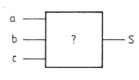

The function of this circuit can also be described by a logic formula. In doing so one may use so-called positive or negative logic, which may result in different formulas. Of the four statements made about this component, one is true. Which one?
 a. The formulas for positive and negative logic are identical.
 b. The formulas for positive and negative logic are each other's dual form and are therefore different.
 c. The formulas for positive and negative logic are each other's negation and are therefore different.
 d. None of the above.

5.4. A circuit is given below.

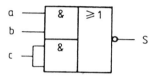

In this diagram *negative logic* has been used.
Which H/L table (note the order!) belongs to the circuit given?

a. a b c	S		b. a b c	S		c. a b c	S		d. a b c	S
L L L	L		L L L	L		L L L	H		L L L	H
L L H	L		L L H	H		L L H	H		L L H	L
L H L	L		L H L	L		L H L	H		L H L	H
L H H	H		L H H	H		L H H	L		L H H	L
H L L	L		H L L	L		H L L	H		H L L	H
H L H	H		H L H	H		H L H	L		H L H	L
H H L	L		H H L	H		H H L	H		H H L	L
H H H	H		H H H	H		H H H	L		H H H	L

5.5. A timing diagram of a circuit with the input signals a, b and c and the output signal S is shown.

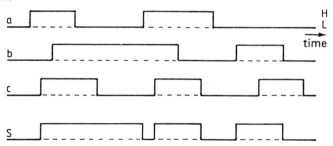

When negative logic is used, the logic function of this circuit is:
a. $S = ab + \bar{a}c$.
b. $S = ac + \bar{a}b$.
c. $S = \bar{a} + b + c$.
d. S is different from the functions specified in a through c.

5.6. A circuit is given below.

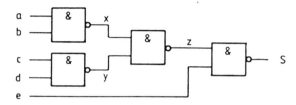

Express the signals on x, y and z in a through d. The circuit realises the following formula on its output S:

a. $S = ab + cd + e$.
b. $S = ab + cd + \bar{e}$.
c. $S = ac + ad + bc + bd + e$.
d. $S = \overline{ac} + \overline{ad} + \overline{bc} + \overline{bd} + \bar{e}$.

5.7. The logic function S is given as:

$S = \bar{x}\bar{y} + yz + xz$.

Besides x through z, $\bar{x}$ through $\bar{z}$ are also available. Minimise S and realise this function with NANDs. This requires at least:
a. two 2-input NANDs.
b. three 2-input NANDs.
c. two 2-input NANDs, and one 3-input NAND.
d. three 2-input NANDs, and one 3-input NAND.

5.8. In the realisation of the logic function S only NANDs are used.

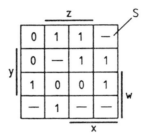

The variables w through z and $\bar{w}$ through $\bar{z}$ are available. The minimal realisation consists of:
a. three NANDs with a total of six inputs.
b. four NANDs with a total of nine inputs.
c. four NANDs with a total of eleven inputs.
d. five NANDs with a total of twelve inputs.

5.9. A circuit is given below.

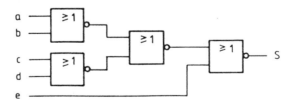

This circuit realises the following formula:
a. $S = \overline{ac} + \overline{ad} + \overline{bc} + \overline{bd} + \bar{e}$.
b. $S = \overline{ab} + \overline{cd} + \bar{e}$.
c. $S = \overline{abe} + \overline{cde}$.
d. $S = \overline{abcde}$.

5.10. Realise the logic function S,

$$S = \bar{y}z + xy\bar{z},$$

with NORs. The variables x through z and $\bar{x}$ through $\bar{z}$ are available. The total circuit then consists of at least:
a. three 2-input NORs.
b. three 2-input NORs, and one 3-input NOR.
c. four 2-input NORs.
d. four 2-input NORs, and one 3-input NOR.

5.11. In the realisation of function S only NOR gates are used. The signals w, x, y, z and $\bar{w}$, $\bar{x}$, $\bar{y}$, $\bar{z}$ are available. The minimal realisation consists of:
a. four NORs with a total of nine inputs.
b. three NORs with a total of eight inputs.
c. three NORs with a total of seven inputs.
d. three NORs with a total of six inputs.

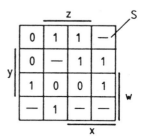

5.12. In the logic function S,

$$S = w + x + \bar{y} + \bar{z},$$

the variable $\bar{y}$ is replaced by a function. How many of the equations below are then correct?

$$S = w + x + \text{NOR}(y,w,x) + \bar{z}.$$

$$S = w + x + \text{NAND}(y,\bar{z}) + \bar{z}.$$

$$S = w + x + \text{NOR}(y,w,x,\bar{z}) + \bar{z}.$$

a. None are equal to $S = w + x + \bar{y} + \bar{z}$.
b. One is equal to $S = w + x + \bar{y} + \bar{z}$.
c. Two are equal to $S = w + x + \bar{y} + \bar{z}$.
d. Three are equal to $S = w + x + \bar{y} + \bar{z}$.

5.13. The circuit given below

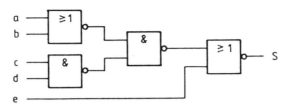

realises the following formula:
a. $S = (a + b)cd + \bar{e}$.
b. $S = (ab + cd)\bar{e}$.
c. $S = (\bar{a} + \bar{b} + \bar{c} + \bar{e})(\bar{a} + \bar{b} + \bar{d} + \bar{e})$.
d. $S = \overline{abce} + \overline{abde}$.

5.14. The logic diagram of a logic circuit to be realised is given in Figure a.

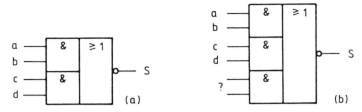

Figure b describes an available standard component, on which the circuit of Figure a can be mapped. One AND gate of Figure b is then superfluous. Which of the statements below applies to this situation?
a. The inputs of the superfluous AND gate do not need to be connected.
b. All inputs of the superfluous AND gate must be set to 1.
c. A zero must be set to at least one input of the superfluous AND gate; the input signal of the other input is then 0 or 1.
d. None of the above.

5.15. S is a function in the variables a through d. For S it holds that:

$$d = 0 \Rightarrow S = f_1(a,b,c);$$
$$d = 1 \Rightarrow S = f_2(a,b,c).$$

It is asserted that S can be written as:

1. $S = \bar{d} \cdot f_2(a,b,c) + d \cdot f_1(a,b,c);$
2. $S = \{\bar{d} + f_2(a,b,c)\} \cdot \{d + f_1(a,b,c)\}.$

Of the formulas for S:
a. the first is correct, and the second wrong.
b. the first is wrong, and the second correct.
c. both formulas are correct.
d. both formulas are wrong.

Chapter 6

6.1. Some gates have so-called Schmitt-trigger inputs, with L → H and H → L thresholds on different levels.

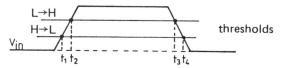

A gate with Schmitt-trigger inputs detects the 'H' level in an L → H → L input pulse between the points:
   a. t_1 and t_3.
   b. t_1 and t_4.
   c. t_2 and t_3.
   d. t_2 and t_4.

6.2. Schmitt-trigger inputs are used to:
   a. improve the AC noise margin (crosstalk).
   b. improve the AC and the DC noise margins.
   c. protect inputs against negative voltages (reflections).
   d. reduce the fanin of an input.

6.3. The 'DC noise margins' of a certain type of TTL ICs and another type of CMOS ICs are given below.

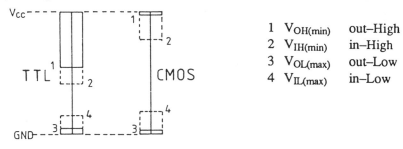

1 $V_{OH(min)}$ out–High
2 $V_{IH(min)}$ in–High
3 $V_{OL(max)}$ out–Low
4 $V_{IL(max)}$ in–Low

When we restrict ourselves to the 'DC noise margins', data transfer is possible from:
   a. TTL → CMOS.
   b. TTL → CMOS and CMOS → TTL.
   c. CMOS → TTL.
   d. No data transfer is possible.

6.4. Sometimes a gate (AND, OR, NAND, etc., part of an IC) has more inputs than necessary. For these inputs one of the four statements below holds. Give the correct statement.
   a. Unused inputs are always allowed to be left open in integrated circuits.
   b. Unused inputs must always be interconnected via a resistor, with the highest level of the supply voltage.
   c. Unused inputs must always be grounded or interconnected to the lowest supply voltage.
   d. None of the above.

6.5. The transition times t_{TLH} and t_{THL} of a signal are measured between 10 % and 90 % of the difference between H and L. The rise time varies between 8 and 27 ns and the fall time between 3 and 14 ns. This signal drives a gate. Its H ↔ L threshold varies somewhat, e.g. as a result of temperature variations. The threshold region lies between 30 % and 65 % of the difference between the H and L levels of the driving signal.

As a result of these effects it is uncertain when an input transition is recognised. What is the maximum uncertainty which this introduces at the input, assuming that all L → H and H → L transitions have the same starting point and increase and decrease, respectively, linearly?
- a. L → H: 18.9 ns; H → L: 10.9 ns (rounded off to one decimal).
- b. L → H: 15.2 ns; H → L: 8.0 ns.
- c. L → H: 9.5 ns; H → L: 5.4 ns.
- d. L → H and H → L transitions introduce other uncertainty intervals.

6.6. For measuring propagation delay times in TTL ICs V_{ref} is:
$$V_{ref} = 1.5 \text{ V.}$$

Measurements show the threshold to lie between 1.65 V and 1.25 V. It is known that the rise/fall time of the driving signals vary between 2 and 25 ns/V. How much uncertainty is introduced by the choice of a fixed reference point, or in other words how much sooner or later, respectively, is the real threshold passed?
- a. $-6.25 \text{ ns} \leq V_t \leq +3.75 \text{ ns}$ (– sooner, + later L → H transition).
- b. $-0.5 \text{ ns} \leq V_t \leq +0.15 \text{ ns}$.
- c. $-10 \text{ ns} \leq V_t \leq +10 \text{ ns}$.
- d. $+0.8 \text{ ns} \leq V_t \leq +10 \text{ ns}$.

6.7. The output driver circuits in TTL technology can be divided into three types:
- totem pole;
- open collector;
- three-state.

With which type of output circuit is one simply allowed to interconnect outputs directly (A), to interconnect outputs under certain conditions (B) or is one not allowed to connect them at all (C)?

	Totem pole	Open collector	Three-state
a.	A	B	C
b.	B	B	C
c.	B	C	A
d.	C	A	B

6.8. A TTL gate has open-collector outputs.
When a number of these outputs are interconnected, this creates a so-called:
- a. wired AND.
- b. wired OR.
- c. active-low wiring.
- d. active-high wiring.

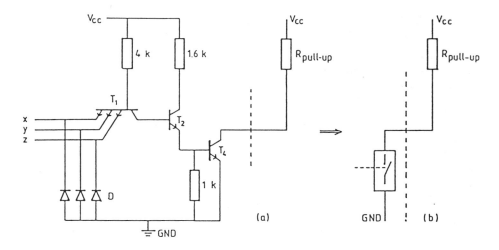

6.9. The major reason the fanout of a gate is restricted, with either TTL or CMOS, is:
  a. TTL: dissipation;   CMOS: dissipation.
  b. TTL: dissipation;   CMOS: propagation delay time.
  c. TTL: propagation delay time; CMOS: dissipation
  d. TTL: propagation delay time; CMOS: propagation delay time.

6.10. The logic function of this network (positive logic) is:
  a. $S = \overline{a \cdot (b + c)}$.
  b. $S = \overline{a + bc}$.
  c. $S = a \cdot (b + c)$.
  d. $S = a + bc$.

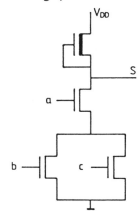

6.11. Four advantages of CMOS over NMOS are listed below. One of these does not apply. Which one?
  a. With CMOS the distance between $V_{IH(min)}$ and $V_{IL(max)}$, is larger than with NMOS.
  b. With CMOS the dissipation in the static state is much smaller than with NMOS.
  c. CMOS circuits are of the ratioless type; either the PMOS part or the NMOS part conducts. This makes the dimensions of the transistors far less critical.
  d. In CMOS, AND/OR gates are simpler than NAND/NOR gates. Therefore, formulas do not need to be converted into NAND-NAND or NOR-NOR form.

6.12. The timing behaviour of gates depends on various physical parameters. For designing with gates it normally suffices to know the propagation delay times $t_{P(min)}$ and $t_{P(max)}$ of the gates. On the basis of these propagation delay times a gate model is made, based on uncertainty intervals in the incoming and outgoing signals.

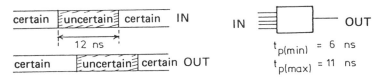

With an uncertainty interval of 12 ns in the input signal IN of the given gate the uncertainty interval at its output is of the following duration:
a. 12 ns.
b. 17 ns.
c. 18 ns.
d. 23 ns.

6.13. A gate circuit is given below.

	$t_{P(min)}$	$t_{P(max)}$	
A	10	15	ns
B	5	12	ns
C	7	13	ns

The total propagation delay time of this circuit lies between:
a. $t_{P(min)} = 5$; $t_{P(max)} = 15$.
b. $t_{P(min)} = 5$; $t_{P(max)} = 40$.
c. $t_{P(min)} = 7$; $t_{P(max)} = 40$.
d. $t_{P(min)} = 22$; $t_{P(max)} = 40$.

6.14. A network in NANDs is given below.

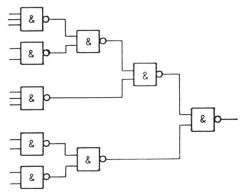

Of the gates used $t_{P(min)}$ and $t_{P(max)}$ are listed below:

2-input NAND: $t_{P(min)} = 10$ ns; $t_{P(max)} = 30$ ns,

3-input NAND: $t_{P(min)} = 15$ ns; $t_{P(max)} = 40$ ns.

For $t_{P(min)}$ and $t_{P(max)}$ of the whole network it holds that:
a. $t_{P(min)} \geq 35$ ns; $t_{P(max)} \leq 120$ ns.
b. $t_{P(min)} \geq 35$ ns; $t_{P(max)} > 120$ ns.
c. $t_{P(min)} < 35$ ns; $t_{P(max)} \leq 120$ ns.
d. $t_{P(min)} < 35$ ns; $t_{P(max)} > 120$ ns.

6.15. A circuit is built with gates of type A and type B.

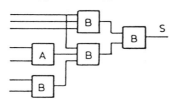

For type A it holds that: $t_{P(min)} = 10$ ns; $t_{P(max)} = 20$ ns.
For type B it holds that: $t_{P(min)} = 5$ ns; $t_{P(max)} = 15$ ns.
The input signals are stable until the moment t_0 and can subsequently change for a period of 30 ns. After this they are stable again. (So the uncertainty interval of the input signals runs from $t = t_0$ to $t = t_0 + 30$ ns.)
The uncertainty interval in the output signal S now ranges from:
a. t_0 to $t_0 + 60$ ns.
b. t_0 to $t_0 + 90$ ns.
c. $t_0 + 15$ to $t_0 + 90$ ns.
d. None of the above.

Chapter 7

7.1. The formulas of sum S_i and transport C_{i+1} of a full adder are given below:

$S_i = (a_i \oplus b_i) \oplus C_i,$

$C_{i+1} = a_i b_i + (a_i \oplus b_i)C_i.$

It is analysed whether the full adder can be made suitable for realising other functions by choosing its inputs and outputs appropriately. The following functions are required:
- Controllable negator: one control input determines whether another input signal is transmitted to the output in negated form or not.
- Controllable AND or OR function: one control input determines whether the AND or the OR function of two other input signals is taken.
- Controllable NAND or NOR function: one control input determines whether the NAND or the NOR function of two other input signals is taken.

A correct choice of inputs and outputs makes the full adder suitable for the realisation of:
a. none of the above-mentioned functions.
b. one of the above-mentioned functions.
c. two of the above-mentioned functions.
d. three of the above-mentioned functions.

7.2. An 8-bit FA is available.

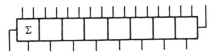

Single FAs are required for a certain application. By connecting the inputs in a certain way these can be obtained from the given 8-bit FA. How many single FAs can be obtained from one 8-bit FA?
a. Two FAs.
b. Three FAs.
c. Four FAs.
d. Eight FAs.

7.3. The structure of a FA circuit is given below.

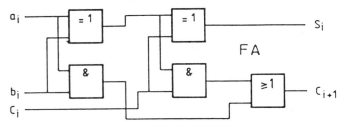

All gates have a propagation delay time $t_{PD} = 12$ ns (fixed). For the entire network it holds (indicate the interval as accurately as possible!) that:

a. $0 \text{ ns} \leq t_{PD} \leq 36 \text{ ns}$.
b. $12 \text{ ns} \leq t_{PD} \leq 36 \text{ ns}$.
c. $24 \text{ ns} \leq t_{PD} \leq 36 \text{ ns}$.
d. $36 \text{ ns} \leq t_{PD} \leq 48 \text{ ns}$.

7.4. A 4-bit full adder is given in which the output signals $C_4, S_3, \ldots, S_0$ are expressed in the P_is and G_is of the sections $0 \ldots 3$ and the incoming carry C_0 ($G_i = a_i b_i$, $P_i = a_i \oplus b_i$). Three statements are made about this 4-bit full adder:
- C_4 can only be 1 if $P_3 = 1$.
- C_0 is only passed to C_4 if $P_3 = P_2 = P_1 = P_0 = 1$.
- Of two neighbouring sections i and i–1 it is stated that section i generates a $G_i = 1$ and section i–1 a $G_{i-1} = 0$. In this case sum bit S_i is always zero.

Of these statements:
a. none are correct.
b. one is correct.
c. two are correct.
d. three are correct.

7.5. The carry generate and the carry propagate functions of a binary adder section are defined as:

$$G_i = a_i b_i$$
$$P_i = a_i + b_i.$$

Then C_{i+1} is:

$$C_{i+1} = G_i + P_i C_i.$$

Two numbers A and B are:

$$A = a_7 a_6 \ldots a_1 a_0 = 11011100_{BIN},$$
$$B = b_7 b_6 \ldots b_1 b_0 = 01000011_{BIN}.$$

These numbers are presented to an adder, with a carry look-ahead generator. The values that G_{4-7} and P_{0-3} will then assume are:

	G_{4-7}	P_{0-3}
a.	0	0
b.	0	1
c.	1	0
d.	1	1

7.6. The carry signal C_4 of an ALU (Arithmetic Logic Unit) is made according to the formula:

$$C_4 = G_{0-3} + P_{0-3} C_0.$$

In this formula:
a. $G_{0-3} = G_3 + P_3 G_2 + P_3 P_2 G_1 + P_3 P_2 P_1 G_0 + P_3 P_2 P_1 P_0$.
b. $G_{0-3} = G_3 + P_3 G_2 + P_3 P_2 G_1 + P_3 P_2 P_1 G_0$.
c. $G_{0-3} = G_3 G_2 G_1 G_0$.
d. $G_{0-3} = G_3 \cdot (P_3 + G_2) \cdot (P_3 + P_2 + G_1) \cdot (P_3 + P_2 + P_1 + G_0)$.

7.7. A circuit to be designed has to compare two non-negative binary encoded numbers $A = a_1a_0$ and $B = b_1b_0$. The output S must be 1 if $A \geq B$. For the circuit a logic specification is made with the help of the propositions

'Bit a_i/b_i is not 0' (this bit has the value of 1 with weight 2^i)

for the input signals and

'$A \geq B$'

for the output signal S.

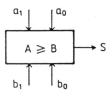

The logic function S is placed in a Karnaugh map, with four variables. This map then contains:
a. ten ones.
b. nine ones.
c. eight ones.
d. another number of ones than the numbers mentioned in a through c.

7.8. Four bits $a_3a_2a_1a_0$ are only permitted to be used in combinations that occur in the BCD code for one digit. To safeguard this a circuit whose function is to check this is designed. Its output S is 1 if a valid combination is found. Otherwise S = 0. If S is written in the *minterm form*, as a function of a_3 through a_0, S contains:
a. two minterms.
b. six minterms.
c. nine minterms.
d. ten minterms.

7.9. The logic diagram of a 2-line-to-4-line decoder/demultiplexer is given below.

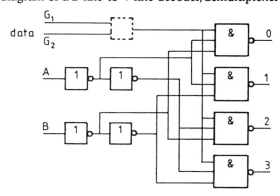

To get the data on the G_2 input onto output 1 the demultiplexer must be set to:
a. $ABG_1 = 011$ and the dashed component is an AND.
b. $ABG_1 = 010$ and the dashed component is a NAND.

c. $ABG_1 = 010$ and the dashed component is a NOR.
d. $ABG_1 = 101$ and the dashed component is a NAND.

7.10. The logic diagram of a 2-line-to-4-line decoder/multiplexer is given below.

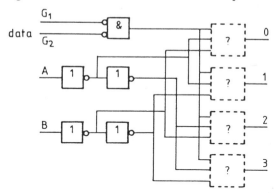

To get the data on the G_2 input onto output 1 (not negated) the decoder must be set to:
a. $ABG_1 = 010$; the dashed gate is an AND and the other outputs are 0.
b. $ABG_1 = 010$; the dashed gate is a NAND and the other outputs are 1.
c. $ABG_1 = 010$; the dashed gate is a NOR and the other outputs are 0.
d. $ABG_1 = 101$; the dashed gate is a NOR and the other outputs are 1.

7.11. For a certain application a 1-out-of-16 selector is required, which is not on hand. However, 1-out-of-4 selectors are available along with some NAND gates, if necessary. The 1-out-of-16 selector can be built with a minimum of:
a. four 1-out-of-4 selectors and one gate.
b. four 1-out-of-4 selectors and four gates.
c. five 1-out-of-4 selectors.
d. five 1-out-of-4 selectors and four gates.

7.12. On a 1-out-of-4 selector the signals w through z have been connected as depicted below.

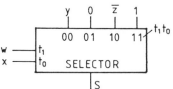

The circuit realises the logic function S, in the variables w through z. In a Karnaugh map of four variables the function S gives the following number of ones:
a. six.
b. seven.
c. eight.
d. nine.

7.13. An eighth bit is added to the signals on the seven data lines in accordance with the 'recipe' of odd parity. Errors can now be detected on the receiver side.

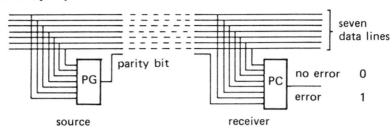

When the circuit denoted as PC is built with EX-OR gates *only* (beware: no error → 0 out), this circuit requires at least
a. four EX-OR gates.
b. five EX-OR gates.
c. seven EX-OR gates.
d. eight EX-OR gates.

7.14.

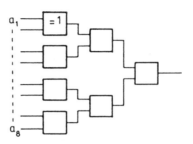

The above circuit can be used as parity generator and also as parity checker. The latter has been indicated below.

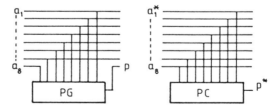

In the generator PG the input a_8 can be used to select between even and odd parity. In the checker PC the output p^* indicates whether or not an error has occurred. The interpretation of the value of p^* (right/wrong) depends on the setting of the generator (even/odd parity). Indicate which value a_8 must have when using even parity and which value p^* then has if one error occurs.
a. $a_8 = 0$; $p^* = 0$.
b. $a_8 = 1$; $p^* = 0$.
c. $a_8 = 0$; $p^* = 1$.
d. $a_8 = 1$; $p^* = 1$.

7.15. An m-out-of-n code has error detecting properties. The following statements are made about this code:
- One error is always detected.
- If m is even two errors are always detected.
- Correction of one error is possible provided that $m \geq 3$.

Of these statements:
a. none are correct.
b. one is correct.
c. two are correct.
d. three are correct.

Chapter 8

8.1. Decimal digits are often represented by 7-segment displays, consisting of seven lamps, denoted as a through g. The decimal digits 0 through 9 can be displayed by driving a certain combination of these lamps. These digits are often recorded in the BCD code internally in digital equipment. To display them, it is necessary to convert the BCD code to the 7-segment code.

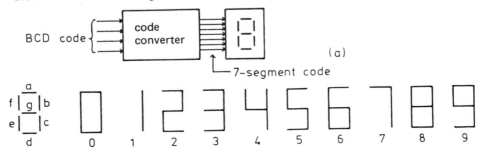

Which line from the truth table of this code converter is represented correctly?

	W X Y Z $2^3\ 2^2\ 2^1\ 2^0$	a b c d e f g *
a.	0 0 0 0	0 0 0 0 0 0 0
b.	0 1 0 0	0 1 1 0 0 1 1
c.	0 1 1 1	1 0 0 0 1 1 1
d.	1 0 0 1	1 1 1 1 1 1 0

* 1 ↔ lamp on

8.2. Combinational circuits cannot only be realised in gates, but also with programmable components such as PROMs and PLAs (Programmable Read Only Memories and Programmable Logic Arrays, respectively). PLAs are generally preferred over PROMs when:

 a. the functions to be realised contain a very large number of product terms.
 b. the functions to be realised are expressed in many variables.
 c. the functions to be realised are expressed in few variables.
 d. there are no programs available to minimise the functions.

8.3. A combinational circuit can be realised in gate circuits or be programmed in a PROM, as desired.

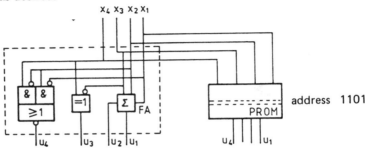

When programming the function of circuit A with inputs x_4 through x_1 and outputs u_4 through u_1 in a PROM the contents of address 1101 have to be:
a. 1110.
b. 1010.
c. 1001.
d. 0110.

8.4. Read Only Memories (ROMs) are components from which, depending on the present address, the contents of the memory location addressed are transported to the outputs.

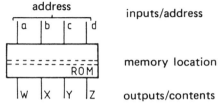

With the circuit as shown four functions are realised:

$W = \overline{ab}c + \overline{cd}$,

$X = ab\overline{cd}$,

$Y = (a \oplus b) \oplus (c \oplus d)$, $\oplus \leftrightarrow$ EX-OR,

$Z = a + b + c + d$.

For the address abcd = 1010 the contents of the memory location addressed are:
a. 1001.
b. 1011.
c. 1000.
d. 0001.

8.5. Read Only Memories (ROMs) are components from which, depending on the present address, the contents of the memory location addressed are transported to the outputs. Figure: see Problem 8.4.

The ROM sketched has *exactly* four inputs and four outputs. When the four functions

$W = ab + \overline{ab}$,

$X = abcd$,

$Y = cd + (\overline{ab} + a\overline{b})\overline{d}$,

$Z = a + b + c + d$,

are realised in this ROM, the memory contains a total of:
a. 16 ones.
b. 24 ones.
c. 32 ones.
d. more than 32 ones.

8.6. Read Only Memories (ROMs) are components from which, depending on the present address, the contents of the memory location addressed are transported to the outputs. Figure: see Problem 8.4.

The ROM sketched has *exactly* four inputs and four outputs. In this ROM four functions are realised:

$W = ab + \overline{abcd},$

$X = ab\overline{c}d,$

$Y = a\overline{b}c + ac\overline{d},$

$Z = a + b.$

The memory then contains a total of:
a. 18 ones.
b. 21 ones.
c. 43 ones.
d. none of the above.

8.7. A given ROM has *eight inputs* and *eight outputs*. Our aim is to realise as many logic functions as possible in this ROM, each function being defined on *four* variables a through d. For that reason a through d are put on the first four inputs. The second group of four inputs is set constant, by which a certain area of the ROM is selected. The function values belonging to the functions defined on a through d are programmed in this part of the ROM.

With the ROM size as given n functions can be realised *simultaneously*. How large is n?
a. n = 1.
b. n = 2.
c. n = 4.
d. n = 8.

8.8. A given ROM has *eight inputs* and *eight outputs*. Our aim is to realise as many logic functions as possible in this ROM, each function being defined on *six* variables a through f. For that reason a through f are put on the first six inputs. The remaining two inputs are set constant, by which a certain area of the ROM is selected. The function values of the functions as defined on a through f are programmed in this area.

By changing the 'constant' inputs another memory area is selected. Other functions can be programmed in this area. With the ROM size given *a total of* m different functions can be realised, k of which are available *simultaneously*. How large are m and k?
a. m = 6; k = 1.
b. m = 32; k = 6.
c. m = 32; k = 8.
d. m = 128; k = 8.

8.9.

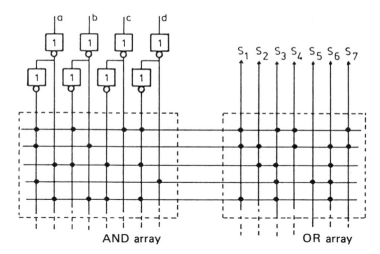

The diagram of a so-called programmable logic array (PLA), consisting of an AND and an OR array, is given above. The programming is as indicated. The dots indicate that there is an interconnection. The output S_3 realises the function:
a. $S = a\bar{d} + ab\bar{c} + \bar{a}cd$.
b. $S = \overline{ab}c\bar{d} + a\overline{bc}\bar{d} + \bar{a}d + ab\bar{c}\bar{d}$.
c. $S = (\bar{a} + \bar{b} + c + \bar{d})(a + \bar{b} + \bar{c} + \bar{d})(\bar{a} + d)(a + b + \bar{c} + \bar{d})$.
d. $S = \bar{a}d + \overline{abc} + a\overline{cd}$.

8.10. A combinational circuit can be realised, as desired, in gate circuits or programmed in a PLA. A PLA consists of a programmable AND and OR array. The programming is shown for the signals $S_1,...,S_4$. The dots indicate that there is an interconnection.

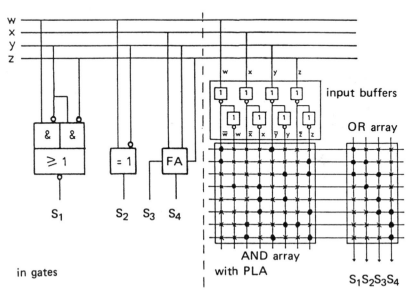

How many of the signals S_1, S_2, S_3, S_4 are programmed correctly?
a. None.
b. One.
c. Two.
d. Three.

8.11. PLAs are combinational circuits, consisting of a programmable AND array and a programmable OR array. When realising the functions

$$S_0 = w\bar{y}z + x\bar{y} + \bar{w}x + \bar{v}z,$$
$$S_1 = w\bar{y}\bar{z} + \bar{v}z + uz,$$
$$S_2 = \overline{wy} + x\bar{y} + \bar{u}xy + \bar{w}x,$$

the *minimum* number of product lines required is:
a. seven.
b. eight.
c. nine.
d. eleven.

8.12. PLAs are combinational circuits, consisting of a programmable AND array and a programmable OR array. Our aim is to realise (= program) three functions S_0, S_1 and S_2 in one PLA.

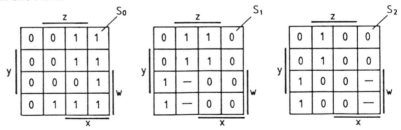

During programming one aims at using as few product terms as possible, in order to keep 'space' free for other functions, if desired. One of the ways of realising this is to use product terms *in common* as much as possible. As a result, the programming of the various functions is not always based on the minimal sum-of-products form. For the three functions given the minimum number of product lines in a PLA is:
a. five.
b. six.
c. seven.
d. eight.

8.13. PLAs are combinational circuits, consisting of a programmable AND array and a programmable OR array. Our aim is to realise (= program) three functions S_0, S_1 and S_2 in one PLA.
During programming one aims at using as few product terms as possible, in order to keep 'space' free for other functions, if desired. One of the ways of realising this is to use product terms *in common* as much as possible. As a result, the programming of the various functions is not always based on the minimal sum-of-products form.

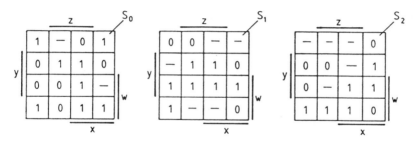

For the three functions given the minimum number of product lines in a PLA is:
a. five.
b. six.
c. seven.
d. eight.

8.14. In PALs the number of product terms per function can be increased by feeding back an output internally. See the figure for an example of the realisation of a function with seven product terms.

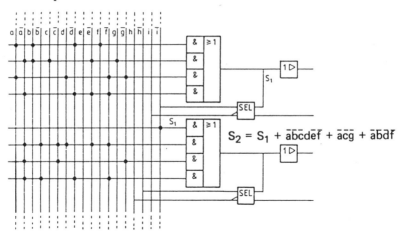

$S_2 = S_1 + \overline{abcdef} + \overline{acg} + \overline{abdf}$

For PROMs, another type of programmable logic, one of the following statements holds. Which one?
a. For a PROM this construction is not necessary / not useful.
b. For a PROM this construction requires an extra address input.
c. For a PROM this construction requires two extra address inputs.
d. For a PROM this construction requires double the number of address inputs.

8.15. Programmable logic can be realised on the basis of PALs. In a PAL the AND array is programmable, while the OR array is not. The figure shows the structure of a certain type of PAL.

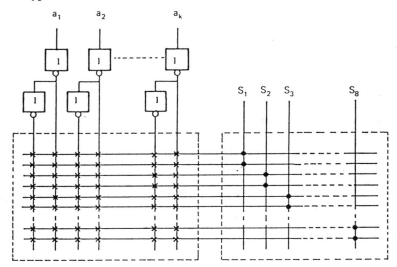

Three statements are made about this PAL:
- In this PAL eight arbitrary functions of maximally k variables can be programmed.
- In this PAL the function

$$S = a_1 a_2 a_3 + a_1 \bar{a}_2 a_3 + a_1 a_2 \bar{a}_4,$$

using one output pin S_i only, cannot be programmed.
- In this PAL it is never possible to program a function in more than two product terms.

Of these three statements:
a. none are false.
b. one is false.
c. two are false.
d. three are false.

Chapter 9

9.1. Consider the following circuit:

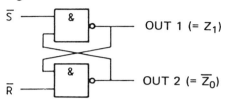

The input signals S and R are both 1 (*take note*: $\overline{S}$ and $\overline{R}$ are used). Which are the corresponding values of the outputs Out1 and Out2?
a. Out1 = 1; Out2 = 0.
b. Out1 and Out2 are not defined.
c. Out1 = Out2 = 1.
d. Out1 = $\overline{\text{Out2}}$ (therefore also $\overline{\text{Out1}}$ = Out2).

9.2. A circuit is given below.

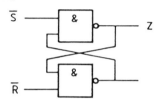

Which formula for Z_{new} belongs to this circuit?
a. $Z_{new} = (\overline{S} + \overline{R}) \cdot Z_{old}$.
b. $Z_{new} = \overline{R} \cdot (S + Z_{old})$.
c. $Z_{new} = S + \overline{R} \cdot Z_{old}$.
d. $Z_{new} = (\overline{S} + R) \cdot Z_{old}$.

9.3. A gated S-R latch is given below.

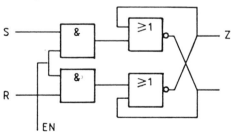

The *minimal* form of the formula that describes the logic operation of this circuit in the static state is:
a. $Z_{new} = \overline{EN} \cdot Z_{old} + \overline{R} \cdot Z_{old} + EN \cdot S \cdot \overline{R}$.
b. $Z_{new} = \overline{EN} \cdot Z_{old} + EN \cdot \overline{R} \cdot Z_{old} + EN \cdot S \cdot \overline{R}$.
c. $Z_{new} = \overline{EN} \cdot Z_{old} + \overline{R} \cdot Z_{old} + EN \cdot S$.
d. none of the above.

9.4. A circuit is composed of four D latches.

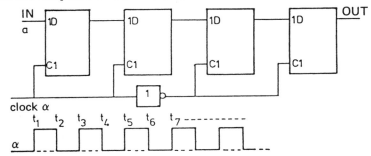

If just before t_1 the signal a is put on IN and a does not change during a large number of clock pulses, what is the earliest moment that signal a is seen on OUT? At or immediately after:

a. t_2.
b. t_3.
c. t_4.
d. a moment later than t_4.

9.5. The following circuit consists of so-called gated S-R latches and a few gates. The propagation delay time of the inverters is considerably shorter than that of the gated latches.

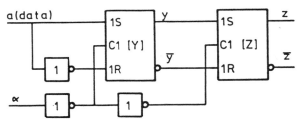

If α and a are as indicated, which timing diagram applies to y and z?

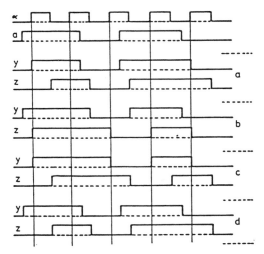

9.6.

S	R
1	0
1	1
0	0
0	1
0	0

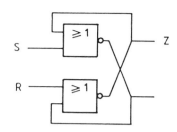

The input signals S and R of the S-R latch successively go through the values of the above table (from top to bottom). At the end S and R both remain 0; output Z has reached its final value. After running down the table and reaching its final value it turns out that:
a. Z *is* continually predictable; the final value is 0.
b. Z is *not* continually predictable; the final value is 0.
c. Z is *not* continually predictable; the final value is *not* predictable.
d. Z *is* continually predictable; the final value is 1.

9.7. The *state of the latch* is understood to be the value of the signal at point Z.

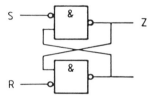

The following input combinations are presented to the S-R latch sketched here:

SR = 10 – 11 – 10 – 00 – 11 – 01 – 10 – 00 – 11 – 10.

With an S-R latch the state after an input transition cannot always be predicted. For the above input row, this turns out to be the case for:
a. zero cases.
b. one case.
c. two cases.
d. three or more cases.

9.8. Of the statements below exactly one is correct. Which one?
a. The state Z_{new} of Z_0, Z_1 and Z_z memory elements is uncertain if, starting from a known state, the input combination SR = 11 is presented.
b. The outputs of Z_0 and Z_1 memory elements always have complementary values with respect to each other.
c. The state Z_{new} of latches is uncertain if the input combination SR = 01 is followed by SR = 10.
d. The state Z_{new} of a Z_1 memory element is uncertain if the input combination SR = 11 is followed by SR = 00.

9.9. In a latch the set command (SET), the reset command (RESET) and the inhibit command (INHIBIT) are encoded as follows:

SET: SR = 10; RESET: SR = 01; INHIBIT: SR = 00.

With respect to this encoding problem one of the following statements is correct. Which one?
a. During the design of a latch these commands can only be encoded in the way indicated above.
b. In principle one is entirely free to choose this encoding.
c. This encoding must be such that no transient phenomena occur between the commands SET → INHIBIT and RESET → INHIBIT, which influence the final state of the latch.
d. This encoding must be such that no transient phenomena occur between the commands SET → INHIBIT, INHIBIT → SET, RESET → INHIBIT and INHIBIT → RESET, which influence the final state of the latch.

9.10. A so-called level mode state diagram is given, i.e. state transitions occur immediately after an input transition.

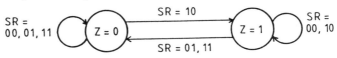

This state diagram describes the behaviour of a:
a. Z_0 memory element.
b. Z_1 memory element.
c. Z_z memory element.
d. none of the above.

9.11. A latch is constructed with two NANDs and has inputs A and B and an output Z. The other output is not considered.

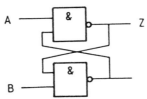

During the design of this latch one is confronted with the query which input combination AB must be used, starting with the present state Z_{old}, to bring the latch into the state Z_{new} as specified. Which of the tables below can be consulted to that end?

a.

Z_{old}	Z_{new}	A	B
0	0	0	0
0	1	1	0
1	0	0	1
1	1	0	0

b.

Z_{old}	Z_{new}	A	B			
0	0	0	0	or	0	1
0	1	1	0			
1	0	0	1			
1	1	0	0	or	1	0

c.

Z_{old}	Z_{new}	A	B			
0	0	1	0	or	1	1
0	1	0	1			
1	0	1	0			
1	1	0	1	or	1	1

d.

Z_{old}	Z_{new}	A	B			
0	0	0	0	or	1	1
0	1	0	1			
1	0	1	0			
1	1	0	0	or	1	1

9.12.

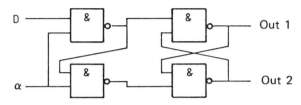

This circuit realises:
a. a gated D latch.
b. an $\overline{S}$-$\overline{R}$ latch.
c. a D flip-flop.
d. no memory element.

9.13.

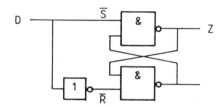

This circuit realises:
a. a gated D latch.
b. an $\overline{S}$-$\overline{R}$ latch.
c. a D flip-flop.
d. no memory element.

9.14. In general the anti-bounce circuit is based on transfer contacts. A make contact with an inverter has been used here instead of a tranfer contact.

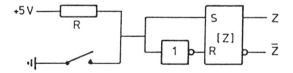

This setup is:
a. always unsuitable as an anti-bounce circuit.
b. suitable as an anti-bounce circuit.
c. suitable as an anti-bounce circuit, on the condition that the contact does not bounce back entirely when switching.
d. suitable as an anti-bounce circuit if the ground connection and the +5 V connection are interchanged.

9.15.

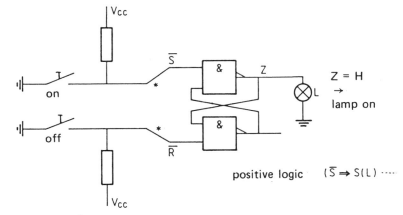

The diagram shows a so-called omnibus circuit. The lamp should switch on when 'on' is pressed, and switch off when 'off' is pressed.
Which of these four statements applies?
a. Inverters (negators) must still be mounted at the points * because of a level conversion to the latch inputs.
b. If both switches are pressed simultaneously the lamp is on.
c. The contacts must first be debounced.
d. If both switches are released simultaneously the lamp will always switch on.

Chapter 10

10.1. The truth table given below specifies how Q^{n+1} of a flip-flop Z depends on Q^n, its previous state, and two input signals A and B.

A^n	B^n	Q^n	Q^{n+1}
0	0	0	0
0	0	1	0
0	1	0	1
0	1	1	0
1	0	0	0
1	0	1	1
1	1	0	1
1	1	1	1

Which formula of Q^{n+1} belongs to this flip-flop?
a. $Q^{n+1} = [AQ + B\overline{Q}]^n$.
b. $Q^{n+1} = [\overline{A}Q + B\overline{Q}]^n$.
c. $Q^{n+1} = [BQ + A\overline{Q}]^n$.
d. $Q^{n+1} = [AB + AQ + BQ]^n$.

10.2. A *clock mode* circuit has three memory elements X, Y and Z. The following sequencing table applies to this circuit:

Step	$[Q_xQ_yQ_z]^n$	$[Q_xQ_yQ_z]^{n+1}$
0	0 0 0	0 1 0
1	0 0 1	1 0 0
2	0 1 0	1 1 0
3	0 1 1	1 1 1
4	1 0 0	0 0 1
5	1 0 1	0 1 1
6	1 1 0	0 0 0
7	1 1 1	1 0 1

From a given initial state the circuit returns to this initial state after a number of clock pulses. The circuit is therefore cyclic. Indicate which statement is true.
a. The circuit has one cycle of eight steps.
b. The circuit has one cycle of three steps and one of five steps.
c. The circuit has two cycles of two steps and one of four steps.
d. The circuit has one cycle of two steps and two of three steps.

10.3. A circuit is given below.

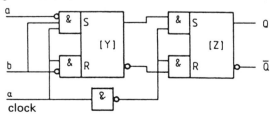

Which formula belongs to this circuit?
a. $Q^{n+1} = [a\overline{Q} + \overline{b}Q]^n$.
b. $Q^{n+1} = [\overline{a}b + bQ]^n$.
c. $Q^{n+1} = [\overline{a} + b]^n$.
d. $Q^{n+1} = [a\overline{b}]^n$.

10.4. A circuit is given below.

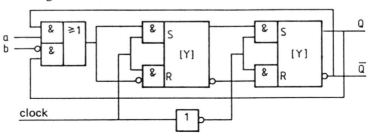

Which formula belongs to this circuit?
a. $Q^{n+1} = [a\overline{Q} + \overline{b}Q]^n$.
b. $Q^{n+1} = [\overline{a}Q + b\overline{Q}]^n$.
c. $Q^{n+1} = [\overline{a}Q + \overline{b}\overline{Q}]^n$.
d. $Q^{n+1} = [a\overline{Q} + bQ]^n$.

10.5. A flip-flop with formula

$$Q^{n+1} = [I_1 + I_2\overline{Q}]^n$$

is realised on the basis of a J-K flip-flop with formula

$$Q^{n+1} = [J\overline{Q} + \overline{K}Q]^n.$$

This is possible, although it may be necessary to add a few extra gates, with:
a. $J = I_1 \cdot I_2$ and $K = 0$.
b. $J = I_1 + I_2$ and $K = \overline{I_1}$.
c. $J = I_1 + I_2\overline{Q}$ and $K = 0$.
d. $J = 0$ and $K = I_1 + I_2\overline{Q}$.

10.6. The logic behaviour of an I_1-I_2 flip-flop has been specified in Table a. Table b is derived from this table.

I_1^n	I_2^n	Q^{n+1}
0	0	$\overline{Q^n}$
0	1	1
1	0	Q^n
1	1	$\overline{Q^n}$

(a) $\longrightarrow$

Q^n	Q^{n+1}	I_1^n	I_2^n
0	0	•	•
0	1	•	•
1	0	•	•
1	1	•	•

(b)

It turns out that a $Q^n \to Q^{n+1}$ transition can sometimes be realised in various ways, depending on the choice of the input combination $[I_1I_2]^n$. The number of combinations that may be used per $Q^n \to Q^{n+1}$ transition is given in the following table. Which column of the table applies?

		Number of combinations $[I_1I_2]^n$			
Q^n	Q^{n+1}	a	b	c	d
0	0	1	2	2	1
0	1	3	2	2	3
1	0	3	2	1	2
1	1	1	2	3	2

10.7. The formula of a flip-flop with input signals I_1 and I_2 is:

$$Q^{n+1} = [I_1\overline{I_2} + I_1Q + \overline{I_2}Q]^n.$$

In the four function tables below, it has been specified which input combination must/can be used for each state transition $Q^n \to Q^{n+1}$. One of these belongs to the given formula of the flip-flop. Which table is the correct one?

a.

Q^n	Q^{n+1}	I_1^n	I_2^n
0	0	0	0
		0	1
		1	1
0	1	1	0
1	0	0	1
1	1	0	0
		1	0
		1	1

b.

Q^n	Q^{n+1}	I_1^n	I_2^n
0	0	1	0
0	1	0	0
		0	1
		1	1
1	0	0	0
		1	0
		1	1
1	1	0	1

c.

Q^n	Q^{n+1}	I_1^n	I_2^n
0	0	0	1
0	1	0	0
		1	0
		1	1
1	0	0	0
		0	1
		1	1
1	1	1	0

d.

Q^n	Q^{n+1}	I_1^n	I_2^n
0	0	0	0
		1	0
		1	1
0	1	0	1
1	0	0	0
1	1	0	0
		0	1
		1	1

10.8. The following circuit consists of an edge-triggered D flip-flop, a gated D latch and a few gates. It may be assumed that the $t_{P(max)}$ of a gate is to be neglected with respect to $t_{P(max)}$ of a memory element.

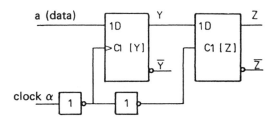

When α and a are as indicated, which timing diagram applies to Q_y and Q_z?

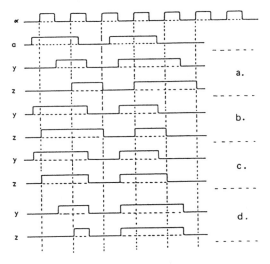

10.9. A circuit consisting of a J-K flip-flop and a NAND is given below. The formula of the J-K flip-flop is $Q^{n+1} = [J\bar{Q} + \bar{K}Q]^n$.

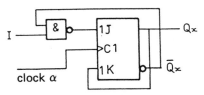

Which statement applies to this circuit?
a. The circuit is a T flip-flop ($Q^{n+1} = [T \oplus Q]^n = [T\bar{Q} + \bar{T}Q]^n$) if I = 0 and is set (goes to 1) if I = 1.
b. The circuit is reset (goes to 0) if I = 1 and subsequently remains in the 0 state, irrespective of I.
c. The behaviour of the circuit does not change if the K input is set to '0' permanently instead of to Q_x.
d. The circuit can be reduced; its behaviour does not change if the NAND is replaced by a negator with I as input signal and a '1' on the K input.

10.10.

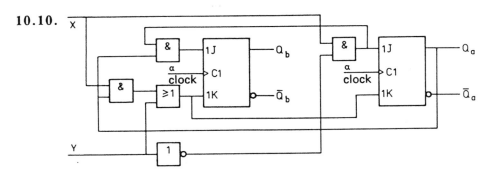

The circuit with input signals X and Y given above is built up of edge-triggered J-K flip-flops ($Q^{n+1} = [J\overline{Q} + \overline{K}Q]^n$). If the states of the flip-flops before the clock pulse correspond to $Q_b^n = 0$ and $Q_a^n = 1$ and the input signals are $X^n = 1$ and $Y^n = 0$, then which states do the flip-flops take *after the next active clock edge*?
  a. $Q_b^{n+1} = 0$ and $Q_a^{n+1} = 0$.
  b. $Q_b^{n+1} = 0$ and $Q_a^{n+1} = 1$.
  c. $Q_b^{n+1} = 1$ and $Q_a^{n+1} = 0$.
  d. $Q_b^{n+1} = 1$ and $Q_a^{n+1} = 1$.

10.11. A circuit is given below.

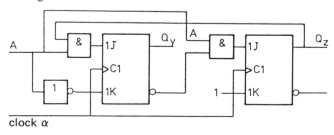

When, starting with the state $Q_y^n Q_z^n = 00$, the input signal becomes $A = 1$ and remains so for the next two clock pulses, which is the state $[Q_y Q_z]^{n+2}$ after two clock pulses?
  a. $[Q_y Q_z]^{n+2} = 00$.
  b. $[Q_y Q_z]^{n+2} = 01$.
  c. $[Q_y Q_z]^{n+2} = 10$.
  d. $[Q_y Q_z]^{n+2} = 11$.

10.12. A circuit is constructed with two flip-flops Y and Z and a number of gates. The circuit furthermore has two input signals A and B, of which it is known that they *cannot* be 1 simultaneously. Flip-flop Y of the circuit has the following behaviour:
  – If AB = 10 it holds that: $Q_y^{n+1} = \overline{Q_z^n}$.
  – If AB = 01 it holds that: $Q_y^{n+1} = \overline{Q_y^n}$.
  – In all other cases, *if any*, it holds that: $Q_y^{n+1} = Q_y^n$.

The behaviour of flip-flop Y is, *as far as* can be determined under the above specifications, specified in a Karnaugh map. Which of the maps below applies?

a.
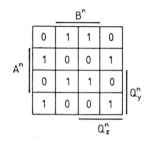

b.

	B^n			
	0	1	1	0
A^n	1	0	0	1
	0	1	1	0
	1	0	0	1
		Q_z^n		

c.

$\overline{B^n}$

	0	1	0	1
A^n	1	—	—	0
	1	—	—	0
	0	1	0	1

Q_y^n

Q_z^n

d. None of the maps.

10.13. For a *clock mode circuit*, consisting of three memory elements X, Y and Z, Q_x^{n+1} as a function of Q_x^n, Q_y^n and Q_z^n is given by the following Karnaugh map:

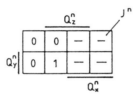

Q^n	Q^{n+1}	J^n	K^n
0	0	0	—
0	1	1	—
1	0	—	1
1	1	—	0

For memory element X a J-K flip-flop is chosen. The function table of this flip-flop is given. For the J and K input signals the following Karnaugh maps are made:

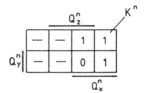

Which statement is correct?
a. The map for J^n is correct, and that for K^n is incorrect.
b. Both maps are incorrect.
c. The map for J^n is incorrect, and that for K^n is correct.
d. Both maps are correct.

10.14. The design of a D flip-flop with input selector is given below. The truth table describes its external behaviour.

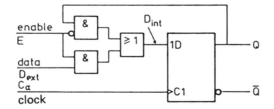

E^n	D_{ext}^n	Q^n	D_{int}^n	Q^{n+1}
0	0	0	0	0
0	0	1	1	1
0	1	0	0	0
0	1	1	1	1
1	0	0	0	0
1	0	1	0	0
1	1	0	1	1
1	1	1	1	1

The circuit is realised on the basis of a J-K flip-flop. The table of this flip-flop is given below.

Q^n	Q^{n+1}	J^n	K^n
0	0	0	–
0	1	1	–
1	0	–	1
1	1	–	0

$Q^{n+1} = [J\overline{Q} + \overline{K}Q]^n$

Karnaugh maps are made for the J and K input signals. The correct formulas for J^n and K^n are:
a. $J^n = [\overline{E} + \overline{D}]^n$; $K^n = [\overline{E} + D]^n$.
b. $J^n = [E + D]^n$; $K^n = [E\overline{D}]^n$.
c. $J^n = [ED]^n$; $K^n = [E\overline{D}]^n$.
d. $J^n = [\overline{E}Q]^n$; $K^n = [\overline{E}\overline{Q}]^n$.

10.15. It is specified in a Karnaugh map how Q_y^{n+1} of a circuit with three T flip-flops depends on Q_x^n, Q_y^n and Q_z^n. The function table of the T flip-flop is given as well.

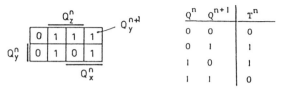

Which of the Karnaugh maps below applies to the T signal of flip-flop Y?

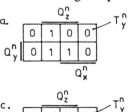

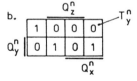

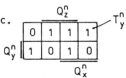

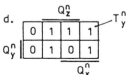

10.16. The function for Q^{n+1} of a D flip-flop as specified in the Karnaugh map is implemented with *NOR gates*. Assume that the negated values of Q^n, a^n, b^n and c^n are also available. To drive the D input signal the following number of gates is *minimally* required:
a. Three gates with a total of nine inputs.
b. Four gates with a total of twelve inputs.
c. Five gates with a total of fourteen inputs.
d. More than five gates.

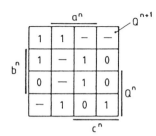

10.17. The figure describes the state diagram of a sequential circuit with four states, of which the state assignment with two flip-flops Y and Z is given. State q_0 is encoded with $Q_yQ_z = 00$, to which the output signal $U = 0$ belongs, etc. All state transitions occur under control of the clock pulse. The flip-flops are positive edge-triggered. Which transition occurs depends on an input signal A, which has been synchronised on the clock pulse.

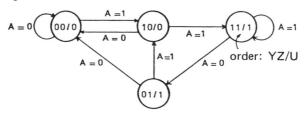

Assume that during t_0 the circuit is in the state encoded with $Q_yQ_z = 00$ and that input signal A assumes the value indicated:

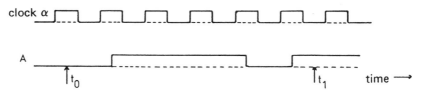

In which state is the circuit at the moment t_1?
a. $Q_yQ_z = 00$.
b. $Q_yQ_z = 01$.
c. $Q_yQ_z = 10$.
d. $Q_yQ_z = 11$.

10.18. A circuit with two edge-triggered D flip-flops and an AND gate is given below.

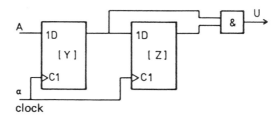

The behaviour of this circuit can be described in a clock mode state diagram such as:

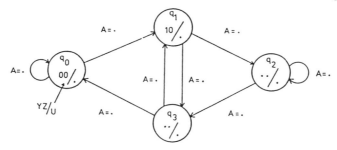

The dot (·) indicates where the diagram still has to be completed. What is the state assignment of state q_3 and which value does input signal A have on the transition of q_3 to q_0?
a. Code Q_3: YZ = 01; transition $q_3 \rightarrow q_0$ if A = 0.
b. Code Q_3: YZ = 01; transition $q_3 \rightarrow q_0$ if A = 1.
c. Code Q_3: YZ = 11; transition $q_3 \rightarrow q_0$ if A = 0.
d. Code Q_3: YZ = 11; transition $q_3 \rightarrow q_0$ if A = 1.

10.19. The state diagram of a sequential circuit with two input signals a and b is given below.

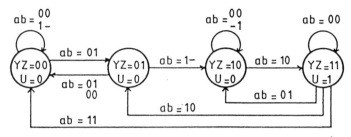

Signals a and b can assume four combinations of values. Output Q_z^{n+1} of memory element Z is specified in a Karnaugh map in four variables. This map contains:
a. three ones.
b. four ones.
c. five ones.
d. more than five ones.

10.20. The state diagram below describes a sequential circuit in the clock mode (state transitions under control of the clock pulse). The state assignment with two D flip-flops Y and Z is indicated. A is the input signal of the circuit.

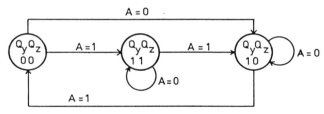

The minimal formula for the signal D_z is:
a. $D_z^n = [A\overline{Q}_y + \overline{A}Q_z]^n$.
b. $D_z = [A\overline{Q}_z + \overline{A}Q_z]^n$.
c. $D_z = [A\overline{Q}_y + \overline{A}Q_y]^n$.
d. none of the above.

Chapter 11

11.1. The diagram of a gated D latch is given below.

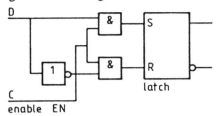

The signal on the D input is meant to determine unambiguously the state of the latch in the inhibit mode (inputs disabled). Which timing diagram specifies the requirements imposed on the signal on the D input with respect to the signal on the C input?

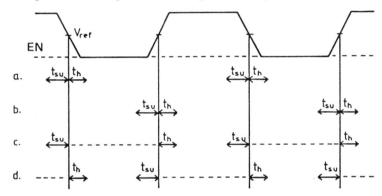

11.2. The timing parameters of a positive edge-triggered D flip-flop are indicated in the timing diagram:

t_{su}: setup time data with respect to the clock,

t_h: hold time data with respect to the clock,

$t_{P(min)}$: minimum propagation delay time Q output with respect to the clock,

$t_{P(max)}$: maximum propagation delay time Q output with respect to the clock.

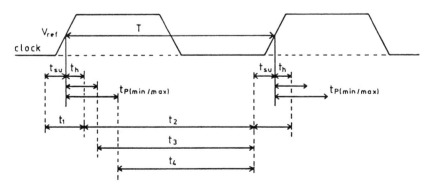

In addition, four intervals t_1 through t_4 are indicated. Data on the D input is only allowed to change during:
a. t_1.
b. t_2.
c. t_3.
d. t_4.

11.3.

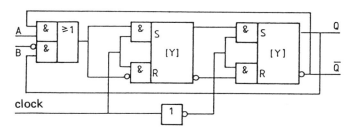

With respect to the clock pulse the flip-flop given above has a:
a. positive edge-triggered timing.
b. negative edge-triggered timing.
c. positive pulse-triggered timing.
d. negative pulse-triggered timing.

11.4. A pulse-triggered flip-flop is given. The clock pulse of this flip-flop is symmetrical and has a repetition frequency of 1 MHz. For this flip-flop the timing parameters are specified as:

$t_{su} = 5$ ns; $t_h = 12$ ns; $t_{P(min)} = 17$ ns; $t_{P(max)} = 25$ ns.

Input data for this flip-flop must be stable in each clock period for:
a. 5 ns.
b. 12 ns.
c. 17 ns.
d. 517 ns.

11.5. A circuit is given below:

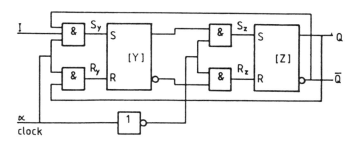

Which timing diagram specifies the requirements to be imposed on the input signal I with respect to the clock pulse α?

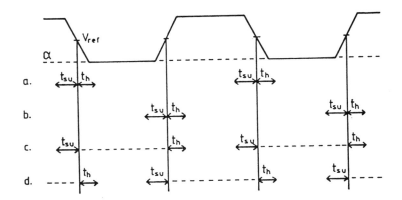

11.6. A circuit is given below:

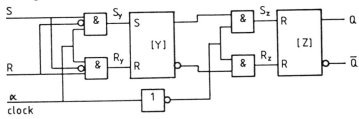

Which timing diagram specifies the requirements to be imposed on the input signals S and R with respect to the clock pulse α.

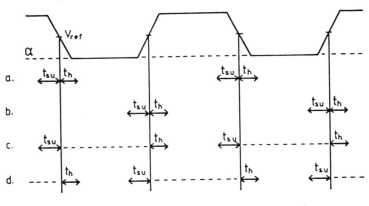

11.7.

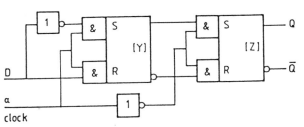

Which symbol belongs to the logic diagram of the flip-flop given above?

a. 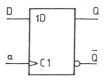   b.

c. 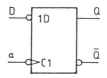   d.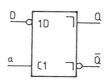

11.8. The timing of flip-flops is characterised by four timing parameters:

$\left.\begin{array}{l}\text{setup time } t_{su} \\ \text{hold time } t_h\end{array}\right\}$ input signals.

$\left.\begin{array}{l}\text{propagation delay time } t_{P(min)} \\ \text{propagation delay time } t_{P(max)}\end{array}\right\}$ output signals.

A certain type of flip-flop has a positive edge-triggered timing. All possible configurations of circuits can be constructed with these flip-flops (clock mode). For a proper operation of the entire system one of the following requirements must hold:

a. $t_{su} > t_h$.
b. $t_{su} \leq t_h$.
c. $t_h < t_{P(min)}$.
d. $t_{P(max)} < t_{su}$.

11.9. A circuit is constructed with two D flip-flops and an AND gate. The circuit has a positive edge-triggered timing.

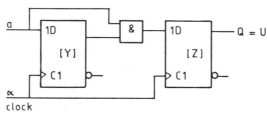

In this circuit the following timing parameters hold:

DFF: $t_{su} = 10$ ns; $t_{P(min)} = 15$ ns;

$t_h = 2$ ns; $t_{P(max)} = 25$ ns.

AND: $t_{P(min)} = 10$ ns;

$t_{P(max)} = 25$ ns.

The clock frequency of this circuit is 5 MHz.
On the basis of the timing parameters specified the input signal a can only be set for a limited part of the period of the clock. This part is:

a. 17.5 % of the period of the clock.
b. 81.5 % of the period of the clock.
c. 85 % of the period of the clock.
d. 94 % of the period of the clock.

11.10. A circuit is constructed with two D flip-flops and an AND gate. The circuit has a positive edge-triggered timing. Figure: see Problem 11.9.
In this circuit the following timing parameters hold:

DFF: t_{su} = 5 ns; $t_{P(min)}$ = 15 ns;
       t_h = 2 ns; $t_{P(max)}$ = 25 ns.

AND: $t_{P(min)}$ = ... ;
       $t_{P(max)}$ = ...

The input signal a is uncertain from 30 ns after the positive clock edge to 85 ns after the positive clock edge.
The clock pulse frequency is set to 10 MHz (clock period is 100 ns). With which values of $t_{P(min)}$ and $t_{P(max)}$ of the AND gate does the circuit operate reliably?
a. $t_{P(min)}$ = 0 ns; $t_{P(max)}$ = 5 ns.
b. $t_{P(min)}$ = 0 ns; $t_{P(max)}$ = 50 ns.
c. $t_{P(min)}$ = 15 ns; $t_{P(max)}$ = 75 ns.
d. With the given clock frequency the circuit cannot operate reliably.

11.11. A circuit consists of two D flip-flops and an AND gate. The circuit has a positive edge-triggered timing. Figure: see Problem 11.9.
In this circuit the following timing parameters hold:

DFF: t_{su} = 5 ns; $t_{P(min)}$ = 15 ns;
       t_h = 2 ns; $t_{P(max)}$ = 25 ns.

AND: $t_{P(min)}$ = 10 ns;
       $t_{P(max)}$ = 25 ns.

The input signal a is uncertain from 30 ns after the positive clock edge to 85 ns after the positive clock edge.
The maximum clock pulse frequency, computed on the basis of these data, is:
a. 8.7 MHz.
b. 10.0 MHz.
c. 18.2 MHz.
d. more than 18.2 MHz.

11.12. The internal organisation of a 'clock buffer' is given below.

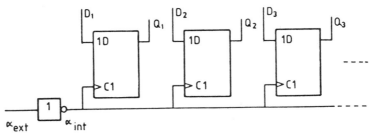

The parameters with which the D signals have to comply, with respect to the internal clock α_{int}, are:

FF: t_{su} = 30 ns;
t_h = 5 ns.

The following holds for the negator: $t_{P(min)}$ = 5 ns; $t_{P(max)}$ = 25 ns.
When the D signals are related to the external clock α_{ext}, they have to comply with:
a. t_{su} = 35 ns; t_h = −20 ns.
b. t_{su} = 25 ns; t_h = 30 ns.
c. t_{su} = 25 ns; t_h = 10 ns.
d. t_{su} = 5 ns; t_h = 10 ns.

11.13. The figure below shows how a D flip-flop can be realised with a J-K flip-flop and a negator.

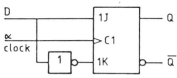

The timing parameters of the components are:

J-K flip-flop: t_{su} = 30 ns; $t_{P(min)}$ = 20 ns;
t_h = 5 ns; $t_{P(max)}$ = 35 ns.

Negator: $t_{P(min)}$ = 5 ns;
$t_{P(max)}$ = 25 ns.

The requirements of the timing of the D input with respect to the clock edge are:
a. t_{su} = 30 ns; t_h = 5 ns.
b. t_{su} = 55 ns; t_h = 5 ns.
c. t_{su} = 55 ns; t_h = 0 ns.
d. none of the above.

11.14. Flip-flop FF1 transfers data to flip-flop FF2. Both flip-flops have the same clock α. The clock pulse is symmetrical; the period is T.

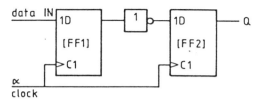

For correct data transfer between flip-flop FF1 and flip-flop FF2 it must hold that:
a. $T > t_{P(min)}(FF1) + t_{P(min)}(NEG) + t_{su}(FF2)$.
b. $T > t_{P(max)}(FF1) + t_{P(max)}(NEG) + t_{su}(FF2)$.
c. $T > 2 \cdot [t_{P(max)}(FF1) + t_{P(max)}(NEG) + t_{su}(FF2)]$.
d. $T > t_{P(max)}(FF1) + t_{P(max)}(NEG) + t_{su}(FF2)$
$T > t_{P(max)}(FF1) + t_{P(max)}(NEG) + t_h(FF2)$.

11.15. A circuit consisting of two flip-flops is given below.

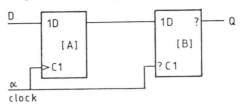

The output of flip-flop A in this figure is interconnected to the D input of flip-flop B. Both flip-flops have the same clock (no clock skew), with a frequency of 5 MHz. The following timing parameters of the flip-flops are given:

$t_{su} = 10$ ns; $t_{P(min)} = 15$ ns;
$t_h = 5$ ns; $t_{P(max)} = 25$ ns.

The name 'positive edge- (pulse-)triggered' means that the input signals have to be stable around the postive edge (pulse) of the clock. Here flip-flop A is of the positive edge-triggered type. Flip-flop B can be of the positive edge-, negative edge-, positive pulse- and negative pulse-triggered type. In how many cases is correct data transfer from flip-flop A to flip-flop B possible?
a. One.
b. Two.
c. Three.
d. Four.

11.16.

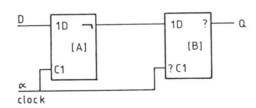

The output of flip-flop A in this figure is interconnected to the D input of flip-flop B. Both flip-flops have the same clock (no clock skew), with a frequency of 5 MHz. The timing parameters of the flip-flops are given:

$t_{su} = 10$ ns; $t_{P(min)} = 15$ ns;
$t_h = 5$ ns; $t_{P(max)} = 25$ ns.

The name 'positive edge- (pulse-)triggered' means that the input signals have to be stable around the rising edge (pulse) of the clock. Here flip-flop A is of the positive edge-triggered type. Flip-flop B can be of the positive edge-, negative edge-, positive pulse- and negative pulse-triggered type. In how many cases is correct data transfer from flip-flop A to flip-flop B possible?
a. One.
b. Two.
c. Three.
d. Four.

11.17.

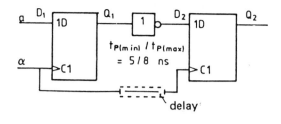

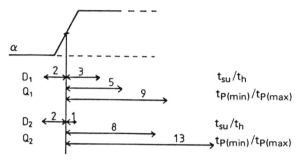

The timing parameters of the circuit are given in the figure. It then holds that:
a. When Q_1 and D_2 are directly interconnected the clock pulse of FF2 may arrive maximally 4 ns later. With a negator in between this is maximally 9 ns.
b. Ditto, with 2 and 10 ns respectively.
c. Ditto, with 5 and 10 ns respectively.
d. Ditto, with 4 and 12 ns respectively.

11.18. A circuit is given below.

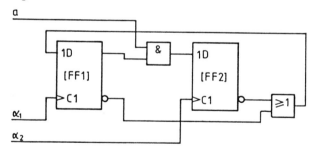

FFs	AND	OR
$t_{su} = 5$ ns; $t_h = 3$ ns;	$t_{P(min)} = 15$ ns;	$t_{P(min)} = 9$ ns;
$t_{P(min)} = 9$ ns;	$t_{P(max)} = 21$ ns;	$t_{P(max)} = 12$ ns;
$t_{P(max)} = 12$ ns.		

It is known that the input signal a is always present 'in time'. Examine first in which interval the signals on the D inputs are uncertain, with respect to the clock pulse $\alpha = \alpha_1 = \alpha_2$. Then check in this diagram to see how many ns sooner or later the clock pulse α_2 of FF2 may be given to make data transfer still just possible.
The limits for clock skew in this circuit are:

a. -13 ns $\leq t_{skew} \leq +19$ ns.   ($-$ means α_2 sooner, and $+$ means later)
b. -19 ns $\leq t_{skew} \leq +13$ ns.
c. -3 ns $\leq t_{skew} \leq +5$ ns.
d. t_{skew} has other limits.

11.19.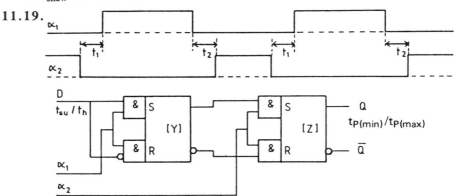

In order to be able to control the 'clock skew' phenomenon a 'double pulse system' is sometimes used as a clock pulse in VLSI designs. Section Y of the flip-flops reads in on α_1, and section Z on α_2. The clock pulses α_1 and α_2 differ mutually in width and in phase, as shown in the figure.

With data transfer, without logic between the flip-flops, the margin for the clock skew is:

a. $t_{skew} = t_{P(min)} - t_h$.
b. $t_{skew} = t_{P(min)} - t_h + t_1$.
c. $t_{skew} = t_{P(min)} - t_h + t_2$.
d. t_{skew} has another formula.

11.20.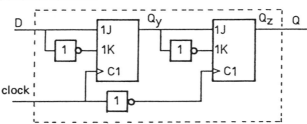

The given circuit consists of two J-K flip-flops and three negators. The pulse-pause ratio of the clock pulse is such that all components have sufficient time to be set.
From an external point of view (D to Q) this circuit reacts as:
a. a positive edge-triggered D flip-flop.
b. a negative edge-triggered D flip-flop.
c. a positive edge-triggered D flip-flop with postponed output reaction.
d. a shift register of two sections, of which the second section only transfers the data on the D input at the falling edge of the second clock pulse.

Chapter 12

The logic diagram below is to be used as a reference to Problems 12.1 through 12.3. The 4-bit shift register has the following facilities:

SH: shift contents from the top downward;
LD: load the data presented in parallel;
INH: store the data just read in.

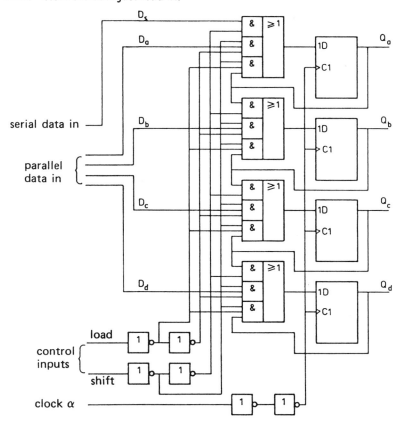

12.1. A function table is given below. One column belongs to the given register. Which one?

Input signals set to		Function of the shift register			
Shift	Load	Column 1	Column 2	Column 3	Column 4
0	0	LD	INH	SH	INH
0	1	INH	LD	SH	SH
1	0	SH	SH	LD	LD
1	1	SH	SH	INH	SH

a. Column 1.
b. Column 2.
c. Column 3.
d. Column 4.

12.2. The given circuit has no possibility to reset the flip-flops. In order to have reset facilities the load/shift inputs are adapted as follows.

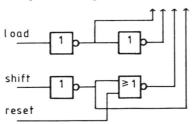

Which of the following statements is correct ?
a. With the adaptation as described a preparatory reset input has been realised. The reset is dominant over the other inputs.
b. With the adaptation as described it is not possible to reset all four flip-flops.
c. To be able to realise a preparatory reset input all four control gates have to be replaced by NORs.
d. With the adaptation as described it is possible to reset all four flip-flops by means of a certain combination of the control signals.

12.3. Our aim is to extend the given register with a *preparatory* reset R. To that end the input circuit is adapted. Which of the circuits below can replace the input circuit, so that the register has a dominant preparatory reset input and the other logic properties do not change (when reset is not active).

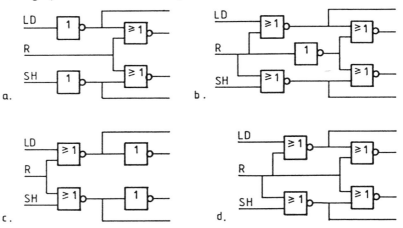

12.4. Many registers have a direct or an asynchronous set and reset. With AND and OR gates a preparatory set and reset can be realised by means of the edge-triggered D input. Indicate for the circuit below which statement is correct.

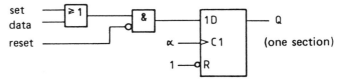

a. The register now has a preparatory set and reset, with dominant data input.
b. The register now has a preparatory set and reset, with dominant set input.
c. The register now has a preparatory set and reset, with dominant reset input.
d. With the circuit given no preparatory set is obtained.

12.5. Many registers have a direct or an asynchronous reset. With a minor adaptation, to be placed between the external data input and the D input of each flip-flop, the register can be equipped with a preparatory (or synchronous) set and/or reset.
We wish to add the following facilities:
- Both set and reset have to be possible;
- With simultaneous 'set' and 'reset' commands, set should dominate.

The following three circuits suggest solutions for this problem.

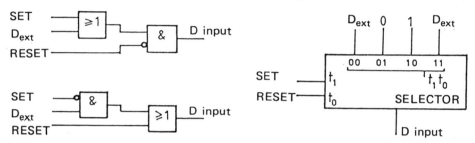

How many of the above circuits give a proper solution?
a. None.
b. One.
c. Two.
d. Three.

The following figure is to be used as a reference for Problems 12.6 through 12.8.

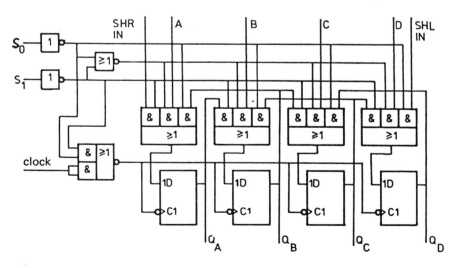

12.6. The given 4-bit shift register has two control inputs S_0 and S_1. The following function table belongs to this register (SHR/L is shift to the right/left, INH is do not change the contents and LD is load):

a.
function	S_0	S_1
LD	1	1
SHR	–	0
SHL	0	–
INH	0	0

b.
function	S_0	S_1
LD	0	0
SHR	–	0
SHL	0	–
INH	1	1

c.
function	S_0	S_1
LD	1	1
SHR	1	0
SHL	0	1
INH	0	0

d.
function	S_0	S_1
LD	0	0
SHR	0	1
SHL	1	0
INH	1	1

12.7. *Data signals* A through D have the following timing with respect to the *external clock pulse* clock:
 a. positive edge-triggered (stable around L → H).
 b. negative edge-triggered (stable around H → L).
 c. positive pulse-triggered (stable from L → H up to H → L).
 d. negative pulse-triggered (stable from H → L up to L → H).

12.8. The *contol signals* S_0 and S_1 have the following timing with respect to the *external clock pulse* clock:
 a. positive edge-triggered.
 b. negative edge-triggered.
 c. positive pulse-triggered.
 d. negative pulse-triggered.

12.9. The clock mode circuit given below consists of a 4-bit shift register and a negator.

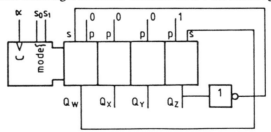

The function table of the shift register is:

S_1	S_0	function
0	0	INH
0	1	SHR
1	0	SHL
1	1	LD

The control inputs S_0 and S_1 and the series inputs of flip-flops W and Z are interconnected as follows:

$S_0 = Q_x$;
$S_1 = \overline{Q_x}$;
series input $D_z = Q_w$;
series input $D_w = \overline{Q_z}$.

After the shift register has loaded the data (WXYZ = 0001) on the parallel inputs, it runs through a sequence. How many clock pulse periods does this sequence last?
a. Two periods.
b. Three periods.
c. Four periods.
d. Five periods.

12.10. The clock mode circuit given consists of a 4-bit shift register and a negator.

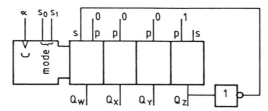

With the control signals S_0 and S_1 the shift register is set to
– parallel loading (register gets contents 0001);
– shift one position to the right. Section W meanwhile reads in the value of $\overline{Q_z}$.
After the shift register has loaded the contents 0001 via $S_0 S_1$ it is set in the 'shift right' mode. Indicate which of the following statements is true.
a. Irrespective of the number of shift pulses, the shift register never returns to its initial state 0001.
b. After four shift pulses the register has returned to its initial state 0001.
c. After eight shift pulses the register has returned to its initial state 0001.
d. After sixteen shift pulses the register has returned to its initial state 0001.

12.11. In a parallel-series converter an 8-bit data word is loaded in parallel and subsequently transmitted serially. With a reset command or while waiting for the next conversion cycle the output S has to be zero. Three configurations of shift registers are given below:

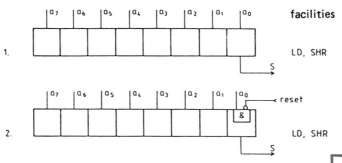

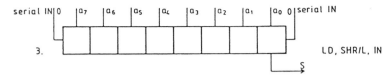

3.

With regard to the problem specification the following configurations meet the specifications:
a. Configurations 1 and 2 and 3.
b. Only configuration 2.
c. Only configuration 3.
d. Both configurations 2 and 3.

12.12. The fact that a shift register can count is sometimes made use of in a parallel-series conversion. This is done by adding one extra 1 and by subsequently letting this 1 shift along. During shifting, the register is then filled up with zeros.

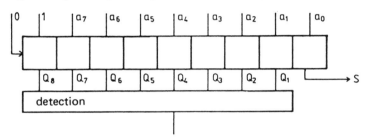

The detection circuit for conversion ready, in its simplest form, consists of

a. 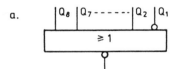   b.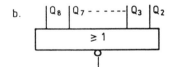

c.   d.

12.13. A modulo-2 feedback shift register of four sections is given below.

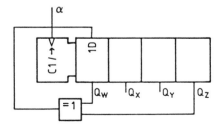

Depending on its initial state this register cyclically goes through:
 a. two sequences.
 b. three sequences.
 c. four sequences.
 d. five sequences.

12.14. A modulo-2 feedback shift register of four sections is given below.

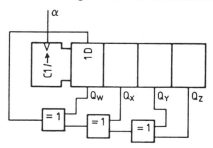

Depending on its initial state this register cyclically goes through
 a. one sequence (length 16).
 b. two sequences (length 0/15).
 c. three sequences (length 0/7/8).
 d. four sequences (length 0/5/5/5).

12.15. A modulo-2 feedback shift register of five sections is given below.

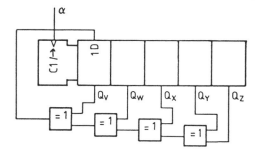

Depending on its initial state this register cyclically goes through:
 a. two different sequences.
 b. three different sequences.
 c. four different sequences.
 d. more than four different sequences.

Chapter 13

13.1. The timing diagram of an asynchronous binary counter is given below.

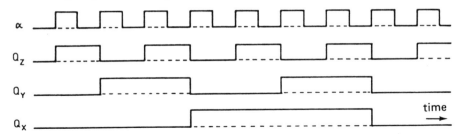

In which one of the following circuits is the behaviour described in the diagram realised?

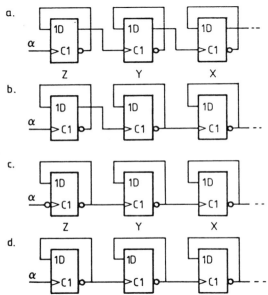

13.2. An asynchronous binary counter is applied within a clock mode environment. The counter consists of toggle flip-flops placed one behind another. The circuit based on this counter, has to give a signal '1' cyclically after each n clock pulses. This has to be given during the next clock period, to be exact. The following is a suggestion for a circuit. The number n can be adjusted.

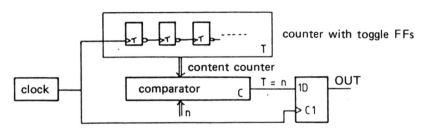

Data:
FFs counter: $t_{su}/t_h/t_{P(min)}/t_{P(max)}$ = 4/3/15/27 ns;
Comparator: $t_{P(min)}/t_{P(max)}$ = 12/35 ns;
D FF: $t_{su}/t_h/t_{P(min)}/t_{P(max)}$ = 8/2/13/23 ns.
The clock frequency is 2 MHz. How large may the number n maximally be?
a. $\leq$ 15.
b. 16.
c. $2^{16} - 1$.
d. This solution does not work for the problem as stated above.

13.3. The logic diagram of an adjustable asynchronous binary U/D counter is given below.

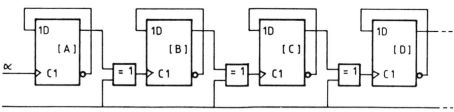

1 UP   0 DOWN

The control signal has the value of 0 and the counter counts n pulses. When the counter has come to a *complete* rest the U/D signal is switched over. There will be no other count pulses α. After some time the contents of this counter are examined. They are:
a. n − 1.
b. n.
c. n + 1.
d. none of the above.

13.4. Design a fully synchronous 4-bit binary down counter, based on J-K flip-flops and gates. The minimal formulas for the J/K signals are:

	a.	b.	c.	d.
$J_a = K_a =$	$\overline{Q_a}$	1	1	$\overline{Q_a}$
$J_b = K_b =$	$\overline{Q_b}\overline{Q_a}$	$\overline{Q_a}$	$\overline{Q_a}$	$\overline{Q_b} + Q_a$
$J_c = K_c =$	$\overline{Q_c}\overline{Q_b}\overline{Q_a}$	$\overline{Q_b}\overline{Q_a}$	$\overline{Q_b}\overline{Q_a}$	$\overline{Q_c} + \overline{Q_b} + \overline{Q_a}$
$J_d = K_d =$	$\overline{Q_d}\overline{Q_c}\overline{Q_b}\overline{Q_a}$	$\overline{Q_c}\overline{Q_b}\overline{Q_a}$	$\overline{Q_c}\overline{Q_b}\overline{Q_a}$	$\overline{Q_d} + \overline{Q_c} + \overline{Q_b} + \overline{Q_a}$

13.5. Design a fully synchronous 4-bit binary up/down counter, based on J-K flip-flops and gates. The minimal formulas for the input signals of the flip-flop D, the most significant bit, are:
a. $J_d = K_d = \text{UP} \cdot Q_c Q_b Q_a + \overline{\text{UP}} \cdot \overline{Q_c}\overline{Q_b}\overline{Q_a}$.
b. $J_d = K_d = \overline{\text{UP}} \cdot Q_c Q_b Q_a + \text{UP} \cdot \overline{Q_c}\overline{Q_b}\overline{Q_a}$.
c. $J_d = K_d = \text{UP} \cdot (Q_c + Q_b + Q_a) + \overline{\text{UP}} \cdot (\overline{Q_c} + \overline{Q_b} + \overline{Q_a})$.
d. none of the above.

13.6. The logic diagram of a 4-bit binary up counter is given below.

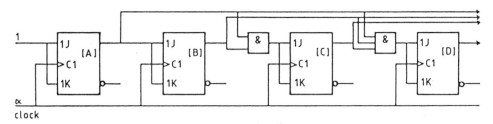

The parameters of the flip-flops and the gates are known.

FFs: $t_{su}/t_h/t_{P(min)}/t_{P(max)} = 5/8/15/22$ ns, $f_{max} \leq 55$ MHz.

ANDs: $t_{P(min)}/t_{P(max)} = 13/21$ ns.

On the basis of these data f_{max}(counter) is computed under stand alone conditions, along with $t_{P(max)}$(clock-to-RC_{out}). We find:

parameter	a.	b.	c.	d.	
$f_{max} \leq$	20.8	23.2	20.8	55	MHz
$t_{P(max)}$(clock-to-RC_{out})	21	21	43	43	ns

13.7. The counter below has a serial enable/ripple carry network.

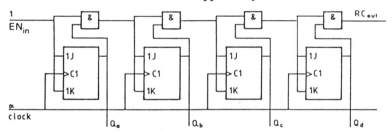

When this principle is extended to a 16-bit counter the maximum propagation delay time is

$$t_{P(max)}(\text{clock-to-}RC_{out}) = t_{P(max)}(\text{FF}) + 16 \cdot t_{P(max)}(\text{AND}) \text{ ns}.$$

To reduce this direct dependence on the number of sections the following 4-bit oriented carry network is designed.

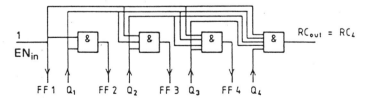

Over the 4-bit sections one can opt for
- a serial structure;
- repeated application of the network.

Examine whether, with the other logic properties of the counter left unchanged, *this* network has advantages or whether a serially oriented network is preferable. The lowest value found for $t_{P(max)}$(clock-to-RC_{out}) is then:

a. $t_{P(max)}(FF) + 2 \cdot t_{P(max)}(AND)$. $(t_{P(max)}(FF) \geq t_{P(max)}(AND))$
b. $t_{P(max)}(FF) + 4 \cdot t_{P(max)}(AND)$. $(t_{P(max)}(FF) \geq t_{P(max)}(AND))$
c. $t_{P(max)}(FF) + 8 \cdot t_{P(max)}(AND)$. $(t_{P(max)}(FF) \geq t_{P(max)}(AND))$
d. $4 \cdot t_{P(max)}(AND)$. $(t_{P(max)}(FF) \geq t_{P(max)}(AND))$

13.8. The logic diagram of a 4-bit synchronous binary counter is given below.

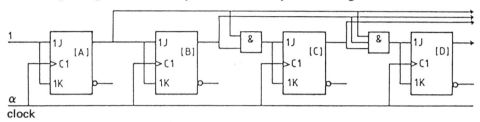

Two statements are made about this counter.
− A 'counter enable' signal can be connected at the point marked 1 near flip-flop A.
− Q_d is the RC_{out} signal of the counter.
For these statements it holds that:
a. statements 1 and 2 are false.
b. statement 1 is false; statement 2 is true.
c. statement 1 is true; statement 2 is false.
d. statements 1 and 2 are true.

13.9. The logic diagram of a 4-bit counter circuit is given on the next page. This counter is:
a. a binary up counter.
b. a binary up/down counter.
c. a BCD up counter.
d. a BCD up/down counter.

13.10. The diagram of a 4-bit counter circuit is given on the next page. Which of the following statements is correct?
a. The ripple carry signal only changes at or around the positive (= upward) edge of the clock pulse.
b. $RC_{out} = 1$ if $Q_aQ_bQ_cQ_d = 1111$.
c. Depending on the application the 'enable T' has different timing requirements from 'enable P'.
d. The load command is dominant over the reset command.

13.11. The diagram of a binary counter circuit is given on the next page. This diagram contains four negators marked ∗. The manufacturer had to incorporate these four negators in the diagram because:
a. for the count logic of sections B through D the negated signals Q_a through Q_d of the previous sections, to name a few signals, are required.

The following figure is to be used as reference for Problems 13.9 through 13.12.

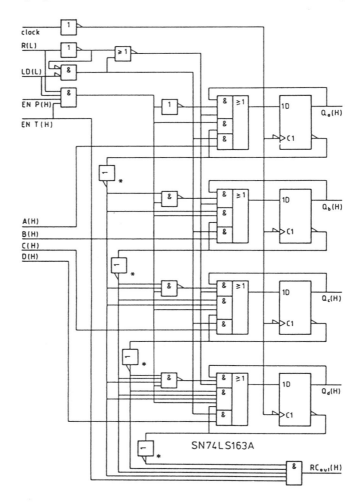

b. for the count logic of sections B through D and for the ripple carry the negated signals $\overline{Q_a}$ through $\overline{Q_d}$ are required.
c. for the count logic of sections B through D and for the ripple carry the signals Q_a through Q_d are required.
d. otherwise the fanout of outputs Q_a through Q_d is/can be too low.

13.12. The logic diagram of a 4-bit counter circuit is given above. The counter has to be loaded with the data on inputs ABCD. The settings *necessary* to that end are:

	R	$\overline{LD}$	EN P	EN T
a.	1	1	–	–
b.	1	0	–	–
c.	1	0	1	1
d.	–	0	–	–

13.13. The diagram of a 4-bit counter section is given below.

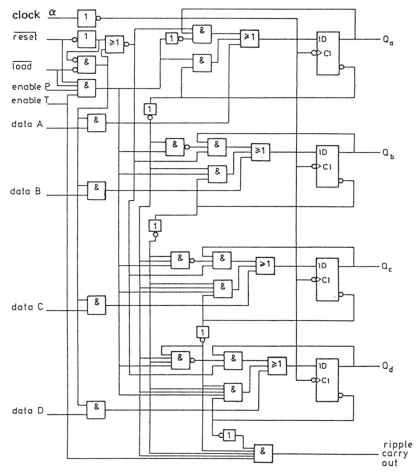

Which statement about this counter section is correct?
a. The ripple carry output = 1 if enable T = 1 and flip-flops A, B, C, D are in state 9 (1001).
b. If $\overline{\text{reset}} = \overline{\text{load}} = 1$ and enable P = enable T = 1 the D input of flip-flop A is presented with its own output signal Q_A.
c. If $\overline{\text{load}} = 0$ data signals A, B, C, D are always presented to the D inputs of the flip-flops independent of the value of enable P, enable T and $\overline{\text{reset}}$.
d. If $\overline{\text{reset}} = 0$ the AND gates with data signals A, B, C, D as input signals are blocked.

13.14. A 4-bit BCD counter section is given. The state of the counter section is denoted by signals W, X, Y and Z with the respective weights 8, 4, 2 and 1.
The outgoing carry for any subsequent BCD counter section is supplied by an AND gate with W and Z as input signals. Due to a disturbance the BCD counter section may end up in one of the (unused) states 10–15.

Indicate which statement is *true*:
a. The carry is only 1 in state 9.
b. The carry is 1 in states 9 and 10.
c. The carry is 1 in states 9 through 15.
d. The carry is 1 in state 9, 11, 13 and 15.

13.15. The detailed logic diagram and the function table of a 4-bit binary up/down counter are given below. The counter is in present state 15 (1111_{BIN}).

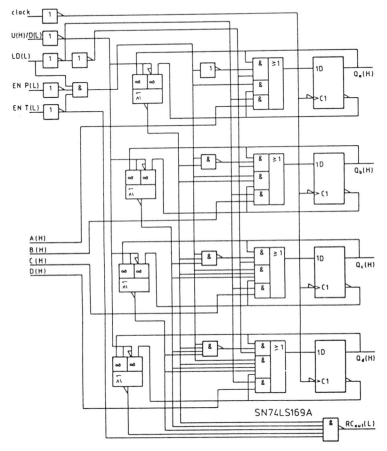

Function	Control in L and H				
	U(H)/D(L)	LD(L)	EN P(L)	EN T(L)	ABCD(H)
Reset	not present				
Load/up mode	H	L	–	L	data
Load/down mode	L	L	–	L	data
Count up	H	H	L	L	----
Count down	L	H	L	L	----
Inhibit/up mode	H	H	H	L	----
Inhibit/down mode	L	H	H	L	----

Function	Control in L and H				
	U(H)/D(L)	LD(L)	ENP(L)	ENT(L)	ABCD(H)
Reset	not present				
Load	–	L	–	–	data
Count up	H	H	L	L	----
Count down	L	H	L	L	----
Inhibit	–	H	H	–	----
Inhibit	–	H	–	H	----

The next state has to be state 14 (1110_{BIN}). When this state is being prepared the RC_{out} signal is not allowed to be suppressed. It has to remain available for test purposes. The possible settings are:

a. CTDN   (L H L L – – – –).
b. LD 14   (– L – – L H H H).
c. LD 14 and CTUP (L L – L L H H H).
d. LD 14 and CTDN (L L – L L H H H).

Answers

0	0.1.	c	4.3.	a		7.14.	c	10.20.	a
	0.2.	b	4.4.	a		7.15.	b	**11** 11.1.	a
	0.3.	d	4.5.	a	**8**	8.1.	b	11.2.	b
	0.4.	d	4.6.	c		8.2.	b	11.3.	b
1	1.1.	d	4.7.	?		8.3.	d	11.4.	d
	1.2.	a	4.8.	?		8.4.	a	11.5.	c
	1.3.	a	4.9.	?		8.5.	c	11.6.	c
	1.4.	c	4.10.	c		8.6.	b	11.7.	c
	1.5.	b	**5** 5.1.	b		8.7.	d	11.8.	c
	1.6.	c	5.2.	c		8.8.	c	11.9.	b
	1.7.	d	5.3.	a		8.9.	a	11.10.	a
	1.8.	b	5.4.	c		8.10.	c	11.11.	a
	1.9.	a	5.5.	a		8.11.	b	11.12.	b
	1.10.	c	5.6.	d		8.12.	a	11.13.	b
	1.11.	b	5.7.	a		8.13.	b	11.14.	b
	1.12.	b	5.8.	b		8.14.	a	11.15.	c
	1.13.	b	5.9.	c		8.15.	d	11.16.	c
	1.14.	a	5.10.	b	**9**	9.1.	c	11.17.	a
	1.15.	d	5.11.	b		9.2.	c	11.18.	a
2	2.1.	b	5.12.	c		9.3.	a	11.19.	c
	2.2.	c	5.13.	d		9.4.	a	11.20.	c
	2.3.	b	5.14.	c (d)		9.5.	b	**12** 12.1.	b
	2.4.	c	5.15.	b		9.6.	b	12.2.	d
	2.5.	c	**6** 6.1.	d		9.7.	a	12.3.	d
	2.6.	b	6.2.	b		9.8.	d	12.4.	c
	2.7.	b	6.3.	c		9.9.	c	12.5.	a
	2.8.	a	6.4.	d		9.10.	a	12.6.	c
	2.9.	c	6.5.	a		9.11.	c	12.7.	a
	2.10.	c	6.6.	a		9.12.	a	12.8.	d
	2.11.	b	6.7.	d		9.13.	d	12.9.	c
	2.12.	c	6.8.	c		9.14.	a	12.10.	c
	2.13.	d	6.9.	b		9.15.	b	12.11.	d
	2.14.	d	6.10.	a	**10**	10.1.	a	12.12.	?
	2.15.	d	6.11.	d		10.2.	d	12.13.	a
3	3.1.	b	6.12.	b		10.3.	b	12.14.	d
	3.2.	b	6.13.	c		10.4.	a	12.15.	d
	3.3.	b	6.14.	d		10.5.	b	**13** 13.1.	d
	3.4.	b	6.15.	d		10.6.	d	13.2.	d
	3.5.	a	**7** 7.1.	c		10.7.	a	13.3.	d
	3.6.	d	7.2.	c		10.8.	a	13.4.	?
	3.7.	d	7.3.	b		10.9.	d	13.5.	?
	3.8.	a	7.4.	b		10.10.	c	13.6.	c
	3.9.	b	7.5.	d		10.11.	c	13.7.	b
	3.10.	a	7.6.	b		10.12.	d	13.8.	a
	3.11.	c	7.7.	a		10.13.	a	13.9.	a
	3.12.	c	7.8.	d		10.14.	c	13.10.	c
	3.13.	b	7.9.	c		10.15.	c	13.11.	d
	3.14.	b	7.10.	b		10.16.	b	13.12.	b
	3.15.	a	7.11.	c		10.17.	b	13.13.	d
4	4.1.	c	7.12.	c		10.18.	a	13.14.	d
	4.2.	c	7.13.	d		10.19.	b	13.15.	c

Symbols and abbreviations

$\oplus$	exclusive OR	0	logic 0
+	logic OR/addition symbol	1	logic 1
·	logic AND/multiplication symbol	AC	alternating current
		A-D	analog-to-digital
$-$	logic NOT	ALU	arithmetic logic unit
$\overline{var}$	negated variable	ALS	advanced low power Schottky
$\overline{name}$	negated 'name'	AND	AND gate
$-$	don't care	AS	advanced Schottky
$\wedge$	logic AND	ASCII	American Standard Code for Information Interchange
$\vee$	logic OR		
$\cap$	intersection	BCD	binary coded decimal
$\cup$	union	BIN	binary
$\in$	belongs to	bit	binary digit
=	equality sign	byte	eight bits (word)
/	NOT (in combination with other symbols)	C	clock/command input
		CMOS	Complementary MOS
≥ 1	qualifying symbol OR	CTDN	count down
&	qualifying symbol AND	CTn	content is n
1	qualifying symbol NOT	CTR	counter
2k	qualifying symbol EVEN	CTUP	count up
2k + 1	qualifying symbol ODD	D	input D(elay) latch/flip-flop
$\neg$	postponed output/NOT	D-A	digital-to-analog
$\rightarrow$	dynamic input indicator	DC	direct current
⊸∣ ⊢	negation indicator	DEC	decimal
⊣∣ ⊢	polarity indicator	DIV	divide by
	open-circuit output	DTL	diode-transistor logic
	open-circuit output	DX	demultiplexer
	open-circuit output	d	don't care/Hamming distance
	open-circuit output	EN P	parallel enable counter
∇	3-state output	EN T	serial enable counter
$\triangleright$	buffered outputs	EXOR	exclusive OR gate
$[...]^n$	clock phase indicator	F	false
[text]	text in symbols	FA	full adder
*	position qualifying symbol/operator	FET	field effect transistor
		FF	flip-flop
$\rightarrow$	shift right/down	f	clock frequency
$\leftarrow$	shift left/up	f_i	function value i
$\Leftrightarrow$	assignment v.v.	$f(x,y,z)$	function in x, y and z
$\Rightarrow$	assignment	G	EN dependency/carry generate function
Δt	propagation delay time		
Ω	Ohm	GND	ground
Π	product	g	base number system/integer
Σ	summation		

H (H)	High (active high, after signal name)	NPN	transistor with p-type base
HEX	hexadecimal	NMOS	n-type MOS
HOLD	hold	NOR	NOT-OR gate
HPRI	highest priority	NOT	negator, negation
Hz	Hertz	n	variabele
I	current	name(H)	name 'true' if High
I_{IH}	input current at H input level	name(L)	name 'true' if Low
I_{IL}	input current at L input level	ns	nanosecond
I_{OH}	output current at H output level	OC	open-collector
I_{OL}	output current at L output level	OCT	octal
IC	integrated circuit	OR	OR gate
IEC	International Electrotechnical Commission	P	enable P/carry propagate function
IEEE	Institute of Electrical and Electronics Engineers	PAL	programmable array logic
		PCB	printed-circuit board
I^2L	integrated injection logic	PLA	programmable logic array
INH	inhibit	PNP	transistor with n-type base
INV	inverter	PROM	programmable ROM
i	input combination/variable	Q	flip-flop output
J	input J-K flip-flop	q	state
K	input J-K flip-flop	R	reset input/resistor
k	variable/kilo(Ω)	RAM	random access memory
L (L)	Low (active low, after signal name)	RC	ripple carry
		REPROM	reprogrammable PROM
LACG	look-ahead carry generator	ROM	read-only memory
LD	load	RTL	resistor-transistor logic
LD0	load zero	R_L	load resistor
LDD	load data	S	set input/circuit output/logic function/sum signal/Schottky
LS	low power Schottky	SH	shift
LSD	least significant digit	SHL	shift left
LSI	large scale integration	SHR	shift right
M	mega	S-R	set-reset
M_i	maxterm i	SRG	shift register
MHz	megaHerz	SSI	small scale integration
MSD	most significant digit	s	second
MOS	metal-oxide semiconductor	s_0, s_1	control inputs shift register
M-S	master-slave	T	input T(oggle) flip-flop/ clock period/true/ enable T
MSI	medium scale integration		
MUX	multiplexer	TOR	teletype-on-radio
m_i	minterm	True	true
max	maximum	TTL	transistor-transistor logic
min	minimum	t	time
ms	millisecond	t_{acc}	access time
NAND	NOT-AND gate	t_c	indicates the period a signal has to be present
Ni	negation dependency		

t_h	hold time
t_i	indicates the period combinatial logic may be set
t_{PD}	propagation delay time
t_{PHL}	ditto, at H→L transition
t_{PLH}	ditto, at L→H transition
$t_{P(min/max)}$	minimum/maximum t_{PD}
$t_{P(typ)}$	typical value t_{PD}
t_{puls}	pulse width
t_{su}	setup time
t_{skew}	clock skew
t_T	transition time
t_{THL}	ditto, H→L transition
t_{TLH}	ditto, L→H transition
U/D	up/down input
V	Volt
VLSI	very large scale integration
Vi	OR dependency
V_{CC}	supply voltage (bipolar)
V_{DD}	supply voltage (MOS)
V_{IL}	input voltage at L input level
V_{IH}	input voltage at H input level
V_{OL}	output voltage at L input level
V_{OH}	output voltage at H input level
V_{ss}	substrate voltage MOS
V_{ds}	drain-source voltage
V_{gs}	gate-source voltage
V_{in}	input voltage
V_{out}	output voltage
V_{ref}	reference level
V_t	threshold voltage
W	Watt
x	input variable
y	state variable/logic variable
Z_{old}	previous state latch/relay
Z_{new}	new state latch/relay
Zi	Z dependency
Z_0	latch with dominating reset
Z_1	latch with dominating set
z	logic variable

Index

1-out-of-n decoder 165
3-state bus line 170
3-state output 118
4-bit counter 319, 326, 328
4-bit full adder 154
4n-bit counter 327, 334

a

absorption law 42
AC noise 105
AC power supply 75
accuracy 12, 16
activating delay 84
active clock edge 258
active high output 120
active low output 120
active low wiring 118, 121, 162
adder, 4-bit full 154
  binary 151
  full 151, 153, 155
  half 152
addition 151
address 45, 179
address decoder 180
address generator 187
adjacent cells 55
alphanumeric character set 23
analog 16
AND 32, 90
  wired 120, 134
AND array 187
AND-NOR, NMOS 134
  TTL 95
AND-OR plane 247
anti-bounce circuit 84, 211
architecture specification 305, 340
arithmetical specification 153
armature 74
ASCII code 25, 173
assignment of truth values 87
assignment of variables 77
associative law 41
asynchronous clear 263
asynchronous counter 308, 310, 335, 340
asynchronous preset 263
autonomous sequential circuit 238

b

base 18, 21, 116
basic logic diagram 290
battery 75
BCD code 20, 306
BCD code, weights 21
BCD counter 61, 324

BCD-to-7-segment decoder 183
bidirectional 82
BIN-DEC conversion 19
BIN-HEX conversion 22
binary 16
binary adder 151
binary code 15, 19
  reflected 15, 28
binary components 17
binary counter 308, 312, 316, 319
binary number system 19
binary variable 39
bipolar 142
bipolar transistor 115, 116, 124, 181
bit line 218
bit partitioning 192
bit string 21
block check 173
Boolean algebra 32
Boolean variable 39
bouncing of contacts 84
branch type 73
break contact 34, 75, 76
buffered output 110
bus, 3-state 171
  open-collector 169
bus driver 170
bus frequency 170

c

canonical form 47
capacitive load 108
capacitor 75
  decoupling 106
  input 125
carry 151, 155
carry generate function 157
carry mechanism 315
carry propagate function 157
carry propagation network 327, 328
CCITT nr. 3 alphabet 24
channel 125
character set 23
circuit, anti-bounce 84, 211
  combinational 73
  omnibus 212
  sequential 197, 238, 245
clipping diode 107, 112
clock 223, 326
clock buffer 272
clock distribution 283
clock frequency 262, 270, 281, 313, 337
  maximum 270, 310, 313
clock input 231

clock interruption 315
clock mode environment 309, 326
clock mode view 223, 225, 245, 307, 335, 343
clock period 249, 257, 269, 335
clock phase 276
clock pulse 258
clock pulse width 262
clock skew 278, 280, 282, 288
CMOS logic 139
CMOS NAND 140
CMOS NOR 141
CMOS, dynamic 141
code, ASCII 25, 173
    BCD 20, 21, 306
    binary 15
    EBCDIC 26
    Gray 15, 28
    m-out-of-n 27
    progressive 15, 27
    reflected binary 15, 28
    telegraph 25
    three-out-of-seven 24, 171
    two-out-of-five 171
code shaft 15
coil 74, 75
collector 116
column decoding 181, 217
column select 181
combinational logic 73, 269, 271
commutative law 41
comparator 161
complementary transistors 131
completeness of specification 60
components, selection 298
composition rules 36
constants, rules for 40
contact 30
    break 34, 75, 76
    make 34, 75, 76
    make-before-break 83
    transfer 75, 76, 78
contact resistance 12, 82
control input 328
control signal 166, 176, 317, 320, 335
control unit 165, 176, 326
convention, mixed logic 89
    negative logic 89
    positive logic 89
conversion, BIN-DEC 19
    BIN-HEX 22
    DEC-BIN 19
    HEX-BIN 22
    repeated division 20
cost function 65
counter 240, 306, 330
    4-bit 319, 326
    4n-bit 327, 334
    asynchronous 308, 310, 335, 340
    BCD 324

    binary 308, 312, 316, 319
    divide-by-n 300
    Johnson 301
    Mobius 301
    ring 300
    synchronous 312, 314, 316
    twisted ring 301
counter-control interaction 339
counter structure 329
counting frequency 311, 337
    maximum 309
counting shift register 304
cover, cost functions in 68
    minimal 65, 67
covering problems 65
critical race conditions 207, 244
cross talk 106, 335, 282
cross-over frequency 142
C switch 139
current source 131
current-mode logic 124
cutoff region 116, 129
cycle length 300, 302, 306
    variable 336
cyclic counter 238

d

data lockout flip-flop 263
data path 176
data signal 176, 320
data transfer 105, 165, 221, 265, 267, 269, 276, 279
    direct 265, 279
    indirect 269, 280
data valid signal 285
DC noise 105
DC noise margin 105, 114, 140
DC power supply 75
DEC-BIN conversion 19
decade counter 307
decimal number system 18, 20
decoder 168, 184
decoder, 1-out-of-n 165
decomposition 78, 97, 98
    Reed-Muller 100
decoupling capacitors 106, 283
delay element 200
DeMorgan law 42
demultiplexer 165, 168
dependent logic variables 234
dependency of logic variables 50
depletion layer 126
depletion-mode MOS transistor 127, 139
depletion-mode pull-up transistor 132
design levels 343
design procedure 176
design rules 156
detailed logic diagram 93, 162, 166, 214, 290, 319

detailed timing model 109, 253
detection circuit 310, 336, 338
D flip-flop 224, 227, 235, 237, 256
digit 18
digital 16
diode 75, 114
    clipping 107, 112
    perfect 115
diode-transistor logic 111
direct acting input 263, 318
direct data transfer 265, 279
direct reset 263, 336
direct set 263
direction information transfer 164
distinctly shaped symbols 90
distributive law 41
divide-by-n counter 300
division 299
dominant input 318
dominant row 67
don't care 62, 80
don't happen condition 62
don't use condition 62
drain 125
dynamic behaviour 126, 128, 136
dynamic CMOS 141
dynamic logic 125, 138, 141
dynamic loop 340
dynamic memory 138
dynamic RAM 217
dynamic state 31, 125, 144

e

EBCDIC code 26
ECL logic 124
edge-triggered flip-flop 257, 282
edge-triggered timing 260, 297
    negative 257
    positive 257
effective propagation delay 337
emitter 116
emitter-coupled logic 124
enable input 162, 231, 237, 315, 316, 328
enable network 327
enable P 315
enable pulse 251
enable signal 204, 307
enable T 315
encoding 15, 164, 207
end-around carry 330
enhancement-mode 139
enhancement-mode MOS transistor 127
enhancement-mode pull-up transistor 131
environment 307, 310, 313
equivalence of logic formulas 200
error correction 171
error detection 24, 171, 185
error propagation 325
essential prime implicant 64, 66

evaluate transistor 141
even parity 173, 175
EX-OR 35, 37, 90, 101
EX-OR gate 123
expander 95
expansion of terms 49
extension of laws 43
external behaviour 236
external timing 271

f

fall time 137, 278
fanin 110, 162, 319
fanout 110, 143, 319
fast carry 155
feedback 228, 260
field programmable logic array, *see* FPLA
field-effect transistor 125
finite state model 305
flip-flop design 287
flip-flop selection 242
flip-flop, edge-triggered 282
    pulse-triggered 271
folding 195
formulas, interpretation of 31, 40
FPLA 188
frame check 173
full adder 151, 1563, 155
    4-bit 154
function 44
function table 209, 210, 227, 235, 293, 318
function value 52
functions, logic operations on 59
fuse 181

g

gain factor 129
gate 125
gate circuit 146
gate model 123, 143, 250
gate symbols 90
gate type 73
gated D latch 205, 250, 253
gated latch 203, 220
gated S-R latch 204, 252, 253
global timing model 109, 253
Gray code 15, 28
ground 75

h

half adder 152
Hamming distance 171, 172
HEX-BIN conversion 22
hexadecimal number system 21, 22
hexadecimal-to-7-segment decoder 185
hidden loop 82
high-impedance state 119
hold time 251, 268, 272
hotel circuit 36

i

I²L logic 121
idempotence law 40
identity comparator 161, 162
IEC system 91
implementation 305, 340
implicant 64
implicants, table with 70
inclusive OR 35
incompletely specified logic function 62
information encoding 15, 164
information transfer 309
inhibit 252, 319
input assignment 207
input capacitor 125
input resistance 126
input signals for flip-flops 234
input transition 256
input value 258
instruction decoding 187
interconnect delay 148
interconnection network 165
    parallel 33
    series 31
internal feedback 260
internal organisation 204
internal structure 155, 275, 292
internal timing 271
interpretation of logic formulas 31, 40, 91
inverter ratio 136
inverter, NMOS 131
    ratio-less 131
    ratioed 130

j

J-K flip-flop 228, 229, 235, 237, 259, 261
Johnson counter 301

k

Karnaugh map 52, 53
    comparison of functions 59
    order of labels 57
    placement 53, 57
    reduction with 57
    reflection axes 55

l

L/W ratio 136
law, absorption 42
    associative 41
    commutative 41
    DeMorgan 42
    distributive 41
    idempotence 40
    modulus 42
    negation 42
leakage current 115
letter/figure shift 24
level mode view 199, 223, 224

line driver 170
line receiver 170
linear region 129
load 319
logic behaviour 225
logic circuits, equivalent 41
    multi-output 58
logic convention 89
logic diagram 214
logic family 105, 111, 121
logic formula 40, 91
logic formulas, equivalent 200
logic function, incompletely specified 62
logic level 86, 104
logic model, detailed 93
    mappable 93
    theoretical 93
logic operations 32, 33, 34
logic operations in I²L 122
logic operations on functions 59
logic specification 153
logic structure 326
logic symbols 75, 89, 119, 226, 262, 319
logic threshold 105, 114, 136, 278
logic value 32
logic variable 39, 40
logically equivalent 41
look-ahead carry 315
look-ahead carry generator 158, 160
loops 82, 321, 339

m

m-out-of-n code 27
magnet 74
magnitude comparator 161, 163
majority carrier 125
make contact 34, 75, 76
    manually operated 75
make-before-break contact 83
mappable logic model 93
mapping problems 215
mask programmable ROM 181
master-slave principle 221, 245
maximum clock frequency 270, 310, 313, 314, 330, 334
maximum likelihood decoding 172
maxterm 47
maxterm form 47, 48
measurement conditions 144
mechanical inertia 84, 211
memory 197
memory array 181, 182
memory element 198
    model 201
    state diagram 199
    state of a 198
    truth table 199
metastability 285
minimal cover 65, 67

minimal form   56
minimal sum-of-products   54
minimum pulse width   251
minterm   45, 47, 64, 181
minterm form   44, 48, 52, 78
mixed logic convention   89
mobility of electrons   125
Mobius counter   301
modular design   164, 168, 314
module   306
modulo-2 feedback   301
modulus law   42
MOS   142
MOS transistor   125, 129
   depletion-mode   127
   enhancement-mode   127
multi-emitter transistor   112
multi-output logic circuits   58
multi-phase clock   301
multiple output specification   102
multiplexer   166
multiplication   299

n

NAND   90, 94, 99, 111, 122
   CMOS   140
   NMOS   133
   recipe for   99
negation indicator   262
negation law   42
negative edge-triggered timing   257
negative logic convention   89
next state   199
NMOS AND-NOR   134
NMOS logic   131
NMOS transistor   126
noise   93, 104, 253, 282, 315, 338
noise reduction   170
noise susceptibility   170, 262, 264
noise, power density   105
NOR   90, 94, 122
   CMOS   141
   NMOS   133
   recipe for   94
normal position   76
   switch   36
NOT   90
NPN transistor   115, 116
number   18
number system, binary   19
   decimal   18
   hexadecimal   21, 22
   octal   21
   range   19

o

observation   12
octal number system   21
odd parity   173, 175

omnibus circuit   212
open input   93, 107
open-collector bus line   169
open-collector output   117, 171, 192
operational condition   144
operator circuit   75
optimisation criteria   176
OR   32, 90
   exclusive   35, 37
   inclusive   35
   wired   120, 134
OR array   187
order of operations   40
order of precedence   37
output, 3-state   118
   active high   120
   active low   120
   open-collector   117
   totem pole   117
output block, universal   247
output driver   114, 117
output enable   118, 169
output symbols   120
output transition   256
output value   258

p

PAL   189, 190, 191
   registered   245
   sequential   245
parallel carry   330
parallel enable   330
parallel interconnection   33
parallel-series converter   302
parameter specification   331
parasitic capacitor   136
parentheses   40
parity bit   26, 172
parity check   172, 174
parity generator   174
pass transistor   137
passive edge   258
pattern generator   338
perfect diode   115
Petrick function   66
phase-splitter   112
physical characteristics of components   200
PLA   188, 191
PNP transistor   115, 116
polarity indicator   262
position, normal   76
position registration   12, 13
positive edge-triggered timing   257, 320
positive logic convention   89
postponed output indicator   262
power circuit   75
power density noise   105
power supply   106, 143
precedence rules   91

precharge transistor 141
preparatory input 263, 318, 339
present input 235
present state 199, 235
prime implicant 64, 68
prime implicant table 66
priority scheme 340
product form 60, 95
programmable array logic, *see* PAL
programmable logic 179, 245
programmable logic array, *see* PLA
programmable shift register 293
progressive code 15, 27
PROM 191
    registered 247
proof of laws 43
propagation delay time 108, 143, 144, 158, 200, 250, 254, 268, 285, 334
proposition 30, 152, 162
pseudo clock mode 307, 335
pull-down resistor 120
pull-up resistor 117, 120, 130, 169
pull-up transistor 131, 137, 140, 217
    depletion-mode 132
    enhancement-mode 131
pulse distortion 220
pulse width 220, 222
    minimum 251
pulse-triggered flip-flop 271
pulse-triggered timing 259, 260, 298
push-button 34, 76

q

quantisation noise 16
quiescent power consumption 140
Quine-McCluskey algorithm 68

r

race condition 232, 207, 244
    critical 207, 244
RAM 216
    dynamic 217
    static 216
Random Access Memory, *see* RAM
random logic 181, 185, 186
rank number 45
ratio-less inverter 131
ratio-typed logic 136
ratioed inverter 130
read-modify-write cycle 218
Read Only Memory, *see* ROM
read/write memory, *see* RAM
realisation 305
realisation memory element 201
recipe for NAND 94, 99
    for NOR 95
reduction with Karnaugh maps 57
redundancy 84
reed relay 74

Reed-Muller decomposition 100
reference level 87, 145, 251, 258
reflected binary code 15, 28
reflection lines 55
refresh cycle 218
register 290
    internal structure 292
registered PAL 245
registered PROM 247
registers, timing of 296
registration method 12, 15
relay, electromagnetic 73
reliability 12
repeated division conversion 20
reset 198, 200, 240, 318, 338
reset gate 228
reset input 264
residu 78
resistor 75
    pull-down 120
    pull-up 117, 120, 130, 169
resistor-transistor-logic 111
restoring logic 137
ring counter 300
ripple carry 155, 318, 324, 329
ripple effect 164
rise time 137, 278
ROM 179
    mask programmable 181
row decoding 181, 217
row select 181
rules for constants 40
rules for logic variables 40

s

S-R flip-flop 231
S-R latch 200
$\overline{S}$-$\overline{R}$ latch 200
sampling 16
saturation mode 129
Schmitt-trigger input 106, 170
segments, number of 14
selector 97, 137, 165, 166
self-timed logic 284
sequential circuit 197, 238, 245
sequential PAL 245
series interconnection 31
series-parallel carry 331
series-parallel converter 302
set combination 198, 200, 264
set gate 228
set/reset encoding 207
setup time 251, 269, 272, 334
shift operation 299
shift register 301
    internal structure 292
    modulo-2 feedback 301
short circuit 119
signal propagation 271

skew time 265
source 125
specifications, completeness of 60
    unambiguous 60
speed-power product 110
spikes 338
SR = 11 input combination 208
stand alone conditions 307
standard component 150
standard form 45
state 225, 226
state assignment 209, 213, 239, 242, 244, 245
state diagram 198, 213, 223, 224, 225, 230, 238
state of a memory element 198
state transition 223
state variable 239
static behaviour 126, 135
static electricity 126
static logic 125, 340
static RAM 216
static state 31, 125
steering logic 137
structure 176, 314
structure of a register, internal 292
subfunction 78
substrate 126
sum form 60
sum-of-products form 54, 94
    minimal 54
summation modulo-2 302
surface charge 126
switch 116
    voltage controlled 128
switching algebra 40
symbol for $\overline{\text{S-R}}$/S-R latch 202, 203
symbolic logic 32
symbols 75, 89, 119, 202, 226, 262, 319
    contact circuit 75
    distinctly shaped 90
    gate 90
synchronisation 283, 285, 342
synchronisation flip-flop 342
synchronous counter 312, 314, 316
system 13

t

table with implicants 70
telegraph code 25
ternary notation 69
testability 176
T flip-flop 230
theoretical logic model 93
three-out-of-seven code 24, 171
threshold 105, 114, 136, 278
threshold voltage level 87, 127, 132, 145
timing 250, 257, 264, 279, 296, 320
timing analysis 109, 275
timing diagram 224, 225

timing external inputs 273
timing model 107, 146, 253, 254
timing of registers 296
timing parameters 107, 328
timing specification 225
timing verification 267
toggle flip-flop 219, 222, 230, 243, 308
top-down design 226
totem pole output 112, 114, 117
transfer characteristic 130, 186
transfer contact 74, 75, 76, 78
transient phenomena 28, 31, 83, 208, 211
transient state 144
transistor, bipolar 115, 116, 124, 181
    depletion-mode MOS 127
    depletion-mode pull-up 132
    enhancement-mode MOS 127
    enhancement-mode pull-up 131
    evaluate 141
    field-effect 125
    MOS 125, 129
    multi-emitter 112
    NMOS 126
    NPN 115, 116
    pass 137
    PNP 115, 116
    precharge 141
    pull-up 131, 137, 140, 217
transistor-transistor-logic 111
transition phenomena 313, 321
transition time 107
transparency 203
transparent latch 221
triode region 129
truth table 31
truth value 30, 32, 87
    assignment of 87
TTL compatible 86
TTL logic family 111, 113, 114
twisted ring counter 301
two-out-of-five code 171
two-phase clocking 287
typical value 258

u

U/D switch 330
ULA 191
unambiguous specification 60
uncertainty 107
uncertainty interval 250, 254, 271
uncommitted logic array, *see* ULA
undefined state 231
universal output block 247
universe 46
unused inputs 93
up/down counter 323, 324, 325

v

variable, binary 39

Boolean 39
  logic 39
variables, assignment of 77
Venn diagram 46
voltage controlled switch 128
voltage level 86

w

weight 18, 20
wired OR 134
wired AND 118, 169
word line 218
working position, switch 36
worst case timing 253

z

$Z_{(0)}$ memory element 207
$Z_{(1)}$ memory element 208
$Z_{(z)}$ memory element 208